Computer Modeling of Water Distribution Systems

AWWA MANUAL M32

Third Edition

American Water Works Association

Manual of Water Supply Practices — M32, Third Edition

Computer Modeling of Water Distribution Systems

Copyright © 1989, 2005, 2012, American Water Works Association
All rights reserved. No part of this publication may be reproduced or transmitted in any form or by any means, electronic or mechanical, including photocopy, recording, or any information or retrieval system, except in the form of brief excerpts or quotations for review purposes, without the written permission of the publisher.

Disclaimer
The authors, contributors, editors, and publisher do not assume responsibility for the validity of the content or any consequences of its use. In no event will AWWA be liable for direct, indirect, special, incidental, or consequential damages arising out of the use of information presented in this book. In particular, AWWA will not be responsible for any costs, including, but not limited to, those incurred as a result of lost revenue. In no event shall AWWA's liability exceed the amount paid for the purchase of this book.

AWWA Publications Manager: Gay Porter De Nileon
Project Manager/Copy Editor: Melissa Valentine
Production Editor: Cheryl Armstrong
Manuals Specialist: Molly Beach

Library of Congress Cataloging-in-Publication Data
Robinson, Laredo.
  Computer modeling of water distribution systems / by Laredo Robinson, Jerry A. Edwards, Lindle D. Willnow.
    p. cm. -- (AWWA MANUAL M32)
  Rev. ed. of: Computer modeling of water distribution systems. 2005.
  Includes bibliographical references and index.
  ISBN 978-1-58321-864-8 (alk. paper)
  1. Water--Distribution. 2. Network analysis (Planning) I. Edwards, Jerry A. II. Willnow, Lindle D. III. Title.
  TD491 .A49 no. M32 2005
  [TD481]
  628.1'44--dc23
                    2012002982

American Water Works Association
6666 West Quincy Avenue
Denver, CO 80235-3098

ISBN 978-1-58321-864-8

Printed on recycled paper

Contents

List of Figures, v

List of Tables, ix

Foreword, xi

Acknowledgments, xiii

Chapter 1   Introduction to Distribution System Modeling 1
- 1.1   Introduction, 1
- 1.2   Purpose of the Manual, 2
- 1.3   Historical Development of Distribution System Modeling, 2
- 1.4   Distribution System Modeling Applications, 4
- 1.5   Hydraulic Models, 8
- 1.6   Distribution System Modeling Within the Utility, 11
- 1.7   Trends, 12
- 1.8   Summary, 15
- 1.9   References, 15

Chapter 2   Building and Preparing the Model 17
- 2.1   Introduction, 17
- 2.2   Planning the Hydraulic Model Construction and Development Process, 19
- 2.3   Data Sources and Availability, 24
- 2.4   Physical Facilities Development, 30
- 2.5   Demand Development, 45
- 2.6   Operational Data, 53
- 2.7   Hydraulic Model Maintenance, 57
- 2.8   References, 62

Chapter 3   Tests and Measurements 65
- 3.1   Introduction, 65
- 3.2   Planning Field Tests and Preparation, 66
- 3.3   Water Distribution System Measurements, 67
- 3.4   Water Distribution System Testing, 76
- 3.5   Data Quality, 83
- 3.6   References, 84

Chapter 4   Hydraulic Calibration 85
- 4.1   Introduction, 85
- 4.2   What is Calibration? 85
- 4.3   Steady-State Calibration, 94
- 4.4   EPS Calibration, 98
- 4.5   References, 102

Chapter 5   Steady-State Simulation 103
- 5.1   Introduction, 103
- 5.2   System Performance Analyses, 104
- 5.3   System Design Criteria, 111
- 5.4   Developing System Improvements, 120
- 5.5   Continuing Use of the Model, 123
- 5.6   References, 123

Chapter 6   Extended-Period Simulation 125
    6.1   Introduction, 125
    6.2   Input Data for Hydraulic EPS Modeling, 126
    6.3   Extended-Period Simulation Setup, 129
    6.4   Extended-Period Model Calibration, 136
    6.5   Types of Extended-Period Simulation Analyses, 136
    6.6   Case Study: City of Fullerton, California, 143
    6.7   References, 146

Chapter 7   Water Quality Modeling 147
    7.1   Introduction, 147
    7.2   Need for Water Quality Modeling, 147
    7.3   Uses of Water Quality Modeling, 148
    7.4   Water Quality Modeling Techniques, 149
    7.5   Governing Principles of Water Quality Modeling, 149
    7.6   Reactions Within Pipes and Storage Tanks, 151
    7.7   Computational Methods, 151
    7.8   Data Requirements, 152
    7.9   Modeling of Multiple Species, 155
    7.10   Objectives of Water Quality Testing and Monitoring, 156
    7.11   Monitoring and Sampling Principles, 156
    7.12   Water Quality Surveys, 158
    7.13   Use of Historical Data, 162
    7.14   Tracer Studies, 162
    7.15   Tank and Reservoir Field Studies, 163
    7.16   Laboratory Kinetic Studies, 164
    7.17   Water Quality Modeling and Testing Case Study, 165
    7.18   References, 170

Chapter 8   Transient Analysis 173
    8.1   Synopsis, 173
    8.2   Introduction, 173
    8.3   Causes of Transients, 176
    8.4   Basic Pressure Wave Relations, 183
    8.5   Governing Equations, 189
    8.6   Numerical Solutions of Transients, 190
    8.7   Methods of Controlling Transients, 191
    8.8   Transient Modeling Considerations, 195
    8.9   Data Requirements, 197
    8.10   Summary, 199
    8.11   Glossary of Notations, 200
    8.12   References, 201

Chapter 9   Storage Tank Mixing and Water Age 203
    9.1   Introduction, 203
    9.2   Types of Tanks and Reservoirs, 204
    9.3   Background, 204
    9.4   Factors Affecting Water Quality , 205
    9.5   Types of Modeling, 207
    9.6   Model Verification, 212
    9.7   Strategies to Promote Mixing and Reduce Water Age, 215
    9.8   References, 217

Index, 219

List of Manuals, 231

Figures

Figure 1-1	The process from model build to analysis	3
Figure 2-1	Basic hydraulic model structures	21
Figure 2-2	Overview of a sustainable modeling process	23
Figure 2-3	Moody Diagram	36
Figure 2-4	GIS detail versus model detail	39
Figure 2-5	Pump curve	40
Figure 2-6	Nodes in close proximity	43
Figure 2-7	Pipe-split candidates	43
Figure 2-8	Intersecting pipes	44
Figure 2-9	Disconnected nodes	44
Figure 2-10	Parallel pipes	45
Figure 2-11	Disconnected pipes	45
Figure 2-12	Diurnal curve	52
Figure 3-1	Chart of pressure logger system pressures	68
Figure 3-2	Hand-held Pitot gauge	69
Figure 3-3	Hand-held Pitot gauge in use	70
Figure 3-4	Three general types of hydrant outlets	71
Figure 3-5	Diffuser with pressure logger	71
Figure 3-6	Traverse positions within a pipe	72
Figure 3-7	Typical velocity profiles at two different gauging points	72
Figure 3-8	Schematic of a strap-on flowmeter	72
Figure 3-9	Schematic of propeller flowmeter and picture of turbine flowmeter	73
Figure 3-10	Existing Venturi tube	74
Figure 3-11	Typical Venturi tube with manometer	75
Figure 3-12	Magnetic meter	75
Figure 3-13	Fire flow test configuration	77
Figure 3-14	Parallel hose method for head loss	79
Figure 3-15	Gauge method for head loss	79
Figure 3-16	Pump tests	81
Figure 3-17	Hydraulic gradient layout	82
Figure 3-18	Hydraulic gradient test	82
Figure 4-1	Steady-state flow calibration	97
Figure 4-2	Steady-state HGL calibration	97
Figure 4-3	EPS hourly peaking factors	100
Figure 4-4	EPS water level calibration	101
Figure 5-1	Idealized maximum day diurnal demand curve	105
Figure 5-2	Pump rating curve versus system head curve	115
Figure 5-3	Multiple pump rating curves	115
Figure 5-4	Pump efficiency curve	116
Figure 5-5	Equalization storage requirements for maximum day conditions	117
Figure 5-6	Storage allocation	118
Figure 5-7	Types of storage and elevation	119
Figure 6-1	Using SCADA data in EPS models	133

Figure 6-2	Examples of typical diurnal demand patterns for different use categories	134
Figure 6-3	Example system diurnal pattern and component patterns	135
Figure 6-4	Example utility demands versus time	137
Figure 6-5	Example of storage versus production for existing conditions, Case 1	139
Figure 6-6	Example of storage versus production with new production, Case 2	139
Figure 6-7	Example of storage versus production with loss of supply, Case 3	139
Figure 6-8	Example of storage versus production with fire fighting, Case 4	141
Figure 6-9	Example of storage versus production with pumping curtailment, Case 5	142
Figure 6-10	Example of storage versus production with supplemental power, Case 6	142
Figure 6-11	Location map for Fullerton case study	145
Figure 7-1	Illustration of water quality model equilibration	153
Figure 7-2	Example results from thermistor study showing temperature variation in tank	164
Figure 7-3	Protocol for chlorine decay bottle test	165
Figure 7-4	Skeletonized representation of Zone I of the North Marin Water District	167
Figure 7-5	Comparison of observed and modeled sodium concentrations in the North Marin Water District	168
Figure 7-6	Average percent of Stafford Lake water in the North Marin Water District	169
Figure 7-7	Comparison of observed and modeled chlorine residual in the North Marin Water District	170
Figure 8-1	Example steady-state transition after a period of rapid transients	176
Figure 8-2	Transient caused by pump shutdown	177
Figure 8-3	Transient caused by pump startup	178
Figure 8-4	Transient caused by rapid valve opening	179
Figure 8-5	Transient caused by rapid valve closure	179
Figure 8-6a	Rupture caused by valve closure (Superaqueduct of Puerto Rico)	179
Figure 8-6b	Damaged pump bowl	180
Figure 8-6c	Broken air admission valve	180
Figure 8-7	Varying pipeline profiles	181
Figure 8-8	Network schematic	182
Figure 8-9	Pressure surge fluctuations (field measurements) following routine pump shutdown	183
Figure 8-10	Pressure wave propagation in a pipe	183
Figure 8-11	Effect of a pipe junction on a pressure wave	186
Figure 8-12	Condition at a control element before and after action	187
Figure 8-13	Wave propagation in a pipe section considering friction	189
Figure 8-14	Flywheels to be installed in a large pump station	193
Figure 8-15	Typical locations for various surge protection devices	195
Figure 8-16	Flowchart for surge control in water distribution systems	196
Figure 8-17	Representative valve closure characteristics	198
Figure 8-18	Typical pump four quadrant characteristics (Suter curve)	199
Figure 9-1	Schematic representation of various types of empirical models	210
Figure 9-2	Tank water age calculated by an empirical model assuming complete mixing	210
Figure 9-3	Effect of thermal differences for tall tank	213

Figure 9-4	Effect of thermal differences for short tank.	213
Figure 9-5	Effect of operational and design changes.	214
Figure 9-6	Water age distribution	214

This page intentionally blank.

Tables

Table 2-1	C-factor values for discrete pipe diameters	33
Table 2-2	Equivalent sand grain roughness for various pipe materials	34
Table 2-3	Typical minor loss coefficients	37
Table 2-4	Operation data required by facility/equipment type	54
Table 5-1	Typical model scenarios	106
Table 6-1	System physical parameters for extended-period simulation analysis	138
Table 8-1	Physical properties of common pipe materials	185
Table 9-1	Example modifications to improve tank mixing characteristics	215

This page intentionally blank.

Foreword

The Engineering Modeling and Applications Committee's (EMAC) mission is to assemble and disseminate information on the use of modeling, geographic information system (GIS), and data management in the design, analysis, operation, and protection of water system infrastructure. The committee was formed in 1982 as the Computer Assisted Design of Water Systems Committee and was renamed the Engineering Computer Applications Committee and eventually modified to its current name. The committee consists of volunteers, a liaison from the Engineering and Construction Division, and an AWWA staff advisor. EMAC develops programs for the AWWA Annual Conference and specialty conferences, manuals, and other documents.

The purpose of M32, *Computer Modeling of Water Distribution Systems*, is to share collective expertise on distribution system modeling so that it is better understood and applied more effectively to benefit water utilities and water customers. The manual is intended to be a basic level or primer reference manual to provide new to intermediate modelers with a basic foundation for water distribution system modeling. M32 is intended to take users through the modeling process from model development through calibration to system analysis. The manual has in-depth discussion on

- Model construction and development
- Field data collection and testing
- Model calibration
- Steady-state analysis
- Extended-period simulation
- Water quality analysis
- Transient analysis
- Tank mixing analysis

M32 is designed to help modelers use water models as effective tools to plan, design, operate, and improve water quality within their water distribution systems.

There have been many advancements in the computer modeling field, and together with emerging issues of the water industry, the main goal of the M32 manual update is to focus on key areas that face the current modeler and utility. Key objectives of the update have been to

- Reorganize the manual for better flow
- Change the manual to address recent changes in the water modeling industry
- Expand the manual to include key topics more relevant to today's modelers

The EMAC is responsible for updating the M32 manual and individuals on this committee have dedicated their time and energy to update the manual to better support the water industry.

This page intentionally blank.

Acknowledgments

The Engineering Modeling Applications Committee acknowledges these individuals for their persistence and dedication as standing subcommittee members assigned to take the lead for updating this manual:

Chair—Larado M. Robinson, P.E., Draper Aden Associates, Blacksburg, Va.
Jerry A. Edwards, P.E., IDModeling Inc., Kansas City, Mo.
Lindle D. Willnow, P.E., AECOM, Wakefield, Mass.

The Engineering Modeling Applications Committee would like to give a special thanks to Laura Jacobsen, committee chair, for her leadership and guidance during this monumental effort.

Authors

Chapter 1   Introduction to Distribution System Modeling
Sharavan V. Govindan, Bentley Systems Inc., Exton, Pa.
J. Erick Heath, P.E., Innovyze, Arcadia, Calif.

Chapter 2   Building and Preparing the Model
Dave A. Harrington, IDModeling Inc., Seattle, Wash.
Paul M. Hauffen, IDModeling Inc., Arcadia, Calif.
Jerry A. Edwards, P.E., IDModeling Inc., Kansas City, Mo.
Rajan Ray, Innovyze, Wakefield, R.I.
Patrick B. Moore, P.E., Bohannan Huston, Albuquerque, N.M.
Nicole M. Rice, Innovyze, Broomfield, Colo.

Chapter 3   Tests and Measurements
Megan G. Roberts, P.E., Hazen and Sawyer, P.C., Greensboro, N.C.
Larado M. Robinson, P.E., Draper Aden Associates, Blacksburg, Va.

Chapter 4   Hydraulic Calibration
Jerry A. Edwards, P.E., IDModeling Inc., Kansas City, Mo.
Nass Diallo, P.E., Las Vegas Valley Water District, Las Vegas, Nev.

Chapter 5   Steady-State Simulation
Scott A. Cole, P.E., Freese and Nichols Inc., Fort Worth, Texas
Jared D. Heller, P.E., Moorhead Public Service, Moorhead, Minn.
Thomas J. Welle, P.E., Apex Engineering Group Inc., Fargo, N.D.
Larado M. Robinson. P.E., Draper Aden Associates, Blacksburg, Va.

Chapter 6   Extended-Period Stimulation
Lindle D. Willnow, P.E., AECOM, Wakefield, Mass.

Chapter 7   Water Quality Modeling
Walter M. Grayman, P.E., Ph.D., W.M. Grayman Consulting Engineer, Cincinnati, Ohio
Lindle D. Willnow, P.E., AECOM, Wakefield, Mass.
Jerry A. Edwards, P.E., IDModeling Inc., Kansas City, Mo.

Chapter 8   Transient Analysis
    Paul F. Boulos, Ph.D., Innovyze, Broomfield, Colo.
    Delbert G. "Skip" Martin, P.E., CH2M Hill, Anchorage, Alaska

Chapter 9   Storage Tank Mixing and Water Age
    Ferdous Mahmood, P.E., Malcolm Pirnie / ARCADIS-US, Dallas, Texas
    Walter M. Grayman, P.E, Ph.D., W.M. Grayman Consulting Engineer, Cincinnati, Ohio

The authors would like to acknowledge the following individuals who provided editorial and technical comments and/or contributed in other ways:

    Xuehua Bai, P.E., Farnsworth Group Inc., Denver, Colo.
    Marie-Claude Besner, Ph.D., Ecole Polytechnique de Montreal, Montreal, Que., Canada
    Linda K. Bevis, P.E., San Antonio Water System, San Antonio, Texas
    Gregory A. Brazeau, P.E., Innovyze, Broomfield, Colo.
    Thomas W. Haster, P.E., Freese and Nichols Inc., Fort Worth, Texas
    Laura B. Jacobsen, P.E., Las Vegas Valley Water District, Las Vegas, Nev.
    Nabin Khanal, P.E., Malcolm Pirnie / ARCADIS-US, Arlington, Va.
    Kathleen M. Khyle Price, P.E., San Antonio Water System, San Antonio, Texas
    Christopher M. Parrish, P.E., American Water, St. Louis, Mo.
    Vasuthevan Ravisangar, Ph.D., P.E., CDM Smith, Atlanta, Ga.
    John E. Richardson, P.E., Ph.D., ARCADIS-US, Blue Hill, Maine
    Larado M. Robinson, P.E., Draper Aden Associates, Blacksburg, Va.
    Vanessa L. Speight, Ph.D., P.E., Latis Associates, Arlington, Va.
    Arnie Strasser, P.E., Denver Water, Denver, Colo.
    Thomas M. Walski, Ph.D., P.E., Bentley Systems Inc., Nanticoke, Pa.
    Z. Michael Wang, Ph.D., P.E., Hazen and Sawyer, P.C., Raleigh, N.C.
    Lindle D. Willnow, P.E., AECOM, Wakefield, Mass.
    Don J. Wood, Ph.D., University of Kentucky, Lexington, Ky.

The following individuals provided peer review of the entire manual. Their knowledge and efforts are gratefully appreciated:

    Antony M. Green, C.Eng, GL Industrial Services, Loughborough, England
    Frank Kurtz, P.E., American Water Works Association, Denver, Colo.
    Saša Tomić, P.E., Ph.D., HDR Engineering Inc., Manhattan, N.Y.
    Thomas M. Walski, Ph.D., P.E., Bentley Systems Inc., Nanticoke, Pa.
    Z. Michael Wang, Ph.D., P.E., Hazen and Sawyer, P.C., Raleigh, N.C.

This manual was reviewed and approved by the Engineering Modeling Applications Committee that included the following personnel at the time of approval:

    Elio F. Arniella, P.E., Halcrow Inc., Marietta, Ga.
    Xuehua Bai, P.E., Farnsworth Group Inc., Denver, Colo.
    Paul F. Boulos, Ph.D., Innovyze, Broomfield, Colo.
    Michael T. Brown, P.E., Gannett Fleming Inc., Harrisburg, Pa.
    Scott A. Cole, P.E., Freese and Nichols Inc., Fort Worth, Texas
    Daniel Creegan, Anchorage, Alaska
    Nass Diallo, P.E., Las Vegas Valley Water District, Las Vegas, Nev.
    Antony M. Green, C.Eng, GL Industrial Services, Loughborough, England

Gary Griffiths, Bentley Systems Inc., Exton, Pa.
Eleni Hailu, Los Angeles County Waterworks, Alhambra, Calif.
Dave A. Harrington, IDModeling Inc., Seattle, Wash.
Thomas W. Haster, P.E., Freese and Nichols Inc., Fort Worth, Texas
Jared D. Heller, P.E., Moorhead Public Service, Moorhead, Minn.
Paul Hlavinka, Gaithersburg, Md.
Paul H. Hsiung, Innovyze, Shawnee Mission, Kan.
Chair—Laura B. Jacobsen, P.E., Las Vegas Valley Water District, Las Vegas, Nev.
Joel G. Johnson, P.E., GL Noble Denton, Houston, Texas
Pranam Joshi, NJBSOFT LLC, Phoenix, Ariz.
Jonathan C. Keck, Ph.D., California Water Service Company, San Jose, Calif.
Carrie L. Knatz, CDM Smith, Carlsbad, Calif.
Douglas J. Lane, MWH, Bellevue, Wash.
Kevin T. Laptos, P.E., Black & Veatch Corporation, Charlotte, N.C.
Foster McMasters, P.E., AECOM, Cleveland, Ohio
Tina Murphy, P.E., HNTB Corporation, Indianapolis, Ind.
Patrick F. Parault, P.E., Malcolm Pirnie, Long Island City, N.Y.
Christopher M. Parrish, P.E., American Water, St. Louis, Mo.
Larado M. Robinson, P.E., Draper Aden Associates, Blacksburg, Va.
Adam Rose, P.E., PMP, Alan Plummer Associates Inc., Fort Worth, Texas
Jeffrey Eric Rosenlund, HKM Engineering, Sheridan, Wyo.
Michael Rosh, Bentley Systems Inc., Sayre, Pa.
Thomas E. Waters Jr., O'Brien and Gere, Louisville, Ky.
Paul West, Newfields, Atlanta, Ga.
Dr. Jian Yang, American Water, Voorhees, N.J.

This manual was also reviewed and approved by the Engineering and Construction Division that included the following personnel at the time of approval:

Mike Elliott, Stearns & Wheler GHD, Cazenovia, N.Y.
Gary L. Hoffman, ARCADIS-US, Cleveland, Ohio
Richard C. Hope, AECOM, Stevens Point, Wisc.
Laura B. Jacobsen, P.E., Las Vegas Valley Water District, Las Vegas, Nev.
David S. Koch, Black & Veatch, Grand Rapids, Mich.
Marlay B. Price, Gannett Fleming Inc., Versailles, Ohio
Michael Stuhr, Portland Water Bureau, Portland, Ore.
Ian P.D. Wright, P.E., Associated Engineering of Canada, Calgary, Alb.

This page intentionally blank.

AWWA MANUAL M32

Chapter 1

Introduction to Distribution System Modeling

1.1. INTRODUCTION

Water utilities seek to provide customers with a safe, reliable, continuous supply of high-quality water while minimizing costs. This water is often delivered through complex distribution systems involving miles of pipe and often incorporating numerous pumps, regulating valves, and storage reservoirs. The performance of these systems is often difficult to understand not only because of their physical size and complexity, but also because of the large amount of system information and data needed to fully grasp how they function. Sometimes, key pieces of information needed to understand a system are missing. One tool that has evolved over time to help water system designers, operators, and managers meet their goals of delivering safe, reliable water supply at a low cost is distribution system modeling.

Distribution system modeling involves the use of a computer model to predict the performance of the system to solve a wide variety of design, operational, and water quality problems. For example, a computer model can predict pressures and flows within a water system to evaluate a design and compare system performance against design standards. Models are also used in operational studies to solve problems, such as evaluating storage capacity, investigating control schemes, and finding ways to deliver water under difficult operating scenarios. Water quality models are used to compute water age, track disinfectant residuals, and reduce disinfection by-products in a distribution system.

Distribution system modeling began with the advent of analog computers and has evolved over time as computer software and hardware advanced to become more powerful and easier to use. Models containing thousands of pipes can now be created and used on readily available personal computers. Models that once took hours to run are now run in seconds or fractions of a second. Originally, models were used only to evaluate system hydraulic grades (and resulting system pressures) and flows. Although this capability remains at the very core of all water distribution modeling work, hydraulic models are now used to calculate water quality, energy costs, and optimize system operations just to name a few.

Historically, model building was an expensive and labor-intensive process. Now models can effectively share data using geographic information system (GIS), computer-aided design and drafting (CADD) systems, supervisory control and data acquisition (SCADA) systems, customer information system (CIS), computerized maintenance management system (CMMS), and asset management system (AMS) software, thus reducing the effort needed to create, update, and maintain a model. Information obtained from a model study can be filtered, organized, and presented in a variety of graphical and nongraphical ways so results can be more easily understood. These advances in technology have broadened the uses of distribution system modeling from just an infrastructure-planning tool to an integrated system used to improve operations, to analyze water quality, and to plan water system security improvements.

1.2. PURPOSE OF THE MANUAL

This manual (M32) was developed by the Engineering Modeling Applications Committee of the American Water Works Association (AWWA). The purpose of this manual is to share collective expertise on distribution system modeling so that it is better understood and applied more effectively to benefit water utilities and water customers everywhere. The manual is intended to be a basic level or primer reference manual to provide new to intermediate modelers with a basic foundation for water distribution system modeling. The manual is intended to take users through the modeling process from model development to system analysis as shown in Figure 1-1. The manual has in-depth discussion on

- Model construction and development
- Field data collection and testing
- Model calibration
- Steady-state analysis
- Extended-period simulation
- Water quality analysis
- Transient analysis
- Tank mixing analysis

M32 is designed to help modelers use water models as effective tools to plan, design, operate, and improve water quality within their water distribution systems.

1.3. HISTORICAL DEVELOPMENT OF DISTRIBUTION SYSTEM MODELING

1.3.1. Pre-1970

Manual engineering calculations for small-pipe systems were used through the 1960s. The Hardy-Cross method was sufficient for single-loop systems, but, without the aid of a computer, this method was impractical for systems having several loops. In 1950,

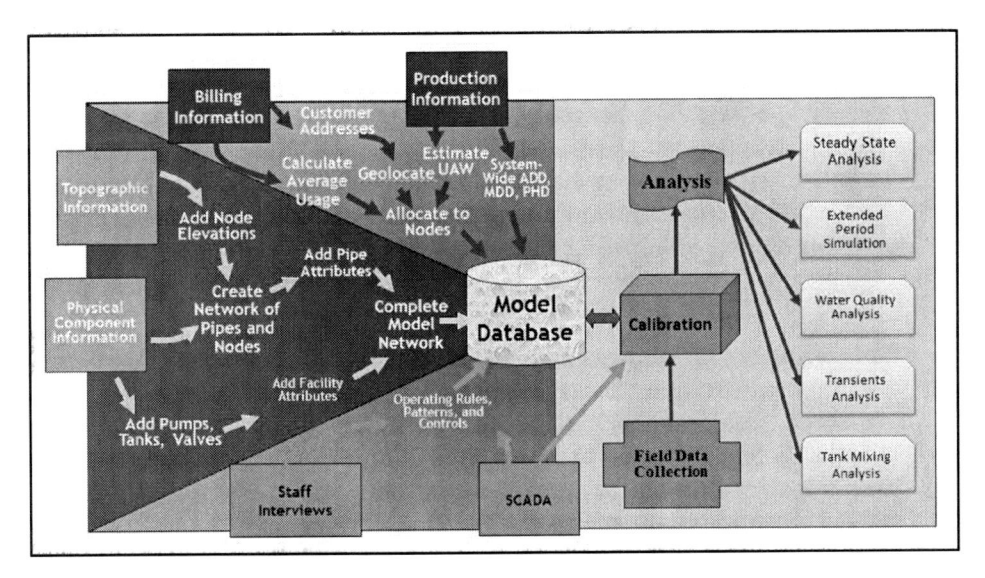

Figure 1-1　The process from model build to analysis

McIlroy simulated the behavior of water distribution systems using electronic circuitry. However, these physical models were large, expensive, and difficult to use. Digital computer models appeared in the 1960s. The FORTRAN programming language was primarily used to develop various models that became available to practicing engineers.

1.3.2. 1970–1990

Modeling software was developed with a variety of features, including extended period simulation and water quality analysis. Graphical user interface capabilities were incorporated for drawing the system and displaying output. Software "packages" were marketed that contained several compatible modular components. Some packages used other specialized software for data entry, display, and reporting of results.

1.3.3. 1990s

The 1990s experienced exponential growth of system modeling capabilities. EPANET, a modeling program developed by the US Environmental Protection Agency (USEPA) to support ongoing research, was made available to the public. Some vendors have taken the EPANET model and added an improved user interface. Software packages were designed to be compatible with other standard software packages, such as AutoCAD by Autodesk® and ArcView by ESRI®. These software packages were developed to integrate with various spreadsheet and database software to improve editing and drawing functions. The result was a familiar user interface and the ability to utilize existing software rather than having to create and update new software. Water quality modeling and extended period simulation became standard features within modeling software packages.

1.3.4. 2000s

Software packages were developed to work more effectively and sometimes were made to run within GIS software environments in response to the adaption of GIS as the asset data management platform by many water utilities. The use of GIS had become more common, and the quality of data was improving, which significantly reduced the effort required to develop models. Automation tools became available for optimizing design and aiding the calibration process. Distribution system security concerns resulted in studies to develop emergency response plans to evaluate the impacts various disasters might have on water distribution systems. Models were used increasingly for water quality analyses, such as evaluating water age and constituent concentrations. USEPA also allowed hydraulic modeling as a means of determining preferred locations for water quality monitoring sites necessary to meet regulatory requirements.

1.3.5. Present

The availability of tremendous amounts of information from various data sources, real-time instrumentation, faster computers, high network bandwidth, and well-funded commercial and academic research and development expanded the applications of water modeling software to newer areas. Water leakage detection and management have helped reduce the amount of nonrevenue water. Pump scheduling optimization tools aid in energy management.

Additionally, transient analysis studies have resulted in designing for transient conditions to reduce potential pipe breaks and water contamination. As concern for water quality within water storage reservoirs increases, modeling of these facilities has become more important to assure proper mixing and meet more stringent quality regulations. Unidirectional flushing programs are being developed to enhance system water quality. Fire flow analysis and capital infrastructure improvement are now among the most popular uses for water modeling. Water quality and system security planning continue to be critical applications because of increased regulations and national threat levels. Software packages now have the sophistication to perform network calibrations as well as help determine where closed pipes exist within the system. The current water models are helping to sustain infrastructure in challenging economic times by allowing system owners to quantify and reduce operating hours, water loss, pipe breaks, energy usage, and a host of other related costs.

1.4. DISTRIBUTION SYSTEM MODELING APPLICATIONS

1.4.1. Benefits of Computer Modeling

To solve hydraulic system problems, there must be one equation for each pipe, pump, and valve, or for each junction, depending on the method used to solve for the unknowns in the hydraulic calculations. The number of equations that must be set up and solved in a system hydraulics problem is very large, even for the most basic water distribution system. The value of a computer model is that tedious calculations are performed much more quickly and more accurately than manual calculations. In addition, the computer is an effective means of managing large amounts of data necessary to analyze a water distribution system. By using computer models, rather than focusing on the procedural mechanics of solving system equations, decision makers can focus more on communicating modeling results and formulating and comparing system design alternatives. Computer models of water distribution systems are not an end in themselves but are tools to help managers, engineers, planners, and operations staff. When properly implemented, models become an integral part of the decision-making process for planning, design, and operation of water distribution systems. Engineers and operators of a water system are still ultimately responsible for decisions based on the results that computer models provide.

Distribution system modeling software generally falls into four categories of application: planning, engineering design, system operations, and water quality improvement.

1.4.2. Planning

A primary planning application of distribution system analysis software is used for assisting in the development of long-range capital improvement plans, which include scheduling, staging, sizing, and establishing preliminary routing and location of future facilities. Other applications include planning for water main rehabilitation and system improvement. Rehabilitation plans identify and prioritize mains that need to be cleaned and/or lined. Distribution system improvement plans identify where installation of new mains, storage facilities, and pump stations are necessary to keep pace with growth and/or new utility standards and regulations. The following are examples of several specific system analysis planning applications.

1.4.2.1. Capital Improvement Program. Water utilities usually have a master plan identifying future capital improvements necessary to respond to projected community growth and replacement of aging infrastructure. These plans typically extend from 5 up to 20 years or more. A model is usually used to identify and schedule these long-term capital improvements.

1.4.2.2. Conservation Impact Studies. Water conservation is desirable for most communities to stretch limited water supplies or to reduce water use so that some capital improvements can be delayed or eliminated. A model is useful to apply expected effects of various conservation measures onto projected system demands to evaluate the potential for success.

1.4.2.3. Water Main Rehabilitation Program. A model is used to identify specific water mains that tend to bottleneck the system, either due to increased demands or tuberculated pipes. The model is used to determine the potential hydraulic effects of replacing, upsizing, or rehabilitating aging mains to evaluate the effectiveness of various alternatives.

1.4.2.4. Reservoir Siting. A reservoir should be sited in a location that optimizes water turnover, that effectively meets peak demands, and that can recharge efficiently during off-peak demand periods. The model is used to explore these scenarios to fine-tune preferred hydraulic solutions.

1.4.3. Engineering Design

Engineering design applications include the sizing of various types of facilities including pipelines, pump stations, pressure regulating valves, tanks, and reservoirs. These facilities are sized using pressures and flows that result from distribution systems modeling. In addition, system performance can be analyzed under fire flow conditions and adjustments made to meet fire demand. The following are examples of engineering design problems that are solved using computer models.

1.4.3.1. Fire Flow Studies. The model is used to simulate fire flow demands at hydrant locations throughout a locality to determine how much water can be delivered to specific fire hydrants within the prescribed fire-flow pressure constraints. Where deficiencies are discovered, distribution system improvements with main reinforcements or looping can also be evaluated with the model. These studies are also used to demonstrate compliance with fire protection standards.

1.4.3.2. Valve Sizing. A distribution system often has pressure-regulating or pressure-sustaining valves to direct flow to different hydraulic zones. Distribution systems may also have throttle valves to direct flow within a zone to different reservoirs or storage locations. The model is used to determine how much flow is required through these valves so the valves are sized appropriately.

1.4.3.3. Reservoir Sizing. Reservoirs are often sized by estimating the total diurnal flow, fire flow, and emergency storage requirements within a particular zone. However, reservoir capacity should also consider the rate of water delivery to the reservoir location and the size of the distribution area. A model is useful to evaluate inflows and outflows to a reservoir to determine an optimal size for a particular location and/or to specify other improvements so that the preferred reservoir site is adequately served by transmission mains and pumping stations.

1.4.3.4. Pump Station/Pump Sizing. Models are used to calculate system curves of distribution systems so that pumps are selected that provide the necessary flow and head. Once sized, the proposed pumps are then used in the model under a wide range of system settings to determine how well they perform under various operating conditions.

1.4.3.5. Calculation of Pressure and Flow at Particular Locations. A water distribution system must provide adequate amounts of water at pressures within a range typically specified by standards used by water utility. A model's core functionality is hydraulic grade and flow calculations. Models are used to predict pressures under specific demand conditions and under a wide variety of scenarios to identify low pressures and to select infrastructure that will improve flow or pressure deficiencies.

1.4.3.6. Zone Boundary Selection. Most water distribution systems deliver water to customers located at a range of different elevations. Distribution systems are separated into pressure zones that follow consistent elevation contours to keep pressures within reasonable ranges. Models are useful to evaluate potential zone boundaries and to determine the adequacy of infrastructure delivering water to each zone.

1.4.4. System Operations

Applications for operations include assisting in the development of operating parameters and strategies, operator training programs, and system troubleshooting guidelines. Operating strategies may be driven by emergency conditions, energy management, water availability, and so on. For example, contingency plans are developed in the event of a key facility component failure, such as a pump station failure. Distribution system modeling is also used to develop operational strategies for energy management and water quality guidelines. Strategies for shifting supply between treatment plants are developed to determine the most efficient use of available water. Optimizing these strategies results in efficient use of pipeline capacities, tank levels, and required treatment plant production, among other things.

1.4.4.1. Personnel Training. Models are used for training personnel that operate the distribution system. System operators can experiment with the model to determine how the system responds to changes in operating parameters and conditions.

1.4.4.2. Troubleshooting. Models are used to troubleshoot potential causes of various problems, such as low pressure, water circulation issues, and events that would otherwise be inexplicable.

1.4.4.3. Water Loss Calculations. In the event of a major main break, the model is used to estimate the amount of water lost through the break as may be required for damage assessments.

1.4.4.4. Emergency Operations Scenarios. Water distribution systems often have critical components; if the components fail, water delivery is interrupted. A model is useful to evaluate the potential impact of a failure and to devise means of reducing the damage or impact of a critical component failure.

1.4.4.5. Source Management. Water treatment plants are sometimes taken out of service for repairs or because the water supply is unavailable for a time. Furthermore, the quality of water at one source may be better at certain times of the year, so the use of the high-quality source can be maximized. The model is useful to devise operating scenarios that best utilize multiple water sources to achieve desired system objectives.

1.4.4.6. Model Calibration. Model calibration is typically thought of as a step in developing a useful model. However, the calibration process is useful to operations staff in discovering anomalies in the distribution system, such as closed valves, tuberculated pipes, leaks, or false or incomplete infrastructure data. This information, once discovered through the calibration process, can explain operational difficulties and identify distribution system problems requiring the development of solutions to resolve and improve system operation.

1.4.4.7. Main Flushing Programs. A hydraulic model is an excellent tool for developing a main flushing program. The model is useful to identify flow paths in the distribution systems so that appropriate flushing locations and sequences can be established.

1.4.4.8. Area Isolation. Water utilities frequently need to isolate a specific area for maintenance or other work. Often, it is helpful to identify those customers whose service will be interrupted by the isolation event. In addition, those planning the event need to know which valves to close in order to minimize impacts of the isolation. Hydraulic models are tools used to accomplish this task.

1.4.4.9. Energy Cost Management. Nationwide, about 4 percent of US power generation is used for water supply and treatment. Electricity represents approximately 75 percent of the cost of municipal water processing and distribution according to the Department of Energy (DOE). Of this, pumping typically accounts for 75 to 80 percent of the power consumed by water utilities. With energy costs being such a high percentage of the overall costs, utilities are trying to find ways to reduce their overall energy consumption. Many of today's models have features to help quantify energy consumption and its related costs, and also have the capability to assist with optimizing pump operation to help reduce electrical usage.

1.4.5. Water Quality Improvement

Water quality regulations in the United States are limiting the level of disinfection by-products (DBPs) in water distribution systems. Standards and expectations for water quality have increased the demand for water quality analysis in these systems. Following are examples of how distribution system modeling is used to improve water quality.

1.4.5.1. Constituent Tracking. If a contaminant enters the distribution system through a treatment plant, well, reservoir, or other location, it can quickly spread throughout the distribution system, affecting water quality of consumers receiving water from that source. The contaminated water may also mix with water from other sources. A model can be used to predict contaminant levels and zones of influence within the distribution system. Customers potentially affected by the contaminant can be identified, and portions of the distribution system that need to be flushed can be identified.

1.4.5.2. Water Source/Age Tracking. Water age is another important water quality parameter in a distribution system. Chlorine levels decay over time, increasing the tendency for DBP levels to increase as chlorine reacts with organic compounds in the water. To maintain water quality, water utilities are striving to minimize water age. This is done by ensuring that water in reservoirs and storage facilities turns over regularly to minimize stagnation and dead ends. When multiple sources serve an area, distribution system modeling helps devise operating strategies to reduce water age where possible.

1.4.5.3. Chlorine Levels. A model is used to predict chlorine decay in a distribution system. This is useful to determine chlorination levels at the treatment plant and to select rechlorination sites where necessary to boost chlorine levels.

1.4.5.4. Water Quality Monitoring Locations. Parts of recent water quality regulations proposed by USEPA include selecting establishing sites within a system to place permanent DBP level monitors and demonstrate compliance with federal regulations. A water model can help identify the most appropriate locations for these water quality monitors.

1.5. HYDRAULIC MODELS

A computer model is composed of two parts: a database and a computer program. The database contains information that describes the infrastructure, demands, and operational characteristics of the system. The computer program solves a set of energy, continuity, transport, or optimization equations to solve for pressure flows, tank levels, valve position, pump status, water age, or water chemical concentrations. The computer program also aids in creating and maintaining the database and presents model results in graphical and tabular forms.

1.5.1. Model Data

A hydraulic model consists of two types of elements: links and nodes. Depending on the model, the major components such as pipes, junctions, pumps, tanks, hydrants, and valves are represented by either a link or a node. Other components such as customer points, rupture disks, or orifices may also be available. The model is a valuable asset to the user and is the result of substantial effort in data collection, entry, and quality control. Such an investment should be protected by maintaining its value over time. One of the keys to maintaining the hydraulic model's value is making sure the model is updated with appropriate changes in infrastructure components, system demands, or operating parameters. To maintain confidence in the results of the model, data within the model should best represent the current system configuration.

1.5.2. Modeling Software

1.5.2.1. Modeling Equations. At the heart of water distribution system modeling software is a system solution algorithm, which solves hydraulic and water quality equations. Depending on the problem, there are several kinds of equations that are solved.

Continuity equations enforce the law of the conservation of mass by keeping track of water flow, making certain that the total flow into a node equals the flow out, plus or minus any changes in storage or demand.

Energy equations enforce the law of the conservation of energy by accounting for energy loss caused by friction in pipes, valves, and fittings, and energy added by pumps. The resulting total energy balance around each closed loop in the system should be zero.

Transport equations in water quality models account for the movement of substances (pollutants or tracers) through a distribution system and any reactions that may occur.

1.5.2.2. Water Model Analysis

1.5.2.2.1. Steady-State Analyses. A steady-state analysis provides a "snapshot" of pipe system conditions at any instant in time. Steady-state analyses are typically used to evaluate maximum day, peak hour, and fire flow conditions.

1.5.2.2.2. Extended-Period Simulation. An extended-period simulation is a series of steady-state simulations at specified intervals performed over a specified time period. This capability may be used, for instance, to model the operation of a water system over a 24-hour period with an analysis run for each hour. Such a simulation is useful in modeling variations in demand, reservoir operations, water quality, and water transfers through transmission pipelines. Extended-period simulation requires that the system package model flow and pressure switches incorporate demand diurnal patterns for nodes and allow for varying tank configurations.

1.5.2.3. Specialized Model Analysis

1.5.2.3.1. Automated Fire Flow Calculation. Some distribution system modeling packages automatically calculate the available fire flow at each node. These calculations are useful in identifying areas having weak firefighting capability.

1.5.2.3.2. Water Quality. Utilities are increasingly interested in modeling the water quality within a distribution system, particularly the decay of chlorine residual and water age. The ability to perform water quality analysis should be a standard part of any modeling package.

1.5.2.3.3. Transients. Transient pressures (water hammer) can cause pipe breaks, contamination, joints to shift and leak, collapse of pipes, and other serious damage to water distribution networks. A transient event is caused by a sudden change in flow velocity that can be created by a valve that is closed quickly, pump failure or a pump simply shutting down, a mishandled fire hydrant, and so on. After identifying a transient condition (valve closure), water-transient analysis software is able to identify where a transient event is likely to happen and evaluate multiple transient protection devices that can help mitigate or prevent damage.

1.5.2.3.4. Energy Analysis. Energy analysis, available in most commercial software, can help identify inefficient pumps and determine better operational strategies. Energy costs are a significant portion of the total expense for most utilities. Some electricity providers have variable rates and electricity costs can vary depending on when the pumps are operated. Energy consumption and energy cost can be quantified through this type of analysis.

1.5.2.4. Model Functionality

1.5.2.4.1. Scenario Generation. Distribution systems with any level of complexity are modeled more easily by applying various combinations of demands, facilities, and operating parameters, such as regulating valve and pumping unit settings. System modeling packages may allow variation and combination of these three types of data in a simulation by keeping them in separate databases for specific or combined model access.

1.5.2.4.2. Selective Reporting of Results. The user is able to specify the results to be reported in tabular form so pages of output need not be generated after each run when the user may only be interested in results in one specific area of the system. This user-specific reporting saves hard disk space and paper while speeding the user's review time.

1.5.2.4.3. Data Management. Modelers are able to export and import data and model results to and from other applications, such as spreadsheets, databases, and GIS systems. These capabilities are widely available and are an important part of any modeling package.

1.5.3. Related Software Systems

Information management trends within utilities are moving toward better information sharing so that decision makers can have as many information resources as possible to make the best decisions. This is often done by using both common databases and files that are shared by a variety of software applications. Distribution system modeling uses a wide variety of information about physical assets, customers, billing data, geographical information, and operational information. Furthermore, modeling activities can benefit a variety of groups within the utility, strengthening the ability to communicate and share information. Brief descriptions of software systems, information systems, and/or corporate-wide databases that are in some way related to distribution system modeling are listed below.

1.5.3.1. Geographic Information System (GIS). A GIS stores and displays geographically referenced information, i.e., information that is easily understood through map display. Spatial relationships between entities are significant for most information stored in GIS. GIS has potential to store vast amounts of information useful for system analysis, including pipe assets, customer meter locations, zoning and land parcel data, aerial photography or other land bases, street locations, digital terrain models (DTMs), digital elevation models (DEMs), and jurisdictional boundaries.

Information in GIS is saved in file formats that most modeling software packages can read. Alternatively, information in a GIS database is translated into a format that can be imported into the model database. Some GIS land-base information and other data layers are displayed directly within some modeling packages. The usefulness of GIS pipe data is often dependent on the way the information is collected and stored in the GIS database. Pipe data in GIS are most useful if the topology and connectivity are already established in the GIS database.

1.5.3.2. Computer-Aided Design and Drafting (CADD). CADD systems are used to manage maps of water distribution systems and are, therefore, a source of pipe information that can be transferred to the model. In addition, CADD systems are useful as a means to display model information and results.

1.5.3.3. Supervisory Control and Data Acquisition (SCADA). SCADA systems are used to remotely control the operation of pump stations, valves, and other system infrastructure. They are also useful to collect data such as pressures, flows, reservoir levels, valve positions, pump status and speed, chlorine levels, and other information useful in monitoring the system. This information is collected at regular intervals and stored for extended periods of time. SCADA is a good source of operational information, as well as calibration data. SCADA data are also used to define the starting point for operational analyses by using the data to define boundary conditions placed in the model. SCADA data usually do not go into the model directly. An interface is usually required that could be as complex as a custom software routine or as simple as importing SCADA data into a spreadsheet via a comma-separated values (CSV) file and formatting the data for import into the model.

1.5.3.4. Customer Information System (CIS). CIS is useful to develop demands based on customer water-use information. Typically, average annual water usage and customer rate classes are extracted from CIS and then linked to the model via GIS, modeling, or customized tools. The specifics of how CIS data can be linked to the model are highly dependent on the data and software used by the utility.

1.5.3.5. Laboratory Information Management System (LIMS). LIMS instruments, protocols, standards, and software are vital components in the monitoring of water quality and ensuring safety and regulatory compliance. Laboratories continuously sample water from various locations and are expected to accurately report any possible water quality issues immediately. Water models can assist with developing flushing routines, chlorination, contaminant tracking, water age, and independently locating the likely source of these water issues.

1.5.3.6. Computerized Maintenance Management System (CMMS). CMMS can act as an access point to post and retrieve system information and costs for operations, maintenance, and rehabilitation. Typically, CMMS is used to store asset inventory, condition assessments, parts inventory, preventative maintenance activities, service requests, and work orders, all of which can be linked to the object by location and imported into models for a variety of applications.

1.5.3.7. Asset Management System (AMS). Similar to CMMS, asset management software is becoming more prevalent as a strategic planning tool for water utilities. This management tool combines long-range planning, life-cycle costing, proactive operations and maintenance, and capital replacement plans based on cost-benefit analyses. These applications are used to meet a wide range of management challenges for water utilities. Water modeling is an integral part of the long-range, decision-making process and can be incorporated into a comprehensive asset management program for achieving system sustainability. AMS seeks to provide informed, timely, and cost-effective decision making for both day-to-day operations and long-term planning. Many of these programs also enable users to integrate many different applications and data sets already in use, including GIS, CMMS, and other work orders or field data systems (SCADA) as well as hydraulic modeling/analysis software.

1.6. DISTRIBUTION SYSTEM MODELING WITHIN THE UTILITY

A successful distribution system modeling program functions best with a team of individuals who can perform system analysis and effectively provide system modeling results to decision makers within the utility. Issues that often need to be addressed when implementing a modeling program are outlined in this section.

1.6.1. In-House Modeling Versus Outside Consultants

The utility should decide whether the model will be developed and maintained by the utility or by an outside consultant.

Usually, a utility understands its system very well and has easy access to model-related information. However, a utility may not have the expertise or resources to develop and maintain the model. Some utilities construct and run their own models, while others hire outside consultants to perform some or all of the modeling work. A model owner who is committed to maintaining the model and to developing expertise is essential for an in-house modeling program. If consultants perform the modeling, ownership rights of the model and data should be clearly delineated in a contract. Regardless of who performs the modeling work, a long-term commitment to maintaining the model and retaining experts to maintain and utilize the model regularly are necessary components of success.

1.6.2. One-Time Versus Long-Term Use

Many decisions made during model development depend on whether the model is used for solving a short-lived problem or for periodic use over an extended period of time. If the model is used for a specific problem, questions regarding the level of detail are easily answered based on satisfying the needs of the problem. If the model is to be used for many purposes, the model should be developed to serve the most demanding applications and simplified, if necessary, for other applications. For example, a model developed to assist in master planning may not contain enough detail for determining available fire flows in subdivisions or for water quality modeling within the system.

Decisions must be made on whether this level of detail should be included in the model or added later, if required. Currently, the power of computers makes it practical to use a complex model for even a simple task. However, many software packages have tools to simplify a model when necessary.

1.6.3. Model Developer Versus Decision Maker

There are two distinct roles in model development: the role of the modeler and the role of the decision maker. The modeler is the person responsible for initial development and running the model. The decision maker is a professional engineer or licensed operator who interprets and makes decisions based on outputs from the model. These two roles could be filled by one individual, two individuals, or even two different companies. The key element is that the decision maker must be satisfied that the modeler has indeed developed a system model that is adequate for the problem or problems being considered. Calibration and sensitivity analyses are two methods available to the decision maker to assure that the model is adequate for the intended purpose.

1.6.4. Modelers Versus Rest of Utility

It is essential that modeling, whether performed in-house or by a consultant, be done with the awareness and cooperation of the rest of the utility. While some individuals may serve as the experts on the model, all interested parties should have input in model development, understand the capabilities and limitations of the model, and appreciate the important role of modeling in decision making. Modeling should be done with thorough consideration of utility operations. For example, utility operators have great insights into the operation of a system, as well as its physical limitations. By working with the operations staff, the modeler can incorporate operators' insights into the model, and operations staff can become sufficiently comfortable with the model to trust its results.

1.6.5. Skeletonized Versus All-Mains Models

All-mains models are becoming increasingly popular with the availability of faster computers and comprehensive GIS data. Choosing to use an all-mains model versus a skeletonized model depends completely on the intended use of the model. All-mains models are better suited for computing fire flow and water quality, while skeletonized models offer quicker results to understand overall system demand and capacity over time. Another popular approach is to model all mains greater than six or eight inches (15.3–20.3 cm) in diameter. Several modelers maintain both skeletonized and all-mains models and choose between them based on the application type.

1.7. TRENDS

Several significant trends in system modeling have become apparent through network modeling surveys, presentations and discussions at conferences, and *Journal AWWA* articles. Some trends are now well established while others are still in their infancy.

1.7.1. Common Databases

In some utilities, computer systems are organized around a type of architecture where common databases are shared by many applications. In such a framework, a distribution systems modeling package extracts data from a large *enterprise* database that is shared by many work groups. A database system that supports multiple work groups gets input from key stakeholders and integrates interdepartmental work flows. Asset management, billing, customer service, work management, facilities, engineering, GIS, CADD, hydraulic modeling, document management, permitting, water quality testing, and other work groups have information that can be linked geospatially. An enterprise database will reduce duplication of efforts, inaccuracies, and outdated information.

1.7.2. Model Sophistication

Model surveys reveal a strong trend toward all-mains models and extended-period simulation to examine water quality issues in distribution systems. The availability of faster processors, memory efficient operating systems, bandwidth, and larger storage capacities has removed several system limitations. The increase in technical resources allocated for modeling, software's ability to connect with multiple data sources, data loggers, and easier model building tools have expanded the power of models.

1.7.3. Demand Allocation

Billing meter information, land use, and population data are used in model loading. This type of information is generally available from GIS and CIS. There is increased interest in demand allocation based on future needs such as estimated land use or population growth in combination with demand density. Demand patterns are generated for day of the week, months, special events, and seasonal variations. This helps with evaluating the system for a wider variety of scenarios.

1.7.4. Information Systems Integration

The near term will see a far greater integration of multiple information and data systems such as GIS, CADD, SCADA, CIS, CMMS, and AMS. This will result in a more accurate model because of demand and distribution network fidelity. In addition, this will facilitate optimization of operational workflows and the road map for future data needs.

1.7.5. Energy Analyses

Water models are increasingly used to improve energy efficiency through better pump scheduling and operations. As energy rates increase, there is greater emphasis on reducing energy consumption and resultant energy costs. Models are used to help identify pumps that need maintenance or replacement or evaluate new energy-efficient pumps from multiple manufacturers. In addition, they are used to evaluate pump combinations that work well, and choose between variable speed and constant speed pumps.

1.7.6. Automated Meter Reading (AMR)

AMR is used for automatically collecting data including demand and status from water meters. This data are then transferred to a database for billing and analysis. Utility providers are able to collect this information in real time (wired or wireless) without having to travel to each physical location to read a meter.

1.7.7. Infrastructure Upgrade (maintenance and rehabilitation versus replacement)

Models are helping with the capital improvement financial analysis and also identifying the critical segments in a water distribution network. Models can help reduce and optimize the overall life-cycle costs of pipes by offering alternatives and can compare potential solutions by cost/benefit ratio, pressure/flow service goals, and available budget.

1.7.8. Transient Analysis

Increased awareness that pipe breaks are not necessarily caused by aging infrastructure and that installing a water hammer protection device is cheaper than replacing a pipe network has caused an increase in transient analysis. There is an improved effort in educating the field crew, contractors, and fire departments that closing a valve or hydrant quickly can result in a water hammer at a nearby location. These days, being able to quantify the magnitude of the transients has helped size and identify the right type of hammer protection device.

1.7.9. Water Quality Analyses

Many system modeling packages have the ability to model water quality parameters in a pipe network and in reservoirs. Utilities find this valuable in response to new water quality regulations and to heightened public interest in water quality. In addition to being able to model water age, source tracing, and constituent concentration (chlorine, chloramine, etc.) in a water distribution network, some models can perform disinfection by-product formation analyses.

1.7.10. Tank Mixing

Driven by the need to improve water quality and reduce storage times, tank mixing is gaining popularity. Tanks can be mixed properly through better design of inflow and outflow ports and control devices, mechanical mixers, and improved operational strategies.

1.7.11. Water Security

System design and operation decisions are increasingly based on assessing risks and vulnerabilities of water supply systems. The ability to track contaminants and isolate and flush potentially affected areas is increasingly important to safe, secure water systems. Water models are used to evaluate multiple scenarios of contamination at susceptible and publicly accessible locations of distribution networks. This aids in the placement of water quality sensors and reduces system vulnerabilities.

1.7.12. Emergency Planning

Emergency planning can range from main breaks to contamination to power failure. Planning ahead may help mitigate potential disasters. Planning protocols including the US government's national incident management system (NIMS) and several state emergency management systems are available to assist with structured emergency planning. System modeling will play an integral part in long-term emergency planning efforts.

1.7.13. Real-Time Modeling

In the past, distribution system modeling packages were typically too slow and unwieldy for system operators to use in generating operating strategies and testing "what-if" scenarios. High-speed processing and data input available from SCADA allow utilities to provide modeling capabilities to their operators. Careful consideration must be given to the user interface in this regard, and simplified models may be required.

1.8. SUMMARY

Water utilities seek to deliver a safe, reliable, continuous supply of high-quality water to customers on a daily basis through a complex distribution network while minimizing costs. Once developed and calibrated, a water distribution model can predict the behavior of a water distribution system, providing an effective tool to help utility service providers meet these goals. This chapter has provided a fundamental overview of hydraulic modeling including a timeline of distribution modeling development, various water modeling applications, essential and desirable features in hydraulic modeling software, and emerging trends. The following chapters provide more detail and guidance in implementing model development and utilizing water models in various applications to analyze, design, and improve the performance of water distribution systems.

1.9. REFERENCES

Anderson, J.L., Lowry, M.V., and Thomte, J.C. 2001. Hydraulic and Water Quality Modeling of Distribution Systems: What Are the Trends in the U.S. and Canada? In *Proceedings of the AWWA Annual Conference*. Denver, Colo.: American Water Works Association (AWWA).

Benedict, R.P. 1910. *Fundamentals of Pipe Flow*. New York: John Wiley & Sons.

Bhave, P.R. 1991. *Analysis of Flow in Water Distribution Systems*. Lancaster, Pa.: Technomics Publications.

Canning, M.E. 2002. Field Report—Field-Testing of Infrastructure for Asset Evaluation. *Jour. AWWA*, 94:12:53–57.

Casey, R. 2001. Integrating Modeling, SCADA, GIS, and Customer Systems to Improve Network Management. In *Proceedings of the AWWA Information Management and Technology Conference*. Denver, Colo.: AWWA.

Cruickshank, J.R. 2010. Hydraulic Models Shed Light on Water Age, *Opflow*, 36:6:18–21.

Draper, T.M. 2008. Automation Works—Leverage Utility Applications With SCADA Data, *Opflow*, 34:5:10–11.

Edwards, J., Koval, E., Lendt, B., and Ginther, P. 2009. GIS and Hydraulic Model Integration: Implementing Cost-Effective Sustainable Modeling Solutions. *Jour. AWWA*, 101:11:34–42.

Eggener, C.L., and Polkowski, L. 1976. Network Modeling and the Impact of Modeling Assumptions. *Jour. AWWA*, 68:4:189–196.

Gessler, J. 1980. Analysis of Pipe Networks. In *Closed Conduit Flow*. Chaudry, M.H., and V. Yevjevich. Littleton, Colo.: Water Resources Publications.

Govindan, S., Walski, T.M., and Cook, J. 2009. Hydraulic Models—Helping You Make Better Decisions. *NRWA Magazine*, First Quarter, 34–40.

Gupta, R., and Bhave, P. 1996. Comparison of Methods for Predicting Deficient Network Performance. *Journal of Water Resources Planning and Management*, 122:3:124.

Hauser, B.A. 1993. *Hydraulics for Operators*. Ann Arbor, Mich.: Lewis Publishers.

Jeppson, T.W. 1976. *Analysis of Flow in Pipe Networks*. Ann Arbor, Mich.: Ann Arbor Science Publishers.

Lindley, T.R., and Buchberger, S.G. 2002. Assessing Intrusion Susceptibility in Distribution Systems. *Jour. AWWA*, 94:6:66–79.

Mays, L.W. 1999. *Water Distribution Handbook*. New York: McGraw-Hill.

Murphy, Brian, and Kirmeyer, Gregg. 2005. Improving Distribution System Security, *Opflow*, 31:10:18–23.

Nayar, M.L. 1992. *Piping Handbook*. New York: McGraw-Hill.

Seidler, M. 1982. Obtaining an Analytical Grasp of Water Distribution Systems. *Jour. AWWA*, 74:12:628–630.

Stephenson, M. 1976. *Pipeline Design for Water Engineers*. Amsterdam and New York: Elsevier Scientific Pub. Co.

US Department of Energy (DOE). December 2006. *Energy Demands on Water Resources—Report to Congress on the Interdependencies of Energy and Water*. Washington, D.C.: US Department of Energy.

Walski, T.M. 1984. *Analysis of Water Distribution Systems*. New York: Van Nostrand Reinhold.

Walski, T.M., Gessler, J., and Sjostrom, J.W. 1990. *Water Distribution—Simulation and Sizing*. Ann Arbor, Mich.: Lewis Publishers.

Walski, T.M. 2006. A History of Water Distribution. *Jour. AWWA,* 98:3:110–121.

Walski, Thomas, Bezts, W., Posluszny, E.T., Weir, M., and Whitman, B.E. 2006. Modeling Leakage Reduction Through Pressure Control. *Jour. AWWA,* 98:4:147–155.

Walski, T.M., Chase, D.V., Savic, D.A., Grayman, W.M., Beckwith, S., and Koelle, E. 2007. *Advanced Water Distribution Modeling and Management.* Exton, Pa.: Bentley Systems.

Water Research Centre Plc. 1989. *Network Analysis—A Code of Practice.* Swindon, UK: Water Research Centre.

Wilkes, D.R. 2010. From Plant to Tap: Optimize Quality, *Opflow,* 36:8:20–23.

Williams, G.S., and Hazen, A. 1920. *Hydraulic Tables.* New York: John Wiley & Sons.

Wood, D.J., and Charles, C.O.A. 1972. Hydraulic Analysis Using Linear Theory. *Journal of the Hydraulics Division, American Society of Civil Engineers (ASCE),* 98:7:1157–1170.

Wood, D.J., and Rayes, A.G. 1981. Reliability of Algorithms for Pipe Network Analysis. *Journal of the Hydraulics Division, ASCE,* 107:10:1247–1248.

AWWA MANUAL M32

Chapter 2

Building and Preparing the Model

2.1. INTRODUCTION

The fundamental goal of hydraulic modeling for water systems is to develop and maintain a model that sufficiently emulates the performance of the distribution system to give the utility confidence in its ability to represent system performance. The principal tasks associated with creating and sustaining a hydraulic model include model construction, calibration, and maintenance. The primary goal of calibration is to achieve an acceptable level of confidence in the model's ability to reasonably represent the performance of the water system. Following its creation and calibration, the model requires periodic maintenance to remain representative of system operations.

Hydraulic models should be created and maintained so that the standards of accuracy, completeness, and level of detail are appropriate for the application. Model development begins with careful planning to define requirements and to investigate the sources of information available. A hydraulic modeling project requires planning by a hydraulic modeler and oversight, at a minimum, by an engineer experienced with utility infrastructure and operation.

The hydraulic model is an ongoing investment for a utility. A maintenance plan should be developed to facilitate periodic updates to the model. These periodic updates will help to keep the model representative of real system conditions and preserve its effectiveness as a decision making tool.

Many types of data are used to build hydraulic models. Documentation during the model construction and calibration tasks is essential to manage these data and provide for effective long term model maintenance. This is especially important when transferring the model to other modelers.

2.1.1. Hardware and Software

A wide array of hydraulic modeling software packages is available, ranging from free public-domain software to commercially available software with features and costs that vary greatly. A software selection process may be needed prior to selecting a software package. Selection criteria may be based on a combination of the following: ease-of-use, features, vendor history, integration options, maintenance policies, costs, technical support, computational engine, vendor location, training options, utility preferences, and current hardware and software requirements.

Software packages used in hydraulic network analysis typically require the same basic system data to set up, refine, and calibrate. Taking the time to gather high quality data for the model may yield a number of benefits to the utility. Better data and a better model typically result in savings elsewhere in the organization as a result of improved system knowledge, leading to better decisions. This investment in data should be protected by ensuring that the data can be easily migrated from one modeling package to a different modeling package if the utility's modeling software requirements change. Most software vendors allow export from their modeling packages to a common, interchangeable file format (EPANET file format is an example of this).

Hardware requirements differ only slightly for the various hydraulic modeling software packages. As computer and server technology improves, the hydraulic modeling software should as well. In general, most modeling packages have a minimum hardware specification and an optimal or recommended specification for processor speed, RAM (random access memory), graphics card, and hard-drive space.

2.1.2. Geographic Information Systems (GIS)

The role of GIS in the water/wastewater industry continues to grow and influence the evolution of hydraulic modeling. Advances in GIS database design, software, and hardware allow more robust interoperability with modeling databases, improved performance with large datasets, and advanced postprocessing tools.

The hydraulic model shares a mutually beneficial existence with GIS. Just as GIS enhances the usefulness of the hydraulic model, the hydraulic model can also extend the functionality of GIS. For utilities that have established a GIS, the added functionality of interfacing with a hydraulic model can prove valuable from an enterprise, customer, and operational perspective. This interfacing can add further value to both the hydraulic model and GIS databases. For example, the process of building a model from GIS and the subsequent calibration may reveal errors within the GIS database such as improperly connected or unconnected pipes, missing attributes, and diameter discrepancies. These adjustments could be identified in the model and should be reincorporated into the utility's GIS.

Integration of the hydraulic model with the GIS requires the highest quality of GIS data. While errors can be identified during the model construction and calibration process, data cleanup efforts should be undertaken within the GIS to make the GIS "model ready." This process may include the creation of a geometric network with predefined topology rules and criteria before model construction.

2.1.3. Enterprise Applications

A hydraulic model can be enhanced with database links to other enterprise-wide applications such as customer information system (CIS), computerized maintenance management system (CMMS), and supervisory control and data acquisition (SCADA) systems. The CIS may contain customer data critical to establishing base demands within the hydraulic model. The CIS also provides the ability to identify customers that may be out of service during a pipe outage given customer identifiers (IDs) or parcel numbers. Computerized maintenance management systems are used to generate field work order requests and can be a valuable source of attribute information for pipes, fittings, valves, and other components of distribution systems. This ground-level utility communication can provide current information on customer complaints, asset maintenance, and field data, and therefore can be used as a support tool with the hydraulic model. Model integration with SCADA also offers a variety of mutually beneficial model applications, such as establishing operating ranges for given facilities, identifying boundary conditions necessary for model calibration, historical event system troubleshooting, and real-time modeling to assist with daily pump scheduling and optimization.

Communicating with and relating to current data at any level of the utility can result in time saved in model development, calibration, and maintenance of the model. This communication allows for advanced reporting on infrastructure assets using model results for pipe replacement prioritization and rehabilitation projects.

GIS and other enterprise applications can make building and maintaining hydraulic models more efficient, and they can support a wide array of analysis and reporting functions. These applications can enhance the value and improve the awareness of the importance of hydraulic modeling across the various departments within a utility. As more accurate data are collected along with further adoption of enterprise applications, the process of hydraulic model building and maintenance will allow for a sustainable model versus a model that needs to be repeatedly rebuilt and recalibrated.

2.2. PLANNING THE HYDRAULIC MODEL CONSTRUCTION AND DEVELOPMENT PROCESS

While most water distribution systems have many of the same types of assets, each water system's size, configuration, and operations are unique. The modeler must consider the system's distinctive structure and specific requirements when developing the plan for the hydraulic model.

2.2.1. Establish Purpose and Need

Before developing a hydraulic model, its current and potential future applications should be identified (see section 1.4). The desired applications will dictate the model's level of detail, its required level of accuracy, and the data that will need to be collected.

After the utility vision of model application has been established, decisions on such topics as model loading and accuracy of calibration will naturally follow. By planning the hydraulic modeling objectives first, a utility can determine any necessary integration plans and what is needed in terms of resources and software.

2.2.1.1. Level of Detail. The degree of detail required in the model is greatly influenced by the intended purpose of the model. On the one hand, a model used for long-range planning, transmission main sizing, and reservoir site evaluation may only require the inclusion of major transmission mains, distributing demand evenly across the system, and the use of theoretical peaking factors for estimating water demands. Where detail is lacking, it is critical that the modeler be aware of the limitations of the model. These limitations should be clearly stated in any report documenting analysis results.

On the other hand, a model used for fire flow capacity evaluations at the neighborhood level, leakage control studies, energy-use management, or water quality modeling will require considerably more detail, up to the possible inclusion of every transmission and distribution pipe—excluding service lines—in the distribution system. These types of models require the modeler to allocate demands in a more detailed manner.

The amount of time and financial resources that are available will also influence the level of detail incorporated in the model. The modeler may be required to conduct a cost–benefit analysis to obtain managerial support for extending extra effort in developing a more detailed model.

2.2.2. Model Structure

It may not be necessary or practical to create the hydraulic model with the level of detail found in the GIS. Models can be categorized based on their level of detail and by the extent to which the model elements (pipes, pumps, valves, and tanks) match the corresponding features in GIS. To understand the various model types, it is necessary to define two model development terms.

- *Skeletonization:* deletion or exclusion of small diameter pipes in the hydraulic model

- *Reduction:* removal of intermediate (hydraulically insignificant) nodes along pipe reaches of similar diameter, material, and construction date and subsequent combining of the adjoining pipe segments into a single pipe segment

Descriptions for various model types and sizes, along with their advantages and disadvantages, are presented in the following sections:

All-Pipes Model. An all-pipes model preserves the same level of detail as the GIS and maintains a one-to-one relationship between individual GIS and model elements. There is no skeletonization or reduction of GIS data to construct an all-pipes model. This one-to-one relationship reduces the effort required to maintain the all-pipes model and establishes data sharing opportunities between the two databases. However, because of the larger size of all-pipes models, they require longer processing time, large computer storage space, and often a more expensive modeling software license. All-pipes models for larger utilities can exceed 100,000 or more pipe segments.

All-Pipes Reduced Model. An all-pipes reduced model is similar to an all-pipes model with the exception that intermediate hydraulically insignificant nodes along pipe reaches of similar diameter, material, and construction date are dissolved (reduced). The entire length of pipe in the GIS is represented in an all-pipes reduced model, but the model pipe segments are comprised of multiple GIS pipe segments (one-to-many relationship). An all-pipes reduced model will require more model maintenance effort and coordination than an all-pipes model due to this one-to-many relationship. However, because of the reduced number of pipes, an improvement in model simulation speed may be achieved. Reduction may decrease the number of pipe segments in an all-pipes model by 50 percent or more.

Skeletonized Model. A skeletonized model includes all pipe segments above a specified diameter, typically 12-in. or 16-in. (30.5 cm or 40.6 cm). Because skeletonized models represent only larger pipes, they are sometimes referred to as *transmission* or *backbone models*. A one-to-one relationship is maintained between pipes in the skeletonized model and GIS. A skeletonized model may include only 10 percent to 20 percent of the pipes found in an all-pipes model of the same system. While system speed is improved in comparison to an all-pipes model, detailed evaluations at a neighborhood level are sacrificed.

Skeletonized Reduced Model. A skeletonized reduced model is the same as a skeletonized model except intermediate (hydraulically insignificant) nodes are dissolved (reduced). Skeletonized reduced models represented the standard modeling practice in the 1980s and 1990s, when models were manually digitized from water system

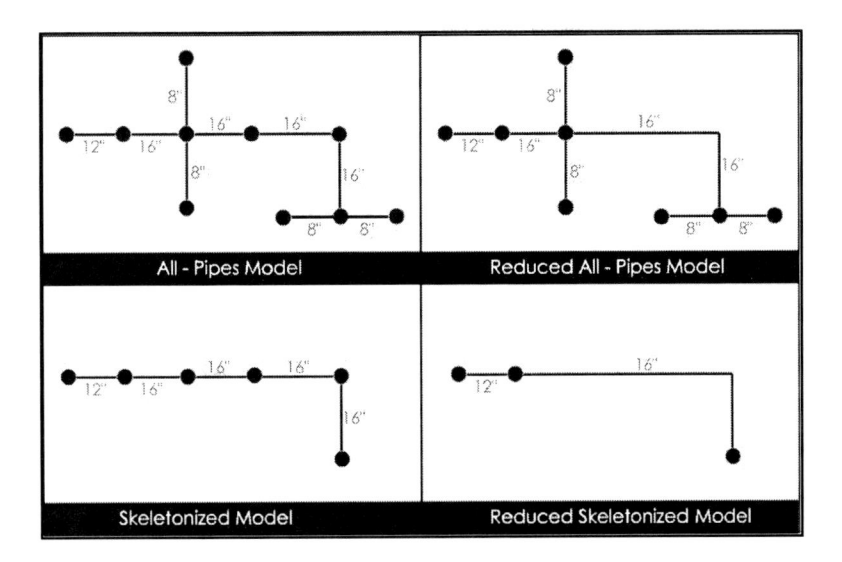

Figure 2-1 Basic hydraulic model structures

paper maps. Many utilities have replaced their skeletonized reduced models with more detailed models of their distribution systems.

Figure 2-1 shows examples of the four basic types of hydraulic model structures.

Variants of the four basic model structures are common. Two of the more common variants are referred to as a *detailed model* or a *hybrid model*. A detailed model is either a skeletonized or a skeletonized reduced model with some of the smaller diameter pipes included (typically some of the 6-in. and 8-in. [15.3 cm and 20.3 cm] pipes). The Stage 2 Disinfection and Disinfectants By-products Rule (Stage 2 DBP Rule), which was promulgated by the US Environmental Protection Agency (USEPA) in 2006, allowed disinfection by-product sample sites to be selected based on modeling of distribution system water age. The rule required that if models were used to select sample sites, they must include at least 50 percent of the total pipe length and 75 percent of the total pipe volume in the distribution system. To satisfy these requirements, some of the 6-in. and 8-in. (15.3 cm and 20.3 cm) pipes in the model would have to be included. A detailed model can effectively evaluate water quality concerns in the distribution system. However, the manual selection of small diameter pipes to include in the detailed model, combined with reducing the pipe data, can result in significant increased model maintenance requirements.

A hybrid model combines large transmission mains (typically with diameters 16-in. [40.6 cm] and larger) throughout the system with all pipes in select pressure zones or regions of the system. In this manner the model can be used to simulate detailed hydraulics and water quality in those portions of the system that are of concern. Hybrid models are typically only used in large distribution systems. Hybrid models improve processing speed compared to all pipes or all-pipes reduced models. However, hybrid models require additional time to configure the mix of systemwide transmission facilities with detailed, localized distribution system facilities.

All types of hydraulic model structures can be constructed using a utility's GIS. However, model structures will dictate model maintenance procedures. For all types of reduced models, a one-to-many relationship must be established between the hydraulic model and GIS in order to conduct automated update procedures. This requires a unique understanding of the related features between the model and GIS to maintain features that are updated, added, or removed. Various methods for working with one-to-many related features to support model maintenance can be used, though they typically require additional development in the GIS (i.e., dynamic segmentation) or programming in order to be effective. All GIS features are maintained in an all-pipes model that facilitates an easier automated transaction of data between the GIS and hydraulic model.

In addition to initial development and ongoing maintenance, other factors must be considered when selecting a model structure. An all-pipes model best facilitates automated maintenance procedures. However, all-pipes models generally require a larger investment in hardware and hydraulic modeling software and may cost more in initial development. The large amount of data in all-pipes models can significantly increase processing and solution time, especially for extended period simulations. Additionally, storage requirements for all-pipes models can significantly exceed the requirements of skeletonized or reduced models. However, processing and storage considerations will become lesser concerns due to the ever increasing advancements in computer processing, storage devices, and software capabilities.

2.2.3. Data Availability and Quality

Building and maintaining a hydraulic model may require collecting data from and collaborating with other departments. Water system asset data for facilities and distribution system piping are frequently provided by GIS or related asset management tools and can be supported by engineering, operations, production, customer billing, planning, and topography data. Information for the pump stations including pump curves, pump tests, and set points may come from the operations department. Demand data may be established from meter records held by the billing department. The utility's field crews may need to be involved if flow testing is required to use for model calibration. Boundary conditions may need to be extracted from the SCADA database. Planning and engineering departments may be involved in validating pipe roughness values in the hydraulic model. With larger utilities, it may be a significant task to navigate through the various departments to locate and retrieve the best available data.

A review of available data sources can also assist in determining the effort and resources required for data collection, data entry, and model calibration. As the data are being located and gathered, it is important to objectively rate the quality of each category of input data. The data quality assessment can provide a long-term plan to improve the quality of each set of data. In addition, these ratings can help in the process to better understand the basis for the hydraulic model and provide a roadmap to guide further refinement of the model during the calibration phase.

2.2.4. Establish Modeling Standards

A crucial aspect of the model planning process is the establishment of modeling standards—how facilities will be graphically represented in the model, what level of detail will be included in the model, and what naming convention will be used for everything from pipes to queries. Procedures should be established for how these standards will be documented, communicated, and reinforced for both current and future users of the model. Consideration should be given to model documentation that includes a log to record changes to the model after initial development and calibration.

2.2.5. Planning for a Sustainable Hydraulic Model

If the model is planned and built with the knowledge and intent that it will be periodically updated and refined, it can provide a sustainable and long-term benefit to the utility. By contrast, if a model is planned and executed as a "one-off," disposable product, it will likely need to be completely rebuilt before its next significant application, resulting in a duplicated effort and a loss of institutional knowledge.

It is important to understand the data structures, development and maintenance procedures, and available tools when developing a sustainable modeling solution. As previously mentioned, an all-pipes model best facilitates automated maintenance procedures and may provide the most cost-effective option for a routine and frequent maintenance program.

The complexity of model development and maintenance from GIS requires that sound procedures be outlined and followed. A number of the development and maintenance tasks can be performed within the GIS environment, using GIS-based tools. Other tasks can be performed using hydraulic model tools within the hydraulic model software. Many tasks can be performed in either environment, and the actual tools used and the order in which they are completed will depend on the type of model constructed, the quality and format of the data source, the skills and availability of staff, and the hydraulic software used.

Figure 2-2 provides a general overview of a sustainable hydraulic model development and maintenance process.

2.2.6. GIS Conversion or Construction

The current GIS stores facilities within a geo-database, which categorizes elements as feature classes and subtypes within the feature class. Feature classes represent element geometry types such as *node*, *line*, or *polygon*, and subtypes of the feature class include more specific facilities or appurtenances of the water system that may be represented as a node, line, or polygon element.

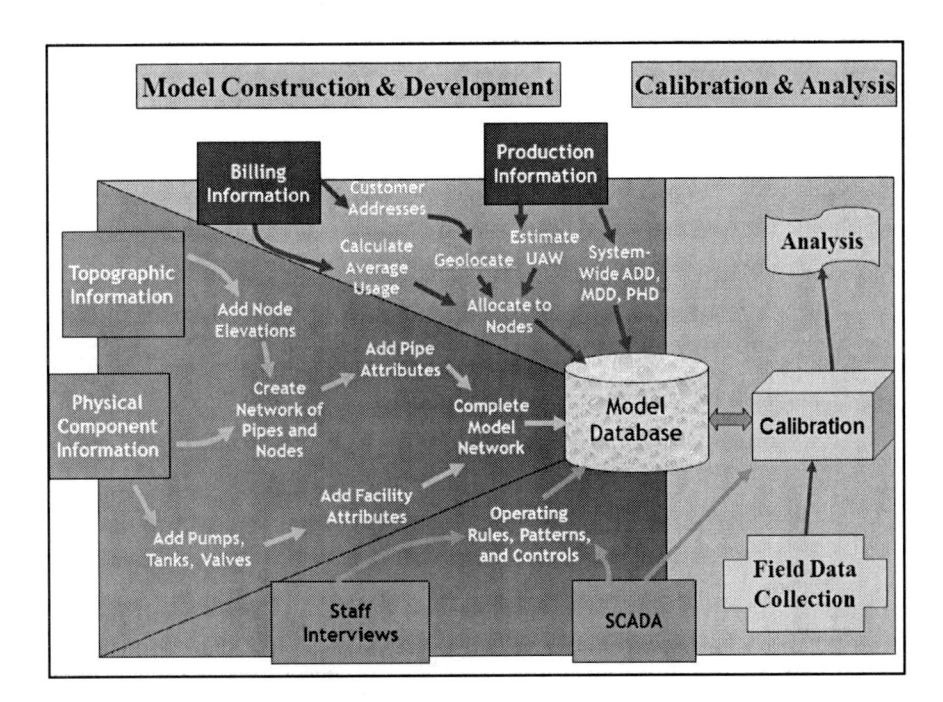

Figure 2-2 Overview of a sustainable modeling process

Tanks, fire hydrants, valves, and customer meters are examples of elements that would be stored as different subtypes within a node (or point) feature class. Distribution system pipes, overflow or drain lines, hydrant laterals, and service laterals are examples of elements that would be stored as different subtypes within a line feature class. Land uses, zoning, demographics, and soil types are examples of features that would be stored as different subtypes within a polygon feature class. Typically, node and line feature classes are those features considered for import to the hydraulic model, with polygon features used as supporting data for a range of hydraulic model considerations and applications.

When importing elements into the hydraulic model, it is important to recognize the different native storage structures of the two databases (GIS and hydraulic model) and plan accordingly. Unique element identifiers (IDs) are required to maintain unique elements within both databases to avoid the duplication of any one or more elements. This represents just one of a number of necessary considerations.

When creating models from GIS, facility and network representation should be clearly defined before model construction begins. The utility's objectives should also be defined and understood before any construction takes place. Defining how each field in the GIS maps to the model elements will greatly assist in the data management plan and effort.

2.3. DATA SOURCES AND AVAILABILITY

The data required to create a model come from a variety of sources, as summarized in the following list. The availability, consistency, quality, and quantity of data received, in combination with the relative size and complexity of each water system, vary significantly with each utility. Engineering judgment, water system experience, and technical software capabilities are essential ingredients to incorporating all the data into a model of the distribution system. Creating a model requires staff do the following to gather data:

1. Identify and collect geographical data:
   - Land use and zoning for existing and future conditions
   - Aerial photographs
   - Elevation data: digital terrain model (DTM), triangulated irregular network (TIN), contour maps, GPS, or other elevation data
   - Street maps and parcel maps: including street names, rights-of-way, and city and township boundaries

2. Establish inventory of existing facilities:
   - Pipes: location, diameter, length, age, material, lining, and connectivity to other pipes
   - Storage facilities: location, capacity, dimensions, base and water surface elevations (minimum and maximum), accessibility, and system connections
   - Water supply facilities: location, elevation, hydraulic grade line (HGL), capacity, operational patterns, and controls
   - Groundwater wells: location, elevation, capacity, pump curves, pumping water levels, recent pump tests, and control settings
   - Connections (including import connections, emergency interties, etc.): location, elevation, capacity, HGL, purpose (take and/or give)

- Pumping and booster stations: location, elevation, capacity, typical suction and discharge pressure, pump curve(s), and related features
- Control valves: Flow control, pressure regulating and sustaining valves (location, elevation, capacities, settings, and operability)
- Other system valves: check valves, isolation valves, and other normally closed valves
- Boundaries: service area and pressure zones

3. Review water consumption records:
   - Water sales (billing records)
   - Service area population
   - Number and type of service connections (current and prior years)
   - Consumption by meter route and by major customers
   - Nonrevenue water (NRW) and water loss records

4. Review water production records:
   - Peaking factors
   - System diurnal demand pattern (A diurnal demand pattern is a dimensionless profile that describes how demands change in a 24-hour period. The diurnal profile value at a particular hour is multiplied by the average daily demand to obtain an hourly demand.)
   - Pressure zone–specific diurnal demand patterns, if available

5. Develop water demands using existing water consumption data, water production data, population data, and land use. Establish current and projected water demands for the total system, pressure zones within the system, and major users on the following basis:
   - Annual average day demand (total water use in a year divided by the days in a year)
   - Maximum day demand (the highest demand occurring in a single day during an entire year)
   - Peak hour demand (the highest hourly demand that could occur during the maximum day)
   - Fire flow demand requirement (typically based on land use and fire authority)
   - Minimum hour demand or maximum storage replenishment rate (occurs during maximum day)
   - Other operationally critical demands within the water system's service area
   - Land use classifications
   - CIS databases with customer addresses and water use records

6. Review system operating records and interview staff:
   - Annual, seasonal, and diurnal water production records
   - Service and booster station flows and pressures
   - Clearwell and storage level variations for various demand scenarios

- System operation criteria and set points for pumping stations, storage facilities, and control valves
- System operating pressures determined from previous testing, monitoring, and fire flow testing
- Observed low-pressure areas indicating system deficiencies
- Observed fire flow capabilities
- Measured roughness characterization

2.3.1. Data Confidence

All data used for the model should be objectively evaluated to determine the level of confidence in each data set provided for model creation, development, and calibration. The level of confidence in the data used, how the data are assigned to the model, how the data are documented, and what efforts are required to find more detailed or reliable information are all important factors in weighing the level of effort required, and the costs, to obtain the qualified data given the model's objectives. Furthermore, the confidence level of the data directly affects the confidence level of the model calibration and results. Particular attention should be given to data that would have a larger impact on model results. For example, a 10-ft (3 m) error in ground elevations could have a much greater effect on resulting pressures than a 10-ft error in pipe length.

2.3.2. Types of Data

Data used for model construction can generally be categorized into four main groups: (1) geographical data, (2) physical facilities data, (3) demand data, and (4) operational data. A detailed water system inventory can reveal much of this information. Certain analyses will require additional information, depending on the type of study. Energy costs, infrastructure costs, water constituent or contaminant concentrations, and demand or customer growth projections are examples of additional data that may be required.

2.3.2.1. Geographical Data. Geographical data includes a land-base aligned to the desired geographical coordinate system, jurisdictional boundaries, street centerlines, aerial imagery, and other information useful to establish the physical location of the model.

2.3.2.2. Elevation or Topographic Data. Elevation or topographic data are used to assign an elevation to the model nodes or junctions, as described later in this chapter. Distribution node elevations are frequently assigned as equal to the ground elevation. Ground elevation data can be obtained from a variety of sources including utility as-builts, planimetric data, or digital terrain models (DTMs). For better accuracy, elevations at pump, control valve, and storage facilities may be based on as-built drawings or field surveying. Elevations for junctions representing pressure monitoring locations must be carefully documented and related to the actual elevation of the pressure-recording device. The model uses the calculated hydraulic grade and elevation to calculate pressure at a given location.

The availability of digital data, such as DTMs, to overlay atop hydraulic model junctions for elevation interpolation is typically a preferred method of elevation assignment. The opportunity to minimize human error and allow computer-based interpolations assists with efficient and accurate performance of this task. The quality of the DTM and confidence level should be evaluated before proceeding with automated elevation assignments. Quality assurance should be performed following the automation of this task to compare hydraulic model junction contours to DTM contours.

2.3.2.3. Physical Facilities Data. Physical data include the geometric descriptions and hydraulic attributes of pipes, pumping facilities, storage facilities, and other control facilities. Physical facilities require proper geometric data and associated hydraulic data. In addition, proper elevation data are required.

Distribution system geometry includes the connectivity, length, and diameter of pipes within the distribution system and within specific facilities. Upon construction of the geometric representation of the water distribution system, hydraulic attributes must be assigned to both facilities and distribution system piping. Hydraulic data for input to the model can include pump type and curve, valve type and characteristics, minor loss characterization, and pipe roughness characterization. Determining reasonable roughness coefficient factors, or C-factors (the friction factor in the Hazen-Williams Friction Headloss Equation most commonly used in the United States), is required to accurately represent the headloss that will occur within the system. Generating a reasonable roughness coefficient typically involves consideration of several factors including pipe diameter, pipe condition, material, year of installation, results of field hydrant testing, and industry standard roughness coefficients for given pipe characteristics. Engineering judgment should be applied to validate the methods, data, and assumptions used to develop this relatively variable hydraulic coefficient.

2.3.2.4. Demand Data. Demand data used for calibration and simulations of a hydraulic model represent the total water supplied to the distribution system at any given time. This may include water supply components such as well production, treatment plant production, and wholesale and/or neighboring agency supply by interconnections. This supply total produced and delivered into the distribution system equals the total water consumption of the system including system water losses.

In recent years, water consumption data (also referred to as *sales data* or *customer billing data*) have been used to obtain a base demand allocation across the distribution system. This base demand is assigned to hydraulic model junctions along the water system piping, using customer billing data addresses as spatial reference. Total water supply recorded as flowing into the system is used to scale the base demand allocation, which results in water demand allocated per water consumption data, with an even distribution of water loss systemwide. There are other approaches for demand generation and allocation to hydraulic model junctions, some of which are summarized in subsequent sections of this chapter.

2.3.2.5. Operational Data. Operating data include parameters such as flow rates, reservoir elevations, pressure-reducing valve set points, valve or pump controls, valve or pump status, and fixed pressures that establish boundary conditions in the model. Operational data applied to the hydraulic model consider performance characteristics of the water system, and would typically be information supplied by the utility operator in connection with the utility SCADA system used to monitor and control system facilities. Operational data required for input to the hydraulic model can include maximum and minimum tank levels, valve position or flow rate, on/off set points for pumps or valves, initial pump status, pump speed, and pump sequencing within the pump station.

2.3.3. Sources of Data

Traditionally, data on distribution systems were paper records maintained by the water system operators, or as in the case of a new distribution system, by the developer and/or their client. Today, many utilities have moved—or are currently moving—their water system records to GIS, so this should be the first place to look for water system data. A GIS provides the opportunity to more efficiently construct a new model or update an existing model.

For utilities without a GIS, the development of a detailed model of their water systems can be a first step in creating a GIS of their water systems. Expanding the model data collection process to include the creation of a GIS will add time to the model building process but ultimately will deliver substantial benefits to the water utility.

2.3.3.1. Paper Records. Paper record sources for distribution system data include the following:

- Water distribution maps
- Contour maps
- Aerial photography
- As-built construction drawings
- Operation and maintenance manuals
- Zone boundary maps
- Water master plans or studies on the water system
- Annual reports
- Planning documents containing information on the age of the water system

The current hydraulic modeling software packages have the ability to graphically present results of a distribution system analysis. These software packages also allow data entry using digitizing techniques, often with aerial photos or other types of a land base that aid in positioning spatial data. In order to do this, however, the input files must include the geographic coordinates (x-y) for each node. The coordinate system used should be verified for spatial compatibility between the two systems. If GIS for the system is not available, the coordinate system of the paper maps or other asset data could be used. A standard coordinate system, such as the United States Geological Survey (USGS), State Plane, or Universal Transverse Mercator coordinate system, is preferred. If the model is on an established coordinate system, the model can be converted to a different coordinate system if the need arises. Using the same coordinate system as the rest of the utility's data simplifies model maintenance and enables the use of model results in conjunction with other utility information.

2.3.3.2. Electronic Records. With the creation of a hydraulic model from electronic data sources, there is an expectation that future model updates will be done quickly in response to infrastructure upgrades or system replacements. Standard procedures should be developed to extract data from electronic sources to allow for timely updates.

The types of electronic data sources typically used as inputs to the model include the following:

- GIS
- SCADA (supervisory control and data acquisition)
- DTM (digital terrain model)
- Maintenance management system (MMS)
- CADD (computer-aided design and drafting) drawings
- Database systems

Depending on the type of structure elected for the hydraulic model creation, a GIS will generally contain more information than is required for building a hydraulic model. As a result, GIS should be filtered before using it as a model creation for updating. Most modeling software packages have routines for model reduction, where pipe segments are combined with the same diameter, age, and roughness coefficient, for example. In addition, there are commercially available GIS utilities that can be used to significantly automate the process of GIS data extraction. The functionality of these GIS utilities typically includes adding a new ID or modifying an existing ID, adding or modifying attributes, modifying topology to prepare it for import to the model, and logging changes in the GIS since the last export. Once the workflow is set up, tested, and proven, future model updates are more efficient and accurate.

SCADA represents the collection of equipment, communications, and software that allows the real time control and monitoring of system operations. SCADA systems are typically used to control the operation of a distribution system and to report on the behavior of equipment, as well as on the distribution system itself. SCADA systems are designed so that pressures, flows, pump status, reservoir levels, disinfectant residuals, valve position, and other information are collected at regular intervals, can be stored in a database, and are usually presented graphically to the operators. The SCADA system therefore provides a wealth of information that is downloadable for use in the model or to compare against model results.

The supervisory control function can be manual or automatic. In manual mode, an operator may manually operate valves, pumps, or other devices via a software interface or at the facility. In automatic mode, a remote telemetry unit (RTU) located at the facility may automate the control of a facility, or SCADA software may control the facilities remotely based on a preset operating strategy. In practice, most SCADA systems have a combination of manual and automatic controls, depending on the field device and the system characteristics. It is recommended that the modeler develop close communication with the system operators to develop a realistic, detailed understanding of how the system operates and how manual system controls are tracked; this information is highly useful during model calibration.

Elevations can be obtained from a variety of electronic sources. A DTM is a type of digital elevation model (DEM) and is a three-dimensional representation of the ground surface. DTMs can be used to assign elevations in a model. DTMs are usually used with CADD or GIS packages.

Utilities that use MMS will usually have a record of repairs, improvements, or other changes to a water distribution system. An MMS is often a source of valuable information regarding the condition of the water distribution network and related facilities and is, therefore, a valuable source of information to construct or update the model.

CADD systems are used to store drawings or maps of a water distribution system. Although more and more utilities use GIS to track utility assets, the CADD system is still a source of valuable asset information. Today's CADD systems can integrate with GIS formats as well.

Utilities will often store information useful for model development in relational databases. Customer billing systems, for example, usually contain information about the water use of each customer. Some utilities also store asset information, repair data, or other information in a database that may be useful to the modeler.

2.3.3.3. Physical Inspections. The physical inspection of a facility can provide information on how the system operates as well as the physical in-ground plumbing (piping) around facilities to improve the modeler's perspective for model facility configurations. Field inspections should be conducted for each system facility to understand how each facility functions. Ideally, the site inspections will be conducted with an operator who can explain how the facility is operated under varying system conditions.

A complete understanding of the function of each facility should be obtained while on-site. In addition, digital photos of the entire site including pumps and valves may prove valuable later when constructing the model away from the facility. These field inspections should include collecting data for the following:

- Number, type, and condition of the water facilities
- Types and age of pipes and overall condition of the piping network
- How system components are controlled and perform over the range of demands
- How the facilities and individual components interface with each other
- Limitations and deficiencies of the facility
- Valve location and position (open or closed)
- Facility site layout and available spacing relative to other potential future facilities that the model and planning efforts may determine

A site visit to water supply sources, pump stations, reservoirs, pressure-reducing valve (PRV) stations, and monitoring points is very useful to understand how the distribution system operates and helps verify that site drawings actually represent current infrastructure.

Observed physical characteristics should be compared against the network and operations data to verify that the piping connections and element locations are accurately represented. The location of all pressure and flow monitors should also be confirmed. Elevations of pressure sensors should be surveyed, if possible, and compared to elevations being used in the model. Control valves should be checked to determine whether or not they are at the correct set points using SCADA or field tests, and that they are fully operational and not stuck in one position. Noises indicating valve or pump cavitation should be noted, as a pump may be operating considerably off its original design point, because of either improper sizing, impeller problems, or incorrect set points.

Where possible, pipes being replaced should be visually inspected to determine the interior condition, assessing how roughness of the pipe has changed in time. This can then be compared with the results of roughness coefficient testing programs (see chapter 3) to determine whether or not the values measured are consistent with observed conditions. A hydrant flow test should be conducted before and after pipeline replacements (and after the installation of pipeline additions) to verify that all system valves are open and that the facilities are functioning as intended.

2.4. PHYSICAL FACILITIES DEVELOPMENT

2.4.1. Model Elements

Hydraulic models use two types of elements to represent the various water facilities: node elements and link elements. Node elements are points in the model that include junctions, tanks, and reservoirs. Link elements are lines in the model that represent pipes. Pumps and valves can be defined as either link or node elements depending on the modeling software (and in some cases the software version).

2.4.1.1. Nodes. The primary functions of a node include connecting links, storing elevation data, and allowing water to enter or leave the network. A node can also represent a boundary condition such as a fixed grade node, tank, pump, or valve. When a node is not a tank or other operational element such as a pump or valve, it may be called a *junction node* or simply a *junction,* especially when used to connect pipes. Nodes may be added to the model to represent meter locations where demands could be placed, or demands may be assigned to nodes that are in the model for other reasons.

Model results report hydraulic grade line (HGL), pressure (HGL minus elevation), fire flow results, water age, and flows (into or out of the network) at node locations.

At a minimum, nodes are required where two or more pipes connect and when pipe data changes, such as diameter, age, or material (which can impact pipe roughness). Hydraulic models that have relatively long pipe runs between node locations and substantial changes in topography may not have nodes at the highest and lowest points in the distribution system. In these cases, additional nodes should be inserted in the network near these high and low points in order to represent these potential locations of extreme pressures. If nodes are not included in the model at hydrant locations, nodes may be added to the pipeline to represent hydrants used during field testing.

2.4.1.1.1. Elevation Data. Elevations used in the hydraulic analysis are separated into two categories: (1) control elevations and (2) ground elevations. There are many different sources available for elevation data including TINs, DTMs, and DEMs, data from USGS, hardcopy maps, and more. The accuracy of the source will directly impact the accuracy of the model.

Facility elevations should be input with a high level of confidence at locations critical to the calibration and operation of the model, such as at boosters or PRV stations. Field data may be gathered at these locations during the model calibration process to determine whether or not the model is performing as expected. Control elevations, if determined as necessary, are often determined by surveying.

Ground elevations are typically used for the remaining model nodes. These elevations are not required for calculating HGL but are necessary for determining available delivery pressures in the water system.

Depending on the application of one's model, elevations should be as accurate as reasonably possible. An elevation difference of 5 ft (1.5 m) will translate to a potential error in model output of approximately ± 2 psi (13.8 kPa). This potential error should be kept in mind during both calibration and system analysis.

2.4.1.1.2. Quality Assurance Tips for Assigning Elevations. High-quality GPS equipment, calibrated to known benchmarks, may be used to provide accurate elevation data. As-built drawings are also used for elevations. Elevations from DTMs and DEMs are of widely varying quality and should be used with caution. Even though the elevations may be printed out to several decimal places, they are frequently extracted from sources that are only accurate to 20 ft (4 m), and hence are only as good as their source. When using automated elevation interpolation methods, the maps should be reviewed and any geographic break lines that limit the use of automation interpolation techniques should be identified.

Due to the significant impact of elevations on model results, it is recommended that the elevations are reviewed once they are in the model. One way of accomplishing this is to generate elevation contours from the model and compare these contours directly to the source file—whether the source is a contour file or DTM. This side-by-side comparison may help identify errors introduced during the interpolation and assignment process.

2.4.1.2. Links. Pipes are critical elements of the hydraulic model. Links are used by the model to represent pipelines or pipe segments. Pipe links are required to convey water from node to node, and to calculate head losses between nodes. In some hydraulic modeling software, links are used to represent pumps and valves in the interface and modeling database. These types of links require additional data not required for pipes.

2.4.1.2.1. Pipe Data and Development. The pipeline data required for hydraulic model calculations consist of diameter, pipe length, and a roughness factor. Pipe roughness factors are typically estimated from the diameter, material, and pipe age.

The hydraulic model considers the available volume of a pipe for hydraulic calculations. During construction of the hydraulic model, nominal pipe size is typically used, as this will produce a reasonable result in the hydraulic loop equations for the purposes of modeling. However, the modeler should be aware of the difference between nominal and true internal diameter when performing detailed calculations, such as for hydrant flow tests or when comparing the economics of two different pipe materials.

For larger pipe sizes, the difference between nominal and actual diameter is significant because increased wall thicknesses are required in larger diameter pipes to maintain sufficient pipe rigidity. For transmission main modeling, the actual internal diameter is recommended over the nominal diameter. When the actual internal diameter is not known, an appropriate roughness factor is determined if the transmission main is monitored during calibration.

The length entered into the hydraulic model is the length of pipe between node locations, including fittings. This length is most typically obtained from the GIS piping layer but can also be achieved from design drawings, scaled maps, or as-builts. In areas with substantial changes in ground elevation, care should be taken to enter the true total length of pipe and not just the horizontal distance between nodes.

2.4.1.2.2. Pipe Roughness Coefficients. Pipe roughness can be estimated by knowing diameter, material, and age (or pipe condition). A testing program is recommended to determine the appropriate correlation of these variables to roughness factors in the local conditions (see chapter 3).

Over the years, new pipe materials have been introduced and manufacturing methods have been largely improved. As a result, many mature water systems are composed of a wide variety of pipe materials with varying hydraulic properties.

Some of the pipe materials found in a water distribution system may include

- Wood stave
- Unlined cast iron
- Lined cast iron
- Ductile iron (cement lined)
- Steel
- Prestressed concrete cylinder pipe
- Asbestos cement (AC, sometimes referred to as Transite)
- Polyvinyl chloride (PVC)
- Polyethylene (PE)

Manufacturers of the different pipe materials typically provide roughness factor information for new, relatively clean pipes. In general, pipes are initially very smooth inside and perform hydraulically as described in the literature for a few years. Over time, however, transformations may occur within the pipes through a number of mechanisms. These mechanisms include corrosion, encrustation, and biofilm development. Corrosion and encrustation rates are dependent on the local water chemistry. These processes effectively reduce the interior diameter of the pipe, which is subsequently reflected in the model by adjusting the roughness coefficient.

Empirical and quantitative formulas were developed over the years to characterize flow of water through pipes. Several of the older formulas were based on tests of pipes manufactured in the early 1900s. These earlier pipes were generally manufactured under less stringent quality controls than present and were connected with many fittings that caused significant disturbance in the flow path. Two formulas, whose derivations were based on smooth pipe, are commonly used in water supply to calculate head loss or volumetric flow rate—they are the Hazen-Williams formula (empirical) and the Darcy-Weisbach formula (quantitative). Both formulas are available for use within hydraulic modeling software packages; however, the Hazen-Williams formula is used predominantly for hydraulic model calculations in the United States. As such, the Hazen-Williams formula will be assumed as referenced in manual discussions unless otherwise indicated.

Table 2-1 lists general roughness factors used in initial model development. The smoother the interior wall of the pipe, the higher the C-factor. This translates into a higher pipe carrying capacity. The inverse is true as well. Though Table 2-1 provides a general guideline for C-factors used, it is important to conduct local roughness factor testing to determine the extent of hydraulic capacity deterioration of the pipes, as hydraulic capacity is influenced by age of pipe, material, and local water quality conditions.

Table 2-2 lists general roughness factors (ε) for the Darcy-Weisbach equation. Unlike the Hazen-Williams equation, the smoother the interior wall of the pipe, the lower the roughness coefficient, which translates into a higher pipe carrying capacity.

The roughness coefficients in Tables 2-1 and 2-2 are provided as general guidelines and do not take into consideration very low values (high values for Darcy-Weisbach) that may be the result of localized conditions. As previously indicated, local testing is recommended to help develop a better estimate of roughness.

Table 2-1 C-factor values for discrete pipe diameters

Type of Pipe	1.0 in. (2.5 cm)	3.0 in. (7.6 cm)	6.0 in. (15.2 cm)	12 in. (30 cm)	24 in. (61 cm)	48 in. (122 cm)
Uncoated cast iron—smooth and new		121	125	130	132	134
30 years old						
Trend 1—slight attack		100	106	112	117	120
Trend 2—moderate attack		83	90	97	102	107
Trend 3—appreciable attack		59	70	78	83	89
Trend 4—severe attack		41	50	58	66	73
60 years old						
Trend 1—slight attack		90	97	102	107	112
Trend 2—moderate attack		69	79	85	92	96
Trend 3—appreciable attack		49	58	66	72	78
Trend 4—severe attack		30	39	48	56	62
100 years old						
Trend 1—slight attack		81	89	95	100	104
Trend 2—moderate attack		61	70	78	83	89
Trend 3—appreciable attack		40	49	57	64	71
Trend 4—severe attack		21	30	39	46	54
Miscellaneous						
Newly scraped mains		109	116	121	125	127
Newly brushed mains		97	104	108	112	115
Coated spun iron—smooth and new		137	142	145	148	148
Old—take as coated cast iron of same age						
Galvanized iron—smooth and new	120	129	133			
Wrought iron—smooth and new	129	137	142			
Coated steel—smooth and new	129	137	142	145	148	148
Uncoated steel—smooth and new	134	142	145	147	150	150
Coated asbestos cement—clean		147	149	150	152	
Uncoated asbestos cement—clean		142	145	147	150	

(continued)

Table 2-1 C-factor values for discrete pipe diameters (continued)

Type of Pipe	1.0 in. (2.5 cm)	3.0 in. (7.6 cm)	6.0 in. (15.2 cm)	12 in. (30 cm)	24 in. (61 cm)	48 in. (122 cm)
Spun cement-lined and spun bitumen-lined—clean		147	149	150	152	153
Smooth pipe (including lead, brass, copper, polyethylene and PVC)—clean	140	147	149	150	152	153
PVC wavy—clean	134	142	145	147	150	150
Concrete—Scobey						
Class 1—Cs = 0.27; clean		69	79	84	90	95
Class 2—Cs = 0.31; clean		95	102	106	110	113
Class 3—Cs = 0.345; clean		109	116	121	125	127
Class 4—Cs = 0.37; clean		121	125	130	132	134
Best—Cs = 0.40; clean		129	133	138	140	141
Tate relined pipes—clean		109	116	121	125	127
Prestressed concrete pipes—clean				147	150	150

Source: Lamont (1981).

Table 2-2 Equivalent sand grain roughness for various pipe materials

Material	Equivalent Sand Roughness, ε (ft)	(mm)
Copper, brass	$1 \times 10^{-4} - 3 \times 10^{-3}$	$3.05 \times 10^{-2} - 0.9$
Wrought iron, stgeel	$1.5 \times 10^{-4} - 8 \times 10^{-3}$	$4.6 \times 10^{-2} - 2.4$
Asphalted cast iron	$4 \times 10^{-4} - 7 \times 10^{-3}$	$0.1 - 2.1$
Galvanized iron	$3.3 \times 10^{-4} - 1.5 \times 10^{-2}$	$0.102 - 4.6$
Cast iron	$8 \times 10^{-4} - 1.8 \times 10^{-2}$	$0.2 - 5.5$
Concrete	$10^{-3} - 10^{-2}$	$0.3 - 3.0$
Uncoated cast iron	7.4×10^{-4}	0.226
Coated cast iron	3.3×10^{-4}	0.102
Coated spun iron	1.8×10^{-4}	5.6×10^{-2}
Cement	$1.3 \times 10^{-3} - 4 \times 10^{-3}$	$0.4 - 1.2$
Wrought iron	1.7×10^{-4}	5×10^{-2}
Uncoated steel	9.2×10^{-5}	2.8×10^{-2}
Coated steel	1.8×10^{-4}	5.8×10^{-2}
Wood stave	$6 \times 10^{-4} - 3 \times 10^{-3}$	$0.2 - 0.9$
PVC	5×10^{-6}	1.5×10^{-3}

Source: Compiled from Lamont (1981), Moody (1944), and Mays (1999).

Hazen-Williams Formula

$$V = 1.318 C r^{0.63} s^{0.54} \qquad (2\text{-}1)$$

Equivalent metric equation:

$$V = 0.849 C r^{0.63} s^{0.54} \qquad (2\text{-}1\text{M})$$

The head loss h_f may be calculated from:

$$h_f = \frac{4.72 Q^{1.852} L}{C^{1.852} D^{4.87}} \qquad (2\text{-}2)$$

Equivalent metric equation:

$$h_f = \frac{10.65 Q^{1.852} L}{C^{1.852} D^{4.87}}$$ (2-2M)

Where:

V = mean velocity, ft/sec (m/sec)
C = Hazen-Williams coefficient
r = hydraulic radius of pipe, ft (m)
S = slope of energy line
h_f = head loss [ft (m)] in pipe length L, ft (m)
Q = discharge, ft³/sec (m³/sec)
L = length of pipe, ft (m)
D = nominal inside diameter of pipe, in. (mm)

From a purely theoretical standpoint, the C-factor of a pipe should vary with the flow velocity under turbulent conditions. However, the difference is usually within the error range of the roughness estimate, and therefore the C-factor is assumed to be constant regardless of flow. If C-factor testing is done at very high velocities (velocities > 10 ft/SEC [3 m/sec]), a significant error can result when the resulting C-factors are used to predict head loss at low velocities (Walski et al. 2007).

Darcy-Weisbach Formula

$$H_L = f\left(\frac{L}{D}\right)\left(\frac{V^2}{2g}\right)$$ (1-3)

Equivalent metric equation:

$$H_L = 0.3048 f\left(\frac{L}{D}\right)\left(\frac{V^2}{2g}\right)$$ (1-3M)

Where:

H_L = head loss [ft (m)] in pipe length L, ft (m)
f = Darcy-Weisbach friction factor
L = length of pipe, ft (m)
V = mean velocity, ft/sec (m/sec)
D = nominal diameter of pipe, in. (mm)
g = acceleration of gravity, 32.2 ft/sec² (9.81 m/sec²)

The friction factor, f, is determined from the Moody Diagram (Figure 2-3), knowing the pipe relative roughness (e/D) and Reynolds number (R), as described in AWWA Manual M11, *Steel Pipe—A Guide for Design and Installation*. Most modeling software packages require that pipe roughness be entered into the model, and then f is calculated automatically.

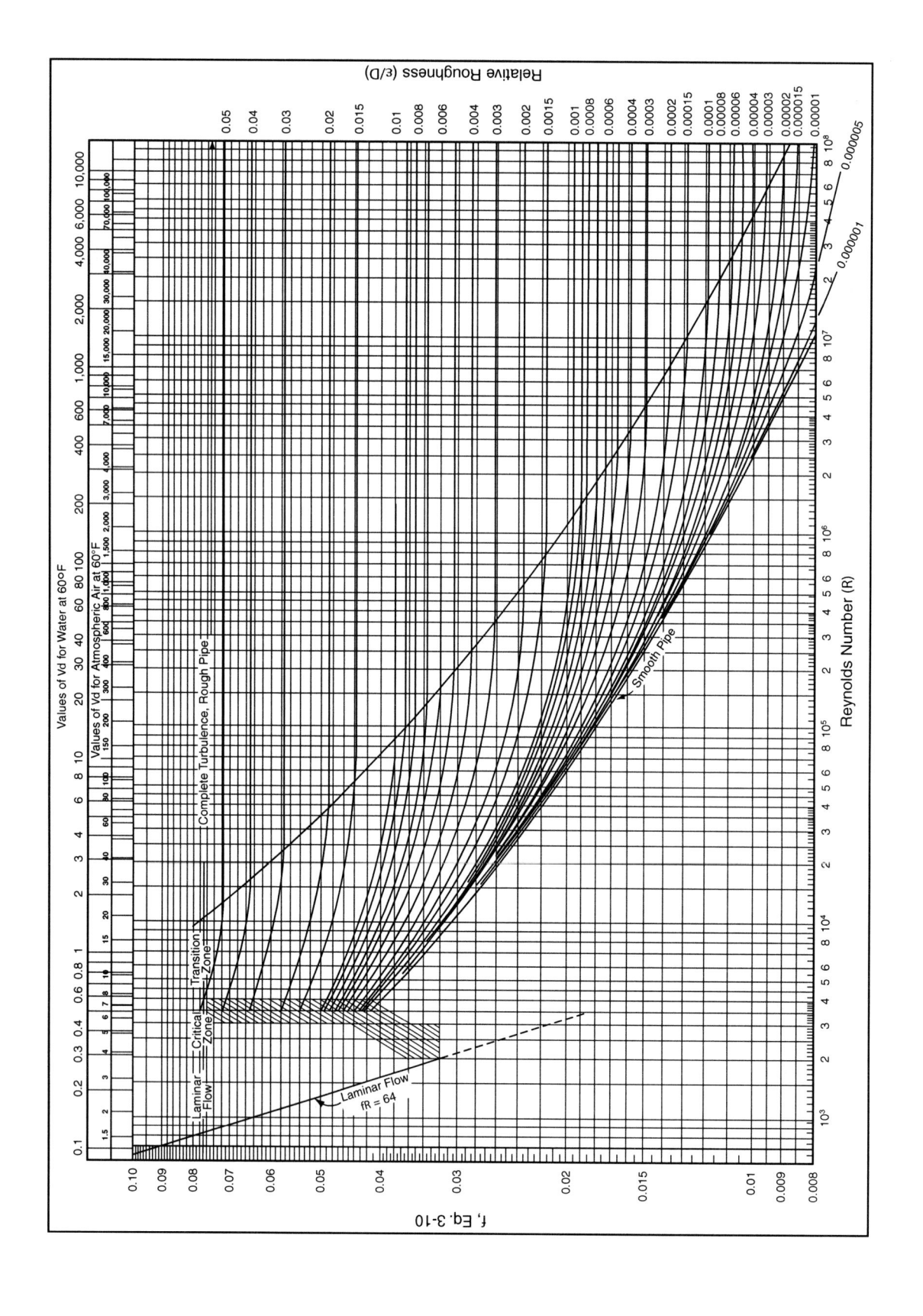

Figure 2-3 Moody Diagram

2.4.1.2.3. Minor Losses. Minor losses occur at valves, tees, bends, and other appurtenances within the distribution system due to turbulence within the flow of the water. Minor losses are computed by multiplying a minor loss coefficient by the velocity head. At low velocities observed during normal conditions—in the 1 to 2 ft/sec (0.3 to 0.6 m/sec) range—the velocity head and resulting minor loss are very low. Minor losses are not typically applied to all fittings in a distribution system model.

If a model is calibrated without explicitly accounting for minor losses in the distribution system, the roughness factors resulting from the calibration effectively incorporate the impact of minor losses. However, minor losses are often input at facility locations such as pump stations and control valves to improve accuracies and account for losses due to the higher velocities at those locations, and at hydrant service laterals (to represent hydrant bends and fittings).

A model with minor losses included will produce more accurate representation of the physical facilities within the model. Minor loss K coefficients are typically determined experimentally and are found in tables from various manufactures or text books. Many times they are expressed in terms of pipe equivalence. Data are available for many different types of fittings and appurtenances. Examples of typical minor loss coefficients are shown in Table 2-3.

2.4.1.2.4. Pipe Direction and Check Valves. In general the pipe direction as defined in the model by the *from node* and the *to node* is not important. However, for pipes that contain check valves, pumps, and controls valves, the direction must be correct and consistent with the direction of flow. Pipes that contain these elements must provide hydraulic representation that reflects the actual direction of the device. Pump and control valve elements are discussed in more detail later in this chapter.

Check valves permit flow through a pipe in one direction only. They are used occasionally in distribution systems to prevent flow from moving across a zone or other type of boundary. Check valves are located on the discharge side of many pumps, but these valves do not necessarily need to be included in the model because most modeling software does not allow reverse flow through pumps with a defined pump curve. However, check valves within the distribution system should be modeled.

Table 2-3 Typical minor loss coefficients*

Type of Component or Fitting	Minor Loss Coefficient
Tee, Line Flow	0.2–0.4
Tee, Branch Flow	0.8–2.0
Elbow, Regular 90°	0.3
Elbow, Long Radius 90°	0.7
Elbow, Regular 45°	0.1
Elbow, Long Radius 45°	0.2
Globe Valve, Fully Open	7–10
Gate Valve, Fully Open	0.15–0.4
Gate Valve, ¼ Closed	0.3–1.5
Gate Valve, ½ Closed	2–4
Gate Valve, ¾ Closed	20–30
Swing Check Valve, Forward Flow	2–4
Ball Valve, Fully Open	0.05
Butterfly Valve, Fully Open	1.0–1.5

* Some software packages have an extensive library of minor loss coefficients.

2.4.2. Modeling of Facilities

Model facilities include reservoirs (water sources), tanks (water storage), pump stations, wells, connections to other systems, and valves such as PRVs, pressure-sustaining valves (PSV), and so on. These elements may or may not exist within a GIS and may require modification or additional detail in the model. GIS facility elements are routinely oversimplified for mapping purposes and do not require hydraulic model detail. For example, the GIS may represent an entire pump station with multiple pumps by a single point feature. This same facility requires significantly more detail within the model. Figure 2-4 illustrates the example of a pump station represented in GIS that includes one point (PS) compared to the multiple elements used for the same facility in the hydraulic model (P representing each pump within the pump station, including upstream and downstream piping).

2.4.2.1. Water Storage Facilities. A water storage facility allows water to be stored from, or drained to, the water system. Hydraulic model elements representing water storage facilities are either tanks or reservoirs. These elements are used to store and release water as demands vary throughout the day within the water system network.

Tank elements require geometric input including diameter, water levels, volume curves, and base elevation. The base elevation plus water levels serve to define the HGL, which may vary based on system demands. The water level indicating an empty tank or the minimum level limit for available storage should also be determined. The tank would be modeled as empty when drained to this minimum level. In a steady-state simulation, initial tank levels are static and are used to define the boundary HGL and total tank volume. In extended-period simulations, tank levels fluctuate in response to system demand and using an accurate tank diameter or tank volume curve becomes a critical part of model validation. A tank can have more than one pipe connected to it, which could represent separate drain and fill lines. In a model, tank elements represent ground level and underground storage facilities, elevated water storage tanks, standpipes, or hydropneumatic tanks.

A reservoir represents a fixed hydraulic grade supply source, an unlimited supply source such as a large lake, a surface water treatment plant, or a groundwater basin. Reservoirs are typically modeled with only one pipe connecting them to the system. These characteristics allow groundwater supplies (such as aquifers) to be modeled as the source of supply for groundwater wells. When modeling a well using a reservoir, the reservoir level should be set to the pumping water level of the aquifer.

For some wells, the pumping water level may vary during the year. In this case, the modeler may wish to select the lower level, the typical level during maximum day demand, or an annual average. Selecting levels that are higher than the actual levels for a specific demand period can result in overestimating the production capacity of a well, causing it to operate at a different point on its pump curve and affect discharge pressures, model calibration, and distribution system recommendations. If modeling water treatment plants that deliver flow to the distribution system without pumping, a flow control valve is required in the model downstream of the reservoir to simulate variable delivery rates.

2.4.2.2. Pumping Stations. Pumps lift water to higher elevations and add energy into the water distribution system. Pumping requirements depend on system demands, the network layout, ground topography, and system losses.

A pump station is typically represented as a single point in GIS for mapping purposes, regardless of how many pumps are located at the station. The use of a single pump element to model an entire pump station should be avoided whenever possible. It is preferred that each pump be modeled as its own pump element, with its own pump curve and controls. There is a continuing trend to accurately represent the true piping configuration at pump stations in the GIS.

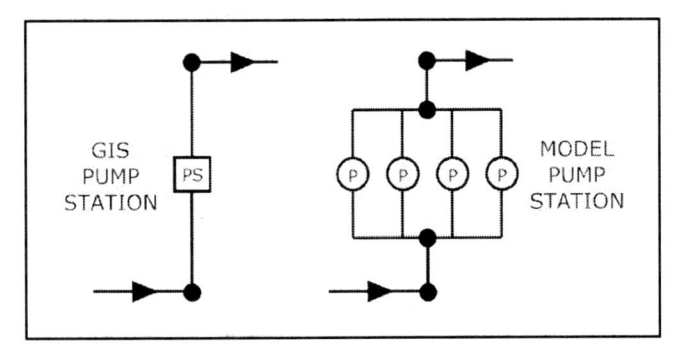

Figure 2-4 GIS detail versus model detail

To simplify identification of total flow through the station, modeled pumping stations may include one dedicated suction pipe into the pump station for each pressure zone that feeds the pump station and one dedicated discharge pipe leaving the pump station for each pressure zone that pump station serves. The dedicated suction and discharge pipes allow the modeler to show the entire flow into or out of the station from and to each pressure zone. In some cases where the model is attempting to reflect the actual station geometry as accurately as possible, total station flow may be available only by adding the flow from two or more pipe elements.

2.4.2.2.1. Pump Curves. Most pumps used in distribution systems are centrifugal or turbine pumps with either a constant or variable speed drive.

A pump characteristic curve is the unique representation of total dynamic head (TDH) delivered (y-axis) versus flow rate (x-axis) for a particular pump as illustrated in Figure 2-5. The size of the pump and the shape of the impeller determine the characteristic curve of a pump. Variable speed pumps have a family of curves, where each curve characterizes the pump at different operating speeds. In most cases, pump characteristic curves are available from the pump manufacturer or supplier. Existing pumps should be tested periodically to determine their characteristic curves, as impeller wear over time affects pump performance.

When designing a water distribution system and selecting pumps, the range of system operating conditions should be considered. A very flat pump curve results in larger changes in flow rate for smaller changes in pressure. In contrast, a very steep curve produces relatively small changes in flow rates over a relatively wide range of pressures.

It is recommended that the most accurate and detailed pump curve data available be used in the model. Compiling the most comprehensive data available to create detailed pump curves is the preferred approach and often starts by locating the manufacturers' pump curves for all pumps and validating these curves through field pump testing. In order of preference and accuracy, the best information for developing pump curves is (1) recent field tests of head versus capacity, (2) performance tests conducted during pump manufacture for initial approval, (3) manufacturer's standard pump curve for that particular unit, and (4) reported pump capacity and head (design point).

Most modeling software allows pump performance information to be entered in multiple ways. Some allow the input of a single design point, head versus flow, and then the software assumes a curve fit intercepting the shutoff head, the design point, and some lower point representing maximum flow at zero head. The curve assumed by the model may not reflect the actual pump performance characteristics. Depending on the type of analysis, it may be best to input actual pump curve information using multiple points of head and flow versus a single point that represent an assumed pump characteristic curve.

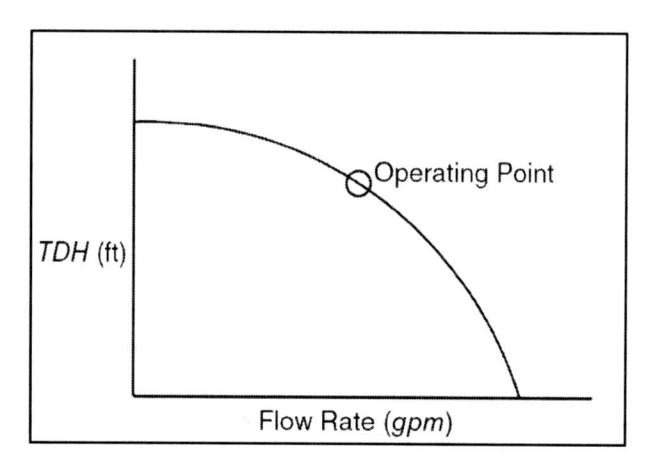

Figure 2-5 Pump curve

For pumps with variable speed drives, most software requires input of the pump performance at a given speed. The software then calculates pump performance at reduced or increased speeds based on distribution system hydraulic characteristics and pump affinity laws. For systems with variable speed drives, it is important to know the pump speed during the calibration period to achieve acceptable results. Pump speed is often not recorded in historical SCADA data and may need to be estimated based on recorded flow rates or discharge pressures.

2.4.2.3. Valves. The valves used in a water distribution system are broadly classified into three main categories: (1) isolation valves, (2) control valves and check valves, and (3) pressure relief and vacuum valves. Isolation valves enable the water system to isolate portions of the system for maintenance or emergency response purposes. Isolation valves are also used in the manual closed position as zone valves to create pressure zone breaks to manage pressures within the system. Isolation valves are discussed further in section 2.4.2.3.1. Check valves allow flow in only one direction and are discussed in section 2.4.1.2.4. Control valves apply a hydraulic restriction or setting to the water system, impacting either hydraulic gradient or flow rate. Control valves are discussed further in section 2.4.2.3.2. Pressure relief and vacuum valves are not typically modeled. However, for transient surge model applications the inclusion of these valves is desired. This valve application is not discussed in this chapter.

2.4.2.3.1. Isolation or Zone Valves. Manual isolation valves are typically either gate valves or butterfly valves. They are located throughout the distribution system and within pumping facilities. In most cases, valves are left either in the fully open or fully closed position and only operated during maintenance or emergency response events. Most isolation valves in the distribution system are normally open. Some valves are fully closed to separate pressure zones. These valves may be commonly referred to as "zone" valves.

2.4.2.3.2. Control Valves. There are five types of control valves typically included in hydraulic models: (1) pressure-reducing valves, (2) pressure-sustaining valves, (3) altitude valves, (4) flow control valves, and (5) throttle control valves. These valves have a sensing component—usually a solenoid valve—that opens and closes the main valve based on pressures, flows, or water levels in the distribution system.

Check valves could be included as another type of control valve. However, they differ in function substantially from the other control valves identified in this chapter. Check valves permit flow through a pipe in one direction only, are not adjustable as are the other types of control valves, and do not require a model setting—only a flow direction.

Pressure-reducing valves (PRVs) detect pressure on the downstream end of the valve and will throttle if necessary to ensure that the downstream pressure does not exceed a set value. They are used in the distribution system where there is a substantial drop in topographical elevation. Without the PRV, excessive pressures can develop in the distribution system. When multiple PRVs service a single area, the potential exists for the valves to work against each other as pressures oscillate between the valves within the zone to reach equilibrium between the valve settings. Setting the pressure points to a slightly different grade line, making one of the valves predominate, sometimes alleviates this problem.

PSVs measure the pressure on the upstream end of the valve and throttle to maintain a setback pressure in the system. They are often used as control valves for reservoirs where flow to refill the reservoir, after depletion from pumping, is throttled. The reservoir feed is regulated to ensure that customers serviced along the main have adequate supply pressure.

Altitude valves are used in conjunction with storage facilities to automate the filling cycle of the reservoir while preventing tank overflow. A level sensor opens and closes the altitude valves based on two (high and low) level set points. Altitude valves are either fully open or fully closed, so they can impose a large demand on a water distribution network when they open—for this reason, they are often used in conjunction with PSVs that throttle flow.

Flow control valves are used to regulate the rate of flow of water. Valve position is adjusted either automatically or by operator control through the SCADA system, based on the desired flow rate through the valve. Typical applications of a flow control valve in the water system may include limiting the supply to a customer at a predefined or not-to-exceed flow rate, or the delivery of water from one pressure zone to another at a rate that allows for adequate fluctuation of water level in storage facilities.

Throttle control valves are used to induce head losses in the distribution system. Flow control valves are in a strict sense throttle control valves because a valve is throttled to deliver the desired flow. However, for this discussion and for modeling purposes, throttle control valves are different from flow control valves. Throttle control valves may be used in the model to simulate a partially closed valve in the distribution system. The known occurrence of partially closed valves is not very common but occasionally exists. A throttle control valve may be used to restrict flow in one of a set of transmission mains, thereby forcing flow into one of the pipes and creating better hydraulic conditions within a pressure zone. Or they may be used to simulate a partially open valve between two areas in the distribution system where such a situation exists.

2.4.2.3.3. Valve Modeling Tips. In certain cases, the use of a single valve element to model multiple valves can provide acceptable hydraulic results while also increasing the stability of the model. Two or more valves should not be directly connected in series. Instead, a pipe should be used to separate the two valves. It should be noted that this may cause the model to become unstable. A single valve should never have more than two pipes connected to it (one upstream pipe and one downstream pipe). Valve stations should be modeled with one suction pipe and one discharge pipe to make it easier to record the entire flow into or out of the station.

Many utilities keep a record of the settings of the pressure reducing valves in their systems. During model construction, these may be taken at face value and entered directly into the model without the realization that the settings may be estimates, that the actual pressure setting may be significantly higher or lower than the originally documented number, or that some valves have been adjusted since the last time the records were updated.

The modeler should also keep in mind that minor losses are often input for throttle control valves. Globe valves can have high head losses resulting from the geometry of the valve. Valves that are modeled with a variable position can have a curve input to the model that defines the relationship between minor loss coefficient and the valve position.

2.4.2.4. Hydrants. When planning a model, the modeler should establish how fire hydrants will be simulated in the model. There are several options available:

- Adding a representative junction along the water main at the hydrant lateral to represent the hydrant location.

- Including the detail of the hydrant lateral and the hydrant in the model.

- Not breaking the water main and relying on the nearest junction to represent the hydrant.

The short- and long-term needs of the model should be considered when choosing among these options. For instance, if the model is developed primarily to generate fire flows, the nearest-junction approach may not be suitable. In this case, if the water main is not split in the GIS at the hydrant lateral, it may be valuable to request that the GIS group add a fitting in the GIS to represent the hydrant lateral connection point and splitting the main at this point. This approach will help ease the process of long-term model maintenance.

Hydrants are typically represented by junctions in the model, although some hydraulic model software provides the ability to model a hydrant as its own element type. In the field, hydrants are located at the end points of hydrant service laterals. If a model includes these service laterals, the number of pipes in the model is much greater. Due to this increase, it is recommended that these service laterals not be included in the model.

2.4.2.5. Turnouts. Turnouts are connections to an adjacent system where water can be received or delivered. Most hydraulic modeling software packages do not have a single model element to represent a turnout. Therefore, turnouts must be modeled using other elements that are available. Modeling a turnout connection supplying a system can be defined using two elements: a *reservoir* and a *flow control valve*. The reservoir is used to simulate the water source and the pressure by which the water is served. The flow control valve regulates how much water is provided to the system from the source. Used together, these elements can accurately represent a turnout as a supply delivery point into a system. For turnouts that represent flow out of a system, no special modeling techniques may be required. It may be sufficient to model this situation as a typical node with an assigned demand. Modeling bidirectional flow through turnouts is more complicated but could be accomplished using a combination of reservoirs, demands, and flow control valves in conjunction with operational controls.

2.4.2.6. Wells. Hydraulic modeling software packages do not have a single model element for a groundwater well. Therefore, wells must be modeled using other elements that are available. Most typical approaches to model a well include:

- A negative demand can be assigned to a junction that acts to input water into the system at the well location.

- A reservoir and a pump best represent the actual operation of a well. The reservoir is used to simulate the groundwater basin from which the well is pumping. The pump is used to simulate the well pump and motor. Used together, these elements can accurately represent a well.

When modeling the groundwater basin as a reservoir, it is important to set the HGL of the reservoir to the pumping (not static) water elevation for the well. Usually reported as pumping water level (this is the depth from the wellhead to the water surface while the pump is running), the HGL would be calculated as the wellhead elevation minus the pumping water level. Alternatively, the static water elevation minus draw down would also provide the pumping water elevation. In some cases, the pumping water elevation may be negative, indicating that the water level is below sea level. Care should be taken to define the gradient of the reservoir that represents the groundwater pumping level. These water levels can change dramatically depending on time of year and the groundwater influence of pumping from other tributary wells.

2.4.3. Quality Assurance Steps

2.4.3.1. Topology and Connectivity Review. Topology and connectivity review of GIS data imported into a model is necessary after an import from the GIS. The following issues are specific to nodes and pipes but can also be applied to tanks, pumps, valves, and reservoirs.

Many software packages have built-in tools to identify and resolve the following topology issues. It is important to note and communicate the model building impacts of these topological issues to those supporting groups providing system data layers. In the following connectivity scenarios, flow may or may not be delivered across the intended pipe segments. If this condition exists throughout the model, flow patterns, system losses, and resulting pressures would differ significantly in the model from actual field conditions. Certain types of issues may prevent a model from running until they are resolved. Field flow tests are an excellent way of validating network topology as well as locating closed valves that should be opened.

Nodes in Close Proximity: Nodes in close proximity designate nodes that are within a very close tolerance of each other, and in most cases, overlap as shown in Figure 2-6. For a model to run properly and before demands are allocated, it is imperative to identify and remove unnecessary nodes and connect appropriate pipe segments.

Pipe-Split Candidates: Pipe-split candidates represent separate pipe segments that should be connected by a common node as depicted in Figure 2-7. It is important to identify where these occur and make sure that the pipes are split where necessary.

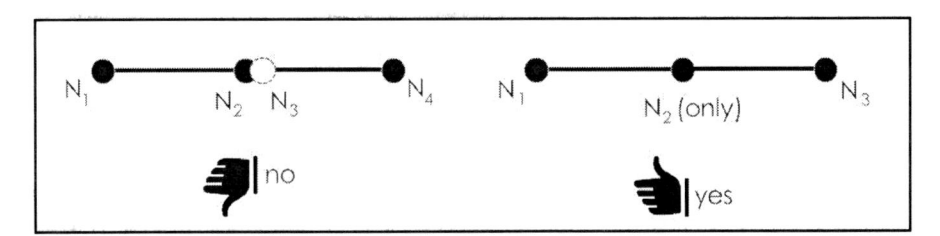

Figure 2-6 Nodes in close proximity

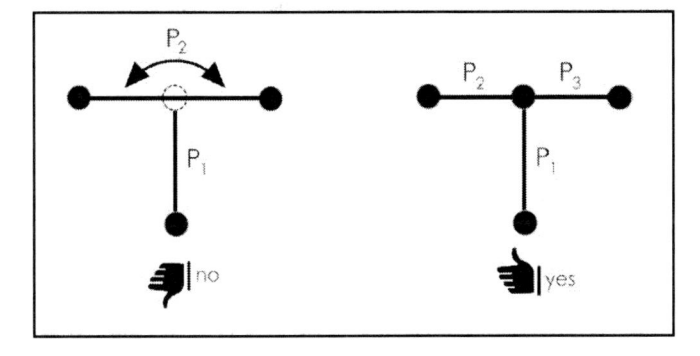

Figure 2-7 Pipe-split candidates

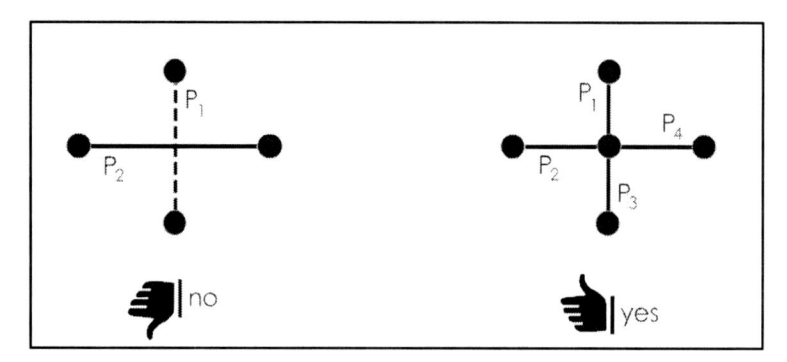

Figure 2-8 Intersecting pipes

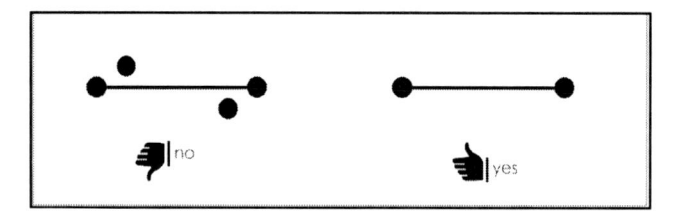

Figure 2-9 Disconnected nodes

Intersecting Pipes: Crossing pipes refer to those pipes that cross but do not intersect at a common node as shown in Figure 2-8. These pipes should be identified and examined carefully to determine whether or not they should be split at intersections. It may be helpful to consult an atlas map of the system to assist in determining if pipes should be intersected.

Disconnected Nodes: Disconnected nodes, stray, or orphan nodes (i.e., nodes that are not connected to any pipe) are separated from the piping network as shown in Figure 2-9. For a model to run properly and before demands are allocated, all disconnected nodes must be resolved. Sources of disconnected nodes may include missing pipes, inaccurate node attribution, duplication of nodes, or other.

Diameter Discrepancies: Diameter discrepancies are typically identified when pipes with significantly different diameters are connected in series. For example, a pipe segment identified as a 36-in. (91.4 cm), connected to a 3-in. (7.6 cm), connected to another 36-in. (91.4 cm) pipe is most likely an error. These diameter discrepancies should be identified and consulted with an atlas map or record drawings to determine if their diameters should be modified.

Parallel Pipes (Duplicate Pipes): Duplicate pipes are superimposed pipes that share the same shape. When duplicate pipes have been identified, the user can then determine if a parallel pipe actually exists and, if so, reroute (redraw) the pipe in a noticeable manner, create an equivalent pipe representing the same volume as the two parallel pipes, or remove/delete it from the model if such a duplicate pipe does not actually exist as shown in Figure 2-10.

Disconnected Pipes: Disconnected pipes or orphan pipes are pipe segments that have no end (terminal) nodes and are therefore not connected to the distribution system. Once disconnected pipes have been identified, junctions should be added to their end points as appropriate. Many software packages will allow the user to automatically add junction nodes at their end points as shown in Figure 2-11.

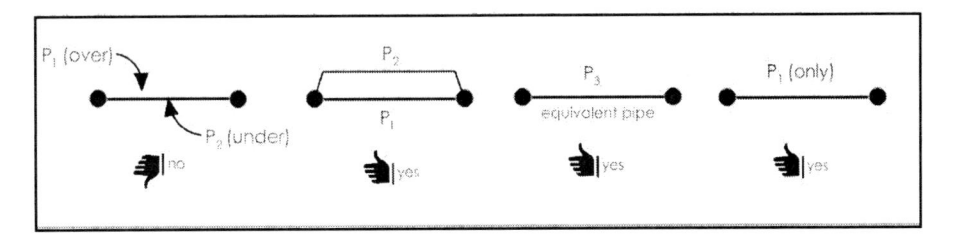

Figure 2-10 Parallel pipes

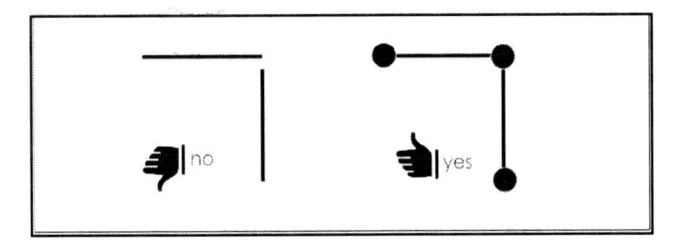

Figure 2-11 Disconnected pipes

2.5. DEMAND DEVELOPMENT

Demand data are added to the model after facility information has been attributed and distribution system topology has been established. There are four steps typically considered when allocating demands to the model: (1) determine the types of water demand, (2) determine the sources of water demand, (3) allocate base demands to the model, and (4) adjust base demands for planning, operational, or design scenarios. There are multiple methods for obtaining, reconciling, and determining demands, then loading these demands into the model. Several of these methods are presented in this section. The modeler will need to evaluate the modeling objectives, budget, and available data in order to select the most appropriate demand calculation and allocation method.

2.5.1. Determine Types of Water Demand

The first step in determining the types of demand within a given water system is to review unique characteristics and land use designations of that system. The information collected should include a listing of major customer types in the community. Irrigation practices and special industries should also be considered, as they may have unique usage rates and/or patterns that require special consideration in the model.

The five customer usage types or sources of water demand that are most often seen in hydraulic analysis include: residential, commercial, industrial, wholesale demands, and water loss. In some communities, additional customer classes, such as multifamily, government, and institutional may be required. Large facilities, such as military bases, prisons, universities, and hospitals, should be kept separate from the other customer classes, because each of these facilities will have its own consumption pattern related to the nature of its business. The modeler will need to select the customer classifications most appropriate for the water system being modeled and the data available to develop the demands.

2.5.1.1. Residential Demands. Residential demands consist of domestic and irrigation consumption.

Domestic consumption consists of indoor water used in flushing toilets, washing, cooking, and drinking. Conservation efforts—as well as changing building codes—have resulted in reduced average day consumption.

If using general per-capita, parcel size, or land-use methods to calculate consumption, the age of a neighborhood and the renovation rates in a community may be important factors to consider when developing demands.

Irrigation consumption consists of outdoor water used on lawns and gardens. This component of consumption is community specific and dependent on yard sizes and climate. A number of communities have water conservation programs that specifically target outside water use. These conservation programs often have a significant effect on the magnitude of irrigation demand. Again, this may be a consideration if using per-capita, parcel size, or land-use methods to develop demands. It is less of a consideration if the demands are developed directly from customer service information.

2.5.1.2. Commercial Demands. Commercial customer usage types typically consist of demands by stores, restaurants, gas stations, offices, and so on. Commercial establishments may have fairly consistent daily usage rates and predictable fluctuations over any given day. The hours of operation and the type of building influence the demand pattern. Depending on the proportion of commercial developments in the community, it may be necessary to separate the commercial category into two groups, depending on general hours of operation—for example, delineate offices and shopping centers that may have different hours of operation or laundromats and car wash businesses.

Historical commercial flow data should be used with caution in forecasting future demands. A number of changes in the design of air conditioners, toilets, and other water fixtures have resulted in significant water cost savings for businesses. Again, the modeler should be aware of renovation changes in the commercial sector and conservation measures initiated by the local utility.

2.5.1.3. Industrial Demands. The industrial customer category consists of all large water customers and is typically comprised of manufacturing plants. Water service to these types of sites may be metered at a location far away from the actual site, so the careful allocation of demands to the proper location on the distribution system is important. Each industrial site may have a unique flow pattern based on hours of production and water use characteristics. Field flow measurements may be helpful to determine diurnal patterns of large customers. Seasonal variation is usually not an issue, but variability according to the day of the week may be significant.

A survey of the large customers in a community is generally recommended to determine their water use patterns and demand characteristics. Other departments with the city or utility may have reports or permits from these large industrial users that may be helpful in determining their average demand and diurnal patterns.

2.5.1.4. Wholesale Demands. Wholesale demands consist of water sold to another water utility. Wholesale demands may have unique diurnal patterns for the delivering system. In many cases, the wholesale customer system has its own storage facilities, resulting in a more constant demand on the delivering system over the course of a day. Wholesale customers may be contractually limited to the rate of water delivered on a daily or hourly basis, and delivery may be physically restricted though a pumping station or control valve. Field flow measurements may be necessary to determine the appropriate diurnal pattern to apply to the model.

2.5.1.5. Water Loss. The system loss is the difference between the total water supply to the system and the authorized billed and unbilled water consumption. Water loss in the distribution system must be added to demands in the model so that the total water supplied will equal the total water demand. A good reference for managing water losses is AWWA's Manual M36, *Water Audits and Leak Detection.*

The amount of water loss in a distribution system may vary widely from system to system. Values ranging from 4 to 70 percent of the total water produced have been reported. Also, the amount can vary from year to year within the same system. The higher values are generally associated with communities that are not fully metered, have soil conditions that prevent surfacing of water from main breaks, or communities that do not maintain an active meter testing/replacement program.

2.5.1.6. Sources of Water Loss. Water loss is always present in a water system. Water loss consists of two distinct components: real loss and apparent loss. Real loss is the physical loss of water from the distribution system, including leakage and tank overflows. Apparent losses including meter inaccuracy, billing error, and unauthorized use.

Water loss results from several factors. These factors include unidentified leaks in pipes, main breaks, fire hydrant flushing, reservoir drainage for maintenance, unauthorized use, unmetered services, inaccurate and nonfunctioning meters, and on-site water plant usage.

Water plant site usage consists of water used for filter backwashing, chemical mixing, rinsing tanks, and sanitation purposes. In certain cases, this water is not metered and can represent as much as 5 percent of the total water production.

2.5.1.7. Allocation of Water Loss. Depending on the significance and location of the water loss, there may be different model demand allocation approaches to consider, including but not limited to: a single-point demand on the system representing the estimated location of significant water loss, an even distribution across the entire network model to consider the water-loss demand systemwide, or zone-by-zone specific allocation, if the water loss can be calculated by pressure zone.

The water loss is usually divided equally to all nodes in the network because specific or isolated causes are difficult to determine, unless district zone metering has been created throughout the distribution system. Systemwide district zone metering allows a more accurate allocation of water loss. To increase allocation accuracy, water utilities have used leakage tests in subareas of the distribution system for prorating the water loss to other areas having similar characteristics, such as pipe material, soil type, and age of water mains.

It is important to note that most steady-state analyses are based on the extreme conditions of peak hour or maximum day, plus fire flow, which are multiples of the average day demand. Under these conditions, the influence of water-loss allocation may be negligible when compared to total system average day demand conditions.

2.5.2. Determine the Sources of Water Demand

Water demand within a water system can be determined using multiple sources, each of which may offer varying levels of quality, completeness, and accuracy, and should be evaluated carefully before allocating to the hydraulic model.

Customer information systems provide water demand at the customer meter (commonly referred to as *water consumption data*). Water consumption data may benefit the hydraulic model by providing both the amount and location of usage, which may improve the accuracy of subsequent water system analyses. Additional sources for estimating water demand within the water system include water production by pressure zone, land use layers, and population data. Each data source may reasonably represent water system usage and may be acceptable for the hydraulic model, while providing varying levels of accuracy. Each data source can be used independently or as support to validate the primary data source used.

2.5.2.1. Customer Information System (CIS). Customer information systems (CISs) store records of water sales at the customer meter. CIS is a database of customer water consumption that provides an excellent source of data to determine and allocate water demands to the hydraulic model. CIS provides two unique advantages for customer water usage as compared to other data sources: first, it represents actual usage over a given time period, and second, it provides the spatial location of where water consumption occurred. These are strengths of CIS that make it a primary and preferred source for water demands for the hydraulic model.

CIS typically tracks the following information for each customer:

- Customer name

- Account number

- Address (service location and billing location)

- Meter-read water usage (actual water consumption data)

- Customer type (customer/demand classes)

- Meter number (meter number may also be available in GIS, providing an easy way to develop the demand distribution)

- Date period for each meter reading

- Average water usage (average water consumption)

- Meter information (make, size, etc.)

Most billing systems contain monthly or bimonthly water consumption information, water use categories, meter route numbers, and customer addresses. Certain billing systems contain additional information, such as number of residences for a multifamily account or business type for a commercial account.

Water consumption data from a billing system may be collected as 12 consecutive months of data for the purpose of generating an average day demand. Consumption data may also be collected for a particular season or period of interest for the purpose of calculating a higher or lower demand and peaking factor relative to the average day demand. Some meters may be read on a monthly basis, while others may be read on a bimonthly or quarterly basis. This meter reading frequency should be confirmed and accounted for prior to calculating the average demand for each customer.

2.5.2.2. Other Sources of Water Demands. If adequate consumption data are not available from CIS, other techniques can be used to estimate and assign demands to the model. The three most common demand calculation techniques are based on production data, land-use classifications, and population counts.

Often, all three techniques are used to develop estimated demand allocation. These techniques are also used to model future growth scenarios.

2.5.2.2.1. Water Production Data by Pressure Zone. Water utilities typically maintain records of total daily production supplied to the distribution system. Depending on the availability and accuracy of production records for each facility, water production may be allocated across the hydraulic model or by pressure zone. If allocated across the hydraulic model systemwide, it is useful to reference population and land-use data sources for spatial allocation.

To calculate demands by pressure zone, a record of the flows entering and leaving the pressure zone is required. Flowmeters and changes in reservoir levels are used to estimate the total volumes consumed in each pressure zone by performing a mass balance calculation for each hour of the day within the pressure zone. Care must be taken to identify any unmetered automatic control valves or check valves that could move water across an adjacent pressure zone boundary.

Using additional information such as population counts and land-use characteristics, the calculated water demand by pressure zone can be further validated and refined.

2.5.2.2.2. Land-Use Classifications. Land-use classifications are used to estimate consumption for a variety of land-use types. The most common are residential, commercial, and industrial land uses. Land-use classifications are usually represented on a map as polygons defining an area of a given land use type. Land uses typically follow the local zoning regulations, and in most cases, there are more land-use classifications than required for water use classification. Consideration should be given to the utility's community plan for future land use (zoning) when planning for growth and considering water demand projections.

Development of water consumption using land use typically involves a water-duty factor that is associated to a land-use classification. The water-duty factor is a ratio of water usage per unit area per land use category, and uses dimensions of gallons/day/acre per land-use type. These water-duty factors can be calculated using population data or customer water consumption within a specific land-use area. It is also common that pre-existing water-duty factors have been created for a given service area or for a neighboring service area with similar characteristics for use, reference, or comparison.

Some communities have developed a GIS to track land use and zoning. GIS adds value in its ability to efficiently and accurately calculate the area of any land-use (or zoning) polygon designation (in acres), apply the water-duty factor to the land-use classification acreage (as gallon/day/acre), and generate a total estimated water demand (as gallons/day) for land-use types. The total demand generated for a land-use classification may then be assigned to hydraulic model junctions contained within the land-use polygon.

Water-duty factors per land-use category are also used to estimate future water demands for a given service area to evaluate master planning scenarios where historical data may not be available to perform demand calculations. Water-duty factors may be used to support water demand projections to validate future water demand. There is also an emerging trend of using multivariate regression demand models to predict future water demands because of the lack of historical information regarding new conservation measures or the effects of climate change.

2.5.2.2.3. Population Counts. Population counts are used to estimate consumption in residential areas. Generally, population information is obtained from government census data, and in certain cases, the local transportation department will maintain traffic count data by neighborhood for use in designing collector and arterial road systems.

The modeler assigns a population value to each node and uses a per-capita multiplier to convert the population number to a consumption value. Per-capita water consumption can vary from community to community based on lawn irrigation needs and the age of the area. Historical water consumption should be reviewed to provide a basis for estimating per-capita consumption.

Future water demand is also evaluated by multiplying changes in population densities or new service area populations by per-capita consumption rate.

2.5.3. Demand Allocation Process

Demand allocation is the process of assigning water demands to the appropriate junctions in the model. The base demand may be considered as the water demand allocated to the hydraulic model that serves as the basis for adjustment for alternate demand conditions. The hydraulic model needs realistic water demand within the network, located at or near the location where it occurs, to predict the flow distribution within the network and calculate the resulting hydraulics. Therefore, the objective of allocating base demands is to distribute the most appropriate data source of water demands across the hydraulic model to establish a representative distribution of flow within the water system network for subsequent demand adjustments or scaling.

Allocating base demand can be a highly automated effort, a semi-automated effort, or a manual effort, depending on the base data source being used, the supporting data available, and the objective of the hydraulic model application. The following section discusses the general approach to using CIS water consumption data as the source for allocating base demands to the hydraulic model.

2.5.3.1. Water Consumption Data. Allocating base demand using water consumption data is effective for hydraulic modeling because the actual water consumption—as well as a spatial representation of where that consumption occurred—can be applied. Assuming the utility has customer meters as a GIS point layer and the meters have been linked to the CIS, water consumption from these meters can be allocated to the nearest pipe or node using GIS spatial analysis tools or demand allocation tools provided by hydraulic modeling software. Thiessen polygons may be created around strategic nodes in the GIS or hydraulic model. Water consumption within each polygon can then be aggregated to the node within that polygon.

If meter locations are not available in the GIS, matching the addresses may be used to establish a point location for each customer's meter. This approach cross-references GIS street centerline addresses with the CIS customer billing address and approximates the location of the CIS meter with its associated consumption data. Utilities using GIS to manage city tax parcels can also use parcel addresses to create a spatially enabled customer meter point file from the CIS. The hydraulic modeler may use this information to link the land parcel to customer consumption data in the CIS. With the meter layer created, the water demand allocation methods previously described would apply. The modeler should ensure that the coordinate system used by the street map conforms to the coordinate system of the model nodes.

Large industrial and wholesale customers should be assigned separately to ensure that large customer demands are properly allocated. Care should be taken to ensure that consumption data are allocated to the correct pressure zone, especially near pressure zone boundaries. Thiessen polygons, created to follow pressure zone boundaries, can be useful to help ensure that consumption data from one zone do not get improperly assigned to junctions in another zone. Quality assurance is recommended during and following the customer meter creation process as CIS customer addresses may not always be representative of the actual location of water consumption.

If a high-quality street centerline shapefile or other GIS data source is not available for geo-coding, a postal code boundary map may be used instead. The modeler should ensure that the addresses used are service addresses, not billing addresses, as these may be different.

2.5.3.2. Service and Nonservice Nodes. When base demands have been accurately quantified, it is important to consider how base demands will be applied to the hydraulic model. Prior to importing base demands to the model, it is recommended that the modeler identify *service junctions* and *nonservice junctions*. Nonservice junctions are those junctions that should not be assigned a demand value. Nonservice junctions may include junctions at the suction and discharge points of pump stations, junctions immediately downstream of water storage tanks, junctions along a large transmission main, or mains from one pressure zone that cross but do not serve customers in another pressure zone. Once service and nonservice junctions have been identified and attributed appropriately in the model, the service junctions should be isolated and assigned base demands.

2.5.4. Adjusting Base Demands

With base demands allocated to the hydraulic model, it is common to adjust or scale these demands to reflect different demand conditions for calibration and analysis purposes. Depending on the data source, the demand condition of the data source (average day, maximum day, minimum day, etc.), and the intended application of the hydraulic model, different demand adjustments may be required.

Typically, base demand allocation using water consumption will be reflective of an average-day demand and scaled to reflect water production for the desired demand condition. Due to the potentially comprehensive nature of CIS, however, it may be possible to acquire water consumption data specifically for historical maximum months, which could allow for average maximum-day water consumption to be used as a base demand allocation. In this case, it would be consistent to adjust the maximum-day demand water consumption to reflect a maximum-day water production value.

It is recommended that base demands allocated to the hydraulic model be reflective of average-day demand as a consistent frame of reference for subsequent demand adjustments.

2.5.4.1. Geographic Variations. Just as the demand varies by the type of customer, it also varies depending on the neighborhood. The average size of the dwelling, the average number of water-use fixtures, age of the neighborhood and accompanying efficiency of the fixtures, and the size of lawns can result in large variations in average residential demand between older and newer neighborhoods. If the demand development and allocation process has been based on actual customer billing data, base demand allocation should account for these variations.

2.5.4.2. Calibration. For calibration purposes, it is the goal of the hydraulic model to perform similar to water system operations. As a result, base demands using water consumption data, land-use data, population data, or other sources outside of production values should be scaled to reflect the total water production in the system during the calibration period of interest. Adjusting base demands to reflect water production also generally considers water loss experienced by the distribution system, which also should be accounted for in the network. It may be challenging to accurately adjust water demands to represent the locations of water loss, but if negligible, a uniform scaling of base demands to reflect water production may be acceptable to account for the water loss component. Large customer demands may be adjusted separately to ensure accuracy of flow distribution.

2.5.4.3. Analysis. For analysis purposes, it remains the goal of the hydraulic model to perform similar to water system operations, as well as to maintain a reasonable level of conservativeness when identifying deficiencies and making recommendations. Therefore, demand adjustments for analyses may reflect water production for a given demand condition and may also consider an adjustment to account for added conservativeness, to consider maximum-day demand, peak-hour demand, and minimum-day demand (depending on the intended analyses). The conservativeness of a demand factor is based on system experience, the confidence level of the data used, other agencies with similar characteristics, the intent of the hydraulic model application, and the engineering judgment of the utility and engineer.

Demands may also be adjusted using scaling factors to reflect future demands as supported by land use and/or water demand projections for analysis of future conditions. Depending on the scenario evaluated, it may also be necessary to determine seasonal variations or hourly variations in demand.

2.5.4.4. Seasonal Variations. Seasonal variation is sometimes significant in distribution systems. Typically, seasonal variation is directly related to irrigation. A community with large institutional or tourist demands should consider the effects of these and account for them when developing a hydraulic model.

Local climate variations determine the irrigation and air-conditioning requirements. The modeler should consider the period of extended hot weather to determine when the associated peaking factor adjustment would be applied.

Systems that experience extreme seasonal variations may be analyzed with unique considerations, resulting in a customized approach to demand development and base demand adjustments within the model for analysis purposes.

2.5.4.5. Diurnal Curves. Initially, base demands are allocated to the hydraulic model as steady-state demands. That is, they represent an instant in time or snapshot of a certain demand condition.

Water demands vary over time. Representing these variations in the model requires a diurnal curve, which is a set of factors—usually on an hourly basis—that represents the ratio of hourly demand to the average demand during the selected time period.

Field-testing and/or analysis of system operating data are required to determine the specific diurnal demand curve for a community. Depending on the data available for diurnal curve calculations, a curve may be developed for the entire water system, for major geographic areas, by pressure zone, by customer, by land-use classification, or by season.

To develop a diurnal demand curve for the overall system, water supplied from treatment plants, pump stations, and groundwater wells must be recorded, in addition to storage inflow and outflow. Ideally, this information is readily obtained from the utility's distribution SCADA system. A diurnal curve is developed by summing the inflows and outflows in a discrete area for each hour of the day to calculate the demand throughout the day. This same mass balance technique is used to develop curves for hydraulically isolated pressure zones.

Diurnal demand curves for specific customer types, such as low-density residential, high-density residential, and commercial, are determined by isolating a homogeneous land-use area and monitoring the flow over a specified time period, typically 24 hours. Strap-on portable meters can be used in specific areas to measure flow such that the diurnal curve can be calculated for specific customers or groups of customers. The diurnal demand curve for the service area under consideration is derived by summing all flows into and out of the area, measured by meters surrounding the area. Figure 2-12 shows an example diurnal curve.

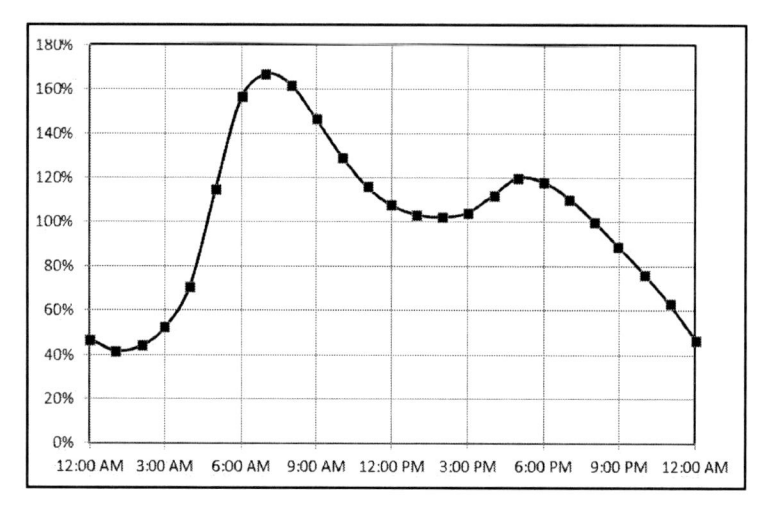

Figure 2-12 Diurnal curve

Diurnal patterns also vary significantly by season. The modeler should consider this factor when developing a sampling program to ensure that the data collected are suitable for the scenario modeled. For example, a winter sampling program is not useful for a hydraulic model assessing transmission main requirements to meet peak-hour summer demands.

2.6. OPERATIONAL DATA

Operational data consist of facility status changes involved in the day-to-day running of a water system to successfully deliver water to utility customers. Understanding, appreciating, and applying operational data are vital to hydraulic model calibration and analyses.

Operational data are generally classified as two main types: (1) automated controls and (2) manual controls. Automated controls are programmed into the utility SCADA system for automatic water system operations and consist of pump and valve control logic and set points for pumps, control valves, pressure-reducing valves, and altitude valves. Most modeling software packages include a control scheme that replicates the behavior of a SCADA system. The model automatically changes valve position or pump status in response to changes in pressure, flow, water level, or time at determined points in the model.

Manual controls are typically used to represent operator intervention, which overrides the automated controls programmed into the SCADA system. It is often challenging to develop hydraulic model control logic for manual controls because the set point to trigger a change in facility status is not predefined. Typically, the modeler will replicate the manual control using time-based controls based on when the SCADA system showed the change of facility status occurring. This approach to emulate manual controls may be suitable for model calibration, but it is typically not recommended for analysis.

Typical operational data required for hydraulic models are listed in Table 2-4. The items are often interrelated, meaning model input changes may automatically result in a recalculation of other model parameters. Operations data are used extensively in the model calibration process so model inputs and outputs agree with actual measured values within a defined calibration tolerance.

There are four main sources of operations data: (1) operations staff, (2) written records, (3) charts, and (4) SCADA.

2.6.1. Operations Staff

The most important source of modeling information is the utility operations staff. The hydraulic modeler should develop a good working relationship with this group while developing hydraulic models. Operations staff can inform the modeler of how the actual system operates as opposed to how it was designed to operate. Operators can identify specific areas to study to determine why the actual operation varies from the design, such as finding closed valves, determining incorrect set points, and identifying distribution system deficiencies. During model calibration, operations staff input helps interpret the data and identify necessary model adjustments.

Involvement of the operations staff in model development greatly enhances the acceptance of analysis results and may motivate more efficient operation of the water distribution system. For these reasons, involving the operations staff in model development results in a more accurate, usable, and sustainable hydraulic model for the utility.

Table 2-4 Operation data required by facility/equipment type

Facility/Equipment	Operation Data Required
Pumps	Pump status (on/off)
	Pump controlling tank, pressure, flow, or other
	Set points for pump control
Pressure-Regulating Valves	Upstream/downstream pressure
	Valve position and setting
	Valve controlling tank, pressure, flow, or other
	Set points for valve
Flow Control Valves	Valve position
	Flow setting
	Valve controlling tank, pressure, flow, or other
	Set points for valve
Node Pressures	Any pressure data collected at pressure-monitoring stations in the distribution system
Pipe Flows	Any in-line flow data measured in the distribution system (includes wholesale supply locations or zone boundary meters)

2.6.2. Written Records

Written records or standard operating procedures (SOPs) consist of the reports and documents that explain how the distribution system is theoretically supposed to operate. Utilities may have what is called an *operations manual* that may provide a schematic for each facility and a narrative of how each facility operates based on varying conditions. The modeler should verify these types of reports against actual operating conditions, which may be different because of the presence of constraints and conditions in the distribution system that may not have been present or recognized when the reports were written.

Typical written reports that should be reviewed are

- Operations and maintenance manuals for each facility
- Design reports and record drawings
- Operator training manuals
- Emergency manuals (to identify how things are expected to operate in an emergency)
- Historical records of deficiencies in the system (i.e., low-pressure complaints, water quality complaints, etc.)
- Maintenance records for all facilities (pump stations, reservoirs, valves, pipes, etc.)
- Valve status reports
- Hydrant flushing reports

Each utility will have a different combination of written records and in many cases the quality of the reports may not be consistent. As stated previously, operations staff offer the best assistance to determine the quality of the data collected.

2.6.3. Charts

Charts are historical records showing time-relative data. Traditionally, these have been paper records and are generally in a circular or strip format. Digital data-logging equipment has reduced the need for paper charts, but software is still required to graphically represent and interpret the collected digital data. Care should be taken to ensure that charts and digital data are calibrated.

Software packages are available from data-logger equipment suppliers that help in analysis of the data. These packages may be worth the cost to reduce the effort required when using standard databases and spreadsheets for analysis.

2.6.4. Supervisory Control and Data Acquisition (SCADA)

SCADA systems are a powerful tool for collecting and maintaining historical information on real-time or historical operation of a water system. The SCADA system database may include flows, pressures, alarms, or tank levels and equipment information such as the on/off status for pumps, valves, tank levels, and more.

The data are collected via the SCADA software and stored in database files for evaluation. The amount of data collected is determined by the polling frequency of the SCADA system. It is also possible to determine total system demand based on a mass balance of flows from the treatment plants and flow in and out of the reservoirs or change in reservoir levels.

2.6.4.1. Importance of SCADA to Modeling. Historically, only the most critical operating points were monitored via SCADA due to the cost of additional field devices and communication lines. These costs have since declined, allowing utilities to increase the monitoring points system. The hydraulic modeler should review the data being collected and possibly coordinate to have additional monitoring points installed at key locations, if budget allows.

SCADA provides a wealth of data on the actual operating conditions of the water distribution system. Differences between the model results and SCADA readings are used to calibrate the model or identify locations where field crew checks can be conducted to locate closed or partially closed valves.

SCADA can generate data files for model calibration and allows the modeler to confirm the accuracy of the model on a regular basis. SCADA software can usually extract facility records to ASCII, delimited text, spreadsheet, or other database formats to facilitate comparison with model results. Models are typically calibrated once they are initially constructed, then every few years thereafter, depending on the frequency and types of changes in the system. Some model software packages will generate log sheets identifying where the model and the system differ.

Developing common interfaces and linkages between the model and the SCADA system improves the ability of the modeler to consider operational scenarios. This is accomplished by entering current operating conditions from SCADA into the model and allowing the operator to perform "what if" analyses to test different scenarios. For example, different pump and valve combinations and the results of those combinations can be tested from the same operational starting point.

2.6.4.2. Data Requirements From SCADA. In general, SCADA data can be used two ways in modeling:

- Model input—these include boundary condition data and system control data. Boundary conditions include information on which pumps are operating, flow and pressure control points, valve status, reservoir levels, water production, and demand loading (if available). This information is required to perform steady-state analyses and to set the initial conditions for extended-period simulations. System control data are related to the changes in boundary conditions that occur over time and includes information on the change in pump or valve status and the variation in water levels or pressures at system boundaries. This information is required to perform extended-period simulation (EPS) analyses. See chapter 6 for more information on how SCADA data is used in EPS models.

- Verification or reference data—these include information collected by flow, pressure, and level monitors and may be presented in a spreadsheet alongside the model output to confirm that the model is sufficiently representing actual operations. These data may be used to provide guidance on areas of the model that should be adjusted for calibration purposes but are not used directly in the hydraulic calculations. These data are used for both steady-state and extended-period calibration.

It is important to note that SCADA data can include transient or surge data (transients are rapid changes in pressure and flow) that can provide a misleading perspective of the average steady-state conditions of the network. Transient data are typically generated during pump startup and shutdown or during changes in valve status. The use of average data instead of instantaneous data values will help ensure that the data placed into the model are a true representation of the pressures and flows in the system. A review of data values in adjacent time periods is also useful to identify anomalies in the data.

Data errors also occur during a SCADA system's communication failure with the facility monitoring point. These should be flagged in SCADA. When assembling data for the model, it may be necessary to interpolate values if some measured values appear to be missing or are not readily available.

2.6.4.3. SCADA Issues. There are several main issues that should be addressed when using a SCADA system in coordination with a hydraulic model for setup and/or calibration purposes:

- Data format—Most field instruments record data in binary format but convert to digital format in the local RTU or the PLCs. When exporting data from a SCADA system, it is helpful to generate this data in a format suitable for use in spreadsheet or database programs rather than as a simple text file. Depending on the software used, this may necessitate using an ASCII file, Microsoft (MS) Excel file, or MS Access database as a common data interchange format. Delimiters between columns of data, such as tabs or commas, are helpful.

- Units—Often SCADA data are presented in reports without units. The user must take care to know what units are being reported and make the appropriate conversions when using this data for modeling.

- Measured versus calculated values—It is important to note whether the data being used are measured directly (such as pressures) or are calculated. Flow measurements are generally calculated based on pressure differences between two points, but these data points are normally calibrated. In many instances, flow rates from multiple flowmeters (such as in a pumping station) are summed to provide a single data point. In these cases, the degree of error in the data could be higher and must be noted when comparing with modeling results.

- Links to model IDs—By far the most time-consuming aspect of relating SCADA data to models is the relationship between IDs in the SCADA system and those used in the model. There are many SCADA points that have no direct relation to hydraulic models, and these systems have distinct nomenclatures with unique tag numbers. The IDs tracked in the SCADA system are usually not the same as the IDs used in the model. Models may have simpler naming conventions but will store data in multiple tables. It is helpful to develop separate tables cross-referencing the SCADA points used in the models and the model data.

2.7. HYDRAULIC MODEL MAINTENANCE

An up-to-date hydraulic model, reflective of current operating conditions, can yield many benefits to a range of utility departments, including: engineering, operations, planning, customer service, and management. Depending on the pace of change, the type of changes, and utility objectives, hydraulic model maintenance may require frequent or regular periodic scheduling to maintain current facilities, hydraulic parameters, controls, and a reasonable level of calibration to have confidence in the hydraulic model results.

Model maintenance is an essential task to leverage the initial hydraulic model investment, and sustain and build on the benefits of the hydraulic model. Historically many models were constructed and used with minimal planning for model maintenance. Models were created periodically from scratch and were viewed as a tool used for major planning studies.

Currently, however, hydraulic models are used frequently to evaluate operational, energy, vulnerability, and water quality issues in addition to the previous more traditional uses. As current hydraulic models increase in complexity and level of detail, the development of a new model becomes increasingly more costly.

Performing updates of an existing hydraulic model rather than fully reconstructing a new model can result in significant time and cost savings as much of the water system remains unchanged and typically would not require updating. Previously constructed models may have useful information within them that may be retained and translated to the updated model. Key model parameters such as pump curves, facility connectivity, and facility controls identified and verified in a previous model can be efficiently captured and retained in the updated model.

Model maintenance can range from manually updating pipes and facilities periodically to an automated approach of updating all modeling elements from GIS. Model maintenance activities are varied but will usually fall into one of the following major groups:

- Updating model elements (pipes, nodes)
- Updating system facilities (pump stations, tanks, wells, turnouts, etc.)
- Updating system demands and diurnal curves
- Updating system operational controls and system settings for control valves
- Performing periodic calibration

Depending on the type of data and technology available to the utility, model maintenance may be database driven, spatially driven, or input manually to the hydraulic model working with as-builts and operator input. GIS-based model maintenance (which could serve both data and spatial driven updates) is the current trend as the GIS data model is being refined to recognize the detail, completeness, and data set required for more efficient hydraulic model construction and maintenance.

2.7.1. Update Frequency

The frequency of required model updates depends on the rate at which changes are occurring in the distribution system, the type of changes occurring, the intended use of the hydraulic model, and the update process in place or planned.

Significant number of changes in pipeline repair/replacements and growth or declines in population and employment within the system leading to overall changes in demand distribution and allocation may impact the overall hydraulics of a water system and influence analyses such as fire flows and resulting pipeline sizing.

The type of changes to a water system may trigger a necessary and immediate update to the model as the change substantially impacts hydraulics in a localized area of interest or even possibly systemwide. Changes, such as pump curves, valve set points, a new well, booster pump, tank, or transmission main, in addition to possible operating changes including a pressure zone reconfiguration or added system redundancies, may have a direct impact on the flow and distribution characteristics of the water system, creating a different hydraulic profile than what was seen and planned for previously.

The method and detail used to construct and calibrate the model initially for its immediate and intended purposes will help guide future applications, frequent uses, and maintenance approach. If a model was constructed and calibrated to evaluate operational and water quality scenarios, the level of detail and accuracy required of the model may be greater than a master plan application of the model.

The frequency of update may also be influenced by the type of data, technology, and ease-of-use during the update process. If the process is relatively automated, the utility may decide to update more frequently. However, if the update process is labor intensive, the utility may elect to update less frequently, with only significant hydraulic changes that may have occurred within the water system acting as triggers for a model update.

Two model maintenance options are described briefly in the following sections.

2.7.1.1. Periodic Model Maintenance. The traditional approach has been to update hydraulic models on a periodic or as-needed basis. Traditionally, the model has been updated when a water system master plan is updated. However, with the expanding capabilities and increased practical applications of the hydraulic model, the update process is being undertaken with more regularity. The model update typically includes addition of new facilities constructed since the model was first developed (or last updated), updating of model demands, and a recalibration of the updated model to reaffirm hydraulic results as compared to field performance. GIS and modeling tools support identifying and adding "new" physical facilities to the model. Updating facilities in the model that have been repaired or rehabilitated may require a database link or spatial reference to facilitate the data update. If neither exists, this process may involve the manual task of translating hydraulic and/or geometry attributes from data sources to the hydraulic model.

2.7.1.2. Model Reconstruction. Because of the improved quality of GIS data and the model construction tools available, some utilities may consider reconstruction of the entire model directly from GIS as a viable model maintenance strategy, when warranted. This eliminates the challenges associated with updating repaired/rehabilitated facilities and ensures that the hydraulic model reflects the current GIS data. Depending on the extent of this approach, information specific to the hydraulic model and not otherwise stored within the GIS database (such as roughness coefficients and demand data) may be maintained in the GIS database. This approach to model maintenance requires a thorough planning stage prior to implementation as GIS and engineering departments need to agree on methods of data storage, naming conventions, possible revisions to the GIS data model to comply with hydraulic model requirements, internal updating workflow processes, and other potentially utility-wide impactful decisions.

2.7.2. The Hydraulic Model and GIS Relationship

Currently, utilities are recognizing the importance of quantifying and justifying decisions while offering an enterprise solution to internal departments, managers, board members, and customers. The communications between hydraulic model results and GIS databases offer just that opportunity: an engineering and operational tool to support utility decision-making presented in a format that allows multidepartmental access for transparency and open communications.

Because of this identified need for accessible, supporting information, database communications between the hydraulic model and GIS are growing closer seemingly every day. These two databases, once thought of as too cumbersome, too complex, and too objectively different to work together with any regularity, are now being leveraged and synergized more frequently because of added computer processing power, broadband internet connections, high resolution video cards, and the experience, exposure, and level of data sophistication brought by the current generation of utility engineers. This trend will continue and accelerate, as the need for hydraulic model solutions grows and coincides with the unavoidable extension of technology with generational acceptance of these technologies for the needs of the utility.

There are several different forms and terminologies that describe hydraulic model and GIS communications. Wholesale reconstructions, wholesale geometry updates, wholesale or local data updates, facility additions, spatial connections, shared interfaces, gateway exchanges, database builders, and so on are all varying terms used to explain an approach to sharing data between the two separate but complementary databases. Generally, there are three categories for the method of communications between the hydraulic model and GIS: *interchange*, *interface*, and *integration*. An overview of GIS and hydraulic model communications is also provided in the US Environmental Protection Agency reference guide for utilities: *Water Distribution System Analysis: Field Studies, Modeling, and Management* and is based on terminology originally presented by Shamsi (2005).

The term *interchange* refers to the lowest level of sophistication for transferring data between the hydraulic model and a GIS. The interchange does not use any direct links between the two databases. They communicate as separate, independent data sources, and information is extracted from one database and stored in a third party database for import into the destination database. This process would represent a relatively manual approach to hydraulic model and GIS communications, and may use data formats such as MS Excel (*.xls), and comma separated or comma delimited files (*.csv) files, as third party communicators between the hydraulic model and GIS. As there is no database link between the databases, interchanging may also involve the manual process of digitizing and attributing water system facilities into either or both databases.

The term *interface* refers to a connection between the two databases to transfer information. Similar to interchange, the two systems operate independently, but there is a direct connection so that intermediate files are not necessary. Protocols and structures must be established and be compatible within the two databases to support the interface. Several hydraulic modeling software vendors offer model development and maintenance tools that provide a sophisticated interface between the GIS and the hydraulic model databases. These tools can provide a link between the two databases, allowing updates in the GIS to be transferred to the hydraulic model and vice versa, with minimal steps. Standard interfacing data formats include MS Access (*.mdb) and visual database (*.dbf) files. These files support both hydraulic and geometric data attributes, and thus communicate on both data and graphical levels between the hydraulic model and GIS. It is vital that the hydraulic model level of detail, GIS database design, verification of unique IDs, treatment of deleted elements, and network topology for all participating GIS features are well defined and established prior to implementing a sustainable interfacing solution.

Integration is the most sophisticated approach to hydraulic model and GIS communications. Pure integration would involve the two databases performing as a single entity, as a single database, with one communication that impacts both databases simultaneously. It is important to recognize that current hydraulic model software, which operates within GIS software, is hosting both hydraulic model and GIS database layers separately. Therefore, true integration is not occurring as there are separate databases being maintained, one for the GIS and one for the model. In all cases where a utility maintains a GIS and a separate hydraulic model, and even when the model is operating within the GIS environment, the second level of data transfer and sharing, or interface, is being achieved.

Sustainable GIS and hydraulic model integration is the process by which new, updated, or abandoned elements in a GIS database are incorporated into a hydraulic model; and in turn data modifications made within the hydraulic model can be synchronized back to the GIS database. Typically, the GIS database is updated prior to the hydraulic model as GIS supports a variety of enterprise-wide data applications requiring on-demand services such as mapping, field work orders, customer service, and asset management.

It is important to recognize that updating a model from the GIS is not simply importing new elements into the model that are in the GIS. Anytime model facilities are revised, model demand values, demand allocations, and operational controls may need to be revisited. In addition, care should be taken to identify what model scenarios will be included within the updated model. This will require verification of model data sets and model operation to verify the updated model is operating as expected. It may be necessary to validate or recalibrate the model.

2.7.2.1. Organizational Coordination. The hydraulic model may be maintained by the engineering department, while GIS may be maintained by a planning, information technology (IT), or dedicated GIS department. Consistent and sustainable interfacing between the two databases requires these two primary department stakeholders to coordinate on expectations, database design, construction tolerances, criteria for updates, and the internal workflow to process changes or updates thoroughly and systematically. This may also require upper management approval to authorize and support departmental staffing, priorities, changes, and utility enterprise objectives.

While utilities recognize the importance, benefits, and technological capabilities of establishing a 1:1 model and GIS relationship, few utilities have sustained a long term 1:1 relationship between the two databases as internal priorities change the pace, course, staffing, and therefore communications between the departments and maintained databases, further highlighting the importance of upper management support for resources and work processes. As a result, disconnects and discrepancies between the model and the GIS grow larger and more labor-intensive to resolve. As the gap between the databases grows, it may result in a complete reconstruction of the model. While this allows the model to represent the current distribution system as maintained in the GIS, it often fails to leverage operator knowledge such as facility controls and operational scenarios, which is vital to the successful performance and useful application of the hydraulic model. Additionally, should a reconstruction occur, hydraulic model specific attributes such as diurnal patterns and pump curves, which may not be shared with the GIS, would need to be reestablished to the hydraulic model database, quality control reinitiated, and hydraulic model calibration validated prior to performing analyses.

Achieving a common purpose and priority across the organization with the approval of upper management is essential for the success of any sustainable 1:1 model to GIS interface.

2.7.2.2. Unique GIS ID. Each element in the GIS database to be imported to the hydraulic model must have a unique element identifier (ID). This unique ID is used to tie the hydraulic model database to the GIS database and is essential to maintain a 1:1 GIS to model relationship. This ID should be unique across the entire GIS for a given model element type and should remain unchanged for a given element. The unique ID should never be reindexed or reused even if an element is deleted from the GIS. There are existing programs that can assist GIS staff in maintaining unique feature IDs. Identifying a procedure for assessing a unique ID to GIS elements may require some forethought and planning but is essential to maintaining a 1:1 model to GIS relationship. This is important for updating the model from GIS and for updating GIS with key information found in the hydraulic model.

2.7.2.3. GIS Data Tracking. Tracking fields between GIS and hydraulic model databases can greatly facilitate communications. Data useful or essential to the hydraulic model may take a significant amount of time to develop and populate. This may include data such as water demands, pipeline roughness coefficients (C-factors), fire flow junctions, fire flow requirements, land use types, and strategic designator fields specifically assigned for distribution system facility communications between the GIS and hydraulic model. If this information is stored in the GIS (from a completed model), it can be used to simplify model updates.

The modeler often faces challenges of translating GIS elements into the proper model element type. For example, most system valves are not typically modeled as a "valve" but rather as a junction. However, control valves are often modeled as a model valve. Database rules with an appropriate database structure must be developed to facilitate the maintenance or translation of GIS elements into their proper model element type to maintain a 1:1 hydraulic model to GIS relationship. This may be performed by first establishing unique, static IDs within the geodatabase among feature classes and subtypes (across the entire GIS without duplication). Using unique IDs, hydraulic model software may be used to connect the geodatabase to the appropriate model element type for importing.

2.7.2.4. GIS and Hydraulic Model Topology. *Topology* is defined as a set of rules governing the spatial relationships (connectivity) between elements within various feature classes. Topology rules define the way mains, valves, fittings, hydrants, and storage facilities are connected. Tools are available within GIS software for automatically creating and checking connectivity by applying user-defined tolerances and parameters. The GIS concept of the geometric network is an important feature to maintain, check, and control topology in GIS.

Following initial GIS connectivity checks using the topology rules a GIS geometric network should be constructed. A geometric network consists of edge and point features classes where edges can only be connected to other edges at point features and every edge must have a point feature at its two end points. During construction of a geometric network, a point feature is inserted at the end point of any edge where a point feature does not already exist. In summary, initial construction of a geometric network adds missing pipe end-point features to the GIS. These inserted junctions should then be replaced by the appropriate point feature (fitting, valve, etc.).

Once the geometric network has been created, a topological relationship exists between point and edge features. If a main or point feature is removed, the related topological features are also removed. Conversely, when a main is added to the network, a default or user-specified point feature such as a fitting, hydrant, or valve is automatically added at the end of the main. A series of connectivity rules should also be created in the geometric network to ensure that all edges (pipes) connect to the appropriate type of point. For example, a connectivity rule can be created to ensure that large mains only connect to smaller mains through a reducer feature.

A geometric network supports complex edges. With complex edges, a point feature (i.e., a valve) or line feature (i.e., a service lateral) can connect to a pipe without requiring the pipe to be split. This aspect of the complex edge makes it very suitable for hydraulic model development and maintenance. For example, features in a GIS that are not required for modeling (i.e., nonoperational valves or service lines) can be ignored during construction of a hydraulic model without eliminating network connectivity. This particular aspect of complex edges can enhance hydraulic model development and maintenance as well as reduce hydraulic modeling processing time.

Application of topology rules and construction of a geometric network complete the GIS network connectivity checking. Prior to initiation of model construction, missing GIS attribute data should be entered into the GIS database. Additionally, all GIS features must be checked to ensure that each record has a unique ID.

Once these tasks are complete, GIS data can be transferred to the hydraulic model for model development. The tools available to transfer data between GIS and the hydraulic model will depend on the modeling software used. Most hydraulic modeling software includes the capability to transfer data from ESRI geodatabase, dbf, and shapefile format to the hydraulic model.

Although the GIS topology checking tools and procedures provide a thorough review of data connectivity, a review of connectivity should also be performed using hydraulic model connectivity checks as well as the calibration process to identify errors that may have been inherited during the GIS import. In addition, there are certain types of "logical" connectivity errors that may not be identified by the GIS topology rules or during construction of a GIS geometric network. Most current hydraulic modeling applications include tools to check for these types of connectivity errors.

In the hydraulic model, pipes, valves, pumps, and storage facilities must be accurately connected to simulate real-world distribution system connectivity and performance. Errors in model connectivity can result in significant differences between actual flows and pressures in the field compared to those predicted by the model. Therefore, it is important to resolve connectivity errors in GIS to facilitate transferring data to the hydraulic model.

2.8. REFERENCES

AWWA. 2004. Manual M11, *Steel Pipe—A Guide for Design and Installation*. Denver, Colo.: AWWA.

Bonema, S.R., Elain, C., Mercier, M., and Tiburce, V. 1995. Linking SCADA With Network Simulation. In *Proceedings of the AWWA Annual Conference*. Denver, Colo.: AWWA.

Cesario, A.L. 1995. *Modeling, Analysis and Design of Water Distribution Systems*. Denver, Colo.: AWWA.

Chase, D., and Jones, G.L. 1994. Linking Hydraulic Network Models and SCADA Systems: The ABCs of System Integration. In *Proceedings of the AWWA Computer Conference*. Denver, Colo.: AWWA.

Deagle, G., and Ancel, S.P. 2002. Development and Maintenance of Hydraulic Models. In *Proceedings of the Information Management and Technology Conference*. Denver, Colo.: AWWA.

Edwards, J.A., Koval, E., Lendt, B., and Ginther, P. 2009. GIS and Hydraulic

Model Integration: Implementing Cost-effective Sustainable Modeling Solutions. *Jour. AWWA*, 101(11):34–42.

Edwards, J.A. 2002. Water Demand Allocation Using GIS. In *ESRI User Conference Proceedings*. August.

Green, D., and Montgomery, C. 1998. SCADA Communication Experience at the Detroit Water and Sewage Department. In *Proceedings of the AWWA Annual Conference*. Denver, Colo.: AWWA.

Fortune, D., Campbell, P., and Cavor, R. 2000. Cost of Ownership of Hyrdaulic Models. In *Proceedings of the Infrastructure Conference*. Denver, Colo.: AWWA.

Hutchison, W. 1991. Operational Control of Water Distribution Systems. In *Proceedings of the AWWA Seminar on Computers in the Water Industry*. Denver, Colo.: AWWA.

Japan Water Works Association and Awwa Research Foundation. 1994. *Instrumentation and Computer Integration of Water Utility Operations*. Denver, Colo.: AWWA and AwwaRF.

Maidment, D., Ed. 2000. *Hydrologic and Hydraulic Modeling Support in GIS*. Los Angeles, Calif.: ESRI Press.

Mau, R.E., Boulos, P., Heath, E., and Brennan, W.J. 1996. Advanced Network Modeling Applications: Dynamic Design and SCADA Interface. In *Proceedings of the AWWA Engineering and Construction Conference*. Denver, Colo.: AWWA.

Moore, P.B., Harrington, D.A., Hauffen, P.M., Ray, R., Tejada, R.W. 2008. Avoid the GIS-Model Disconnect: Identifying the Major Hurdles Associated With Hydraulic Model Creation and Updating From GIS. IDModeling Whitepaper. September 2008.

Ray, R., Moore, P.B., Harrington, D.A., and Hauffen, P.M. 2008. The Achilles' Heel of GIS-built Hydraulic Models: Maintaining/Updating a Model from GIS Data. In *AWWA Conference Proceedings*. June 2008.

Rehnstrom, D.J., and Butler, C.L. 2001. Maximizing the Use of Your Hydraulic Model. In *Proceedings of the Information Management and Technology Conference*. Denver, Colo.: AWWA.

Schulte, A.M., and Bonema, S.R. 1995. SCADA Data for Predictive, Training and On-Line Water Distribution System Simulation. In *Proceedings International Water Supply Association, Specialty*. Denver, Colo.: AWWA.

Shamsi, U. 2002. *Tools for Water Wastewater and Stormwater*. Reston, Va.: ASCE.

Shamsi, U. 2005. *GIS Applications for Water, Wastewater, and Stormwater Professionals*. Florence, Ky.: CRC Press.

USEPA. 2005. *Water Distribution System Analysis: Field Studies, Modeling and Management, A Reference Guide for Utilities EPA/600/R-06/028*. Cincinnati, Ohio: US Environment Protection Agency, December 2005.

USEPA. 2006. *Initial Distribution System Evaluation Guidance Manual for the Final Stage 2 Disinfectants and Disinfection By-products Rule, EPA/815/B-06/002*. Washington, D.C.: US Environmental Protection Agency, January 2006.

Walski, T.M., Chase, D.V., Savic, D.A., Grayman, W.M., Beckwith, S., and Koelle, E. 2007. *Advanced Water Distribution Modeling and Management*. Exton, Pa.: Bentley Systems.

WEF. 2004. *GIS Implementation for Water and Wastewater Treatment Facilities*. Alexandria, Va.: Water Environment Federation.

Wilson, L.L., Moshavegh, F., and Bolze, M.A. 2001. Constructing, Maintaining, and Utilizing a Large Water Model. In *Proceedings of the Information Management and Technology Conference*. Denver, Colo.: AWWA.

Wood, P. 1994. Application Integration for Improved Utility Operations. In *Proceedings of the AWWA Computer Conference*. Denver, Colo.: AWWA.

This page intentionally blank.

AWWA MANUAL M32

Chapter 3

Tests and Measurements

3.1. INTRODUCTION

The key objective of hydraulic modeling is to develop and use a model that will predict the performance of an existing water distribution system. Hydraulic tests and measurements (field tests) are important in developing such a model. Hydraulic tests are used to obtain and verify system information and are an integral part of model calibration. At times, information is not available without taking measurements in the distribution system. Data, such as pump curves, pipe friction factors, flow rates, system demands, and reservoir levels, can be obtained from field tests when no other source of this information is available. There are also times when information provided from other sources is incorrect, the model does not match the data provided, or a large variation exists between a theoretical estimate and an actual measurement. In these cases, hydraulic tests and measures can be used to verify and understand such discrepancies in order to determine why the system does not behave as expected. During the model calibration process, field testing is critical in identifying closed valves, incorrect data records, severely tuberculated pipes, and other problems within a distribution system. By using field testing to calibrate the hydraulic model, accuracy of simulations increases, confidence in the model is reinforced, and the model becomes an efficient planning, design, and operational tool. This chapter covers the following areas related to hydraulic tests and measures:

- Planning field tests
- Flow rate measurements
- Meter calibration
- Hazen-Williams C-factor tests
- Hydraulic gradient tests
- Fire flow tests

3.2. PLANNING FIELD TESTS AND PREPARATION

To be effective, field testing should be well planned. Field testing involves additional cost but is necessary to develop a calibrated model that can be used with confidence. The potential for failure is high when planning has been insufficient, parameters affecting the test are poorly understood, incomplete data are collected, and/or test equipment does not function properly. When planning a field test, testing objectives must first be established. These could include (1) calibration of all or a portion of a model, (2) measurement of or estimation of Hazen-Williams C-factors, (3) checking meter calibration, or (4) identifying the cause of anomalies in the distribution system. Personnel experienced with the distribution system, such as field personnel from the utility and local fire department, should be consulted when planning field tests. Their understanding and agreement to tests to be performed are critical.

Prior to gathering any field data, all the equipment used must be calibrated. Any SCADA equipment or permanent meters should also be calibrated if the data are to be used to calibrate the water model. Calibrating equipment prior to field testing will ensure confidence in the data collected. Poorly calibrated equipment can result in gross inaccuracies that can cause problems later with model calibration.

3.2.1. Mapping

A particularly valuable tool for field testing is a map of the water distribution network identifying zone boundaries, key transmission mains, and major system facilities. This map should be annotated with all the important elements necessary for planning and implementing the field testing. For example, the map can be used to indicate locations of fire hydrants to be tested, valves that need to be closed, and/or any system operational changes needed for a particular test. The map can also be used to indicate installation points for a temporary pressure logger (recorder) or flow monitoring equipment. The map should also identify existing system data collection monitoring points and any other information useful for the planning process.

3.2.2. Flow Mass Balance

Where possible, flow monitoring should be performed at points where flow is entering and exiting the system, such that the modeler has the information necessary to calculate a mass balance of the distribution area. If not already equipped, flow monitoring devices can be placed at distribution system entry points, e.g., discharge of water treatment facilities or connections with imported water supplies, pump stations, and/or pressure-reducing valve (PRV) stations. In specific areas of interest, intermediate flow monitoring can be performed. Typically, flow monitoring is performed at locations where waterlines are easily accessible. However, excavation of a waterline may also be necessary to perform flow monitoring. If continuous flow monitoring is performed, diurnal demand patterns can easily be obtained from the data.

3.2.3. Equipment Preparation

After field testing requirements are established, an assessment should be made of equipment necessary to carry out the tests. Some utilities may have pressure loggers available to mount on fire hydrants. Equipment can also be rented or purchased, depending on need. It is very important to verify that all equipment has been calibrated prior to its use in the field. Preliminary testing to ensure all equipment is in proper working order is critical to ensure data obtained are valid for the intended purpose. Batteries should be fresh or fully charged, and digital memory storage must be adequate for the tests being performed.

3.2.4. Other Considerations

The time of day and duration of the tests should be considered to ensure customers are not adversely impacted during testing and to ensure that the equipment and the system will not be adversely impacted by weather conditions. Testing for periods of several days or weeks can be common. In the event that operational problems occur, the entire test may not necessarily be completely invalidated. Data from the days without operational problems may still be utilized, depending on the nature of the problem. Fire flow testing is typically performed on weekdays during daytime hours. However, depending on the situation, it may be necessary to perform system testing on weekends and/or during the night.

It is also best to ensure field personnel from the utility or local fire department are available to assist with testing and that system operators are aware the system testing is occurring. Keeping communication channels open during this time is vitally important to ensure that the utility can adequately continue to serve its customers during the testing period and that firefighting operations are not obstructed. An additional consideration when planning fire flow testing is making sure sufficient head losses can be achieved during testing. Sufficient head loss in the mains is necessary to ensure head losses are greater than the accuracy and precision parameters of the instrumentation. Otherwise, the field test will fail to yield meaningful results. Generally, the greater the head loss, the more accurate the test results will be. To ensure sufficient head losses are achieved, more than one hydrant may need to be opened simultaneously during the test.

3.3. WATER DISTRIBUTION SYSTEM MEASUREMENTS

3.3.1. Pressure Measurements

Measuring pressures in a distribution system is key to understanding its performance. Pressure is typically measured in pounds per square inch (psi) or in metric kilopascal (kPa) with 1 psi (6.89 kPa) being equivalent to 2.31 ft of water (0.704 m). This is based on the characteristics of potable water typically found within a distribution system. Pressure can further be defined as absolute pressure or gage pressure. Absolute pressure is the pressure measured with absolute zero (a perfect vacuum) as its datum, and gage pressure is the pressure measured with atmospheric pressure as its datum (Walski 2007). Because gauge pressure is generally used for water distribution modeling applications, all pressures indicated within this manual are as such.

Two types of devices are typically used for pressure measurement, the *mechanical pressure gauge* and the *pressure logger*. The mechanical pressure gauge has been used for years but can be difficult to read at times, which can introduce human error. Pressure fluctuations common in many water systems due to flow transients and dynamic system changes can make determining pressures difficult, especially during fire flow testing. In any case, the user must make sure gauges have the proper range and accuracy for expected pressures. Unfortunately, pressure gauges capable of reading higher pressures typically are less precise. Pressure loggers provide the advantage of recording pressure over an extended period of time and can fully capture the system's response to changes in flow. Once logged, pressure data are collected and downloaded, and software is available to graph system pressures (Figure 3-1), allowing the user to perform statistical analyses, such as calculating average values, to eliminate noisy readings and easily obtain usable pressures. Regardless of the type of device used, the user must make sure the elevation of the pressure gauge is accurately known. Every 2.31 ft (0.704 m) of error in the elevation of the gauge means a 1 psi (6.89 kPa) error in recorded pressure. Similarly, differences between pressure gauge elevation and model node elevation will also introduce error.

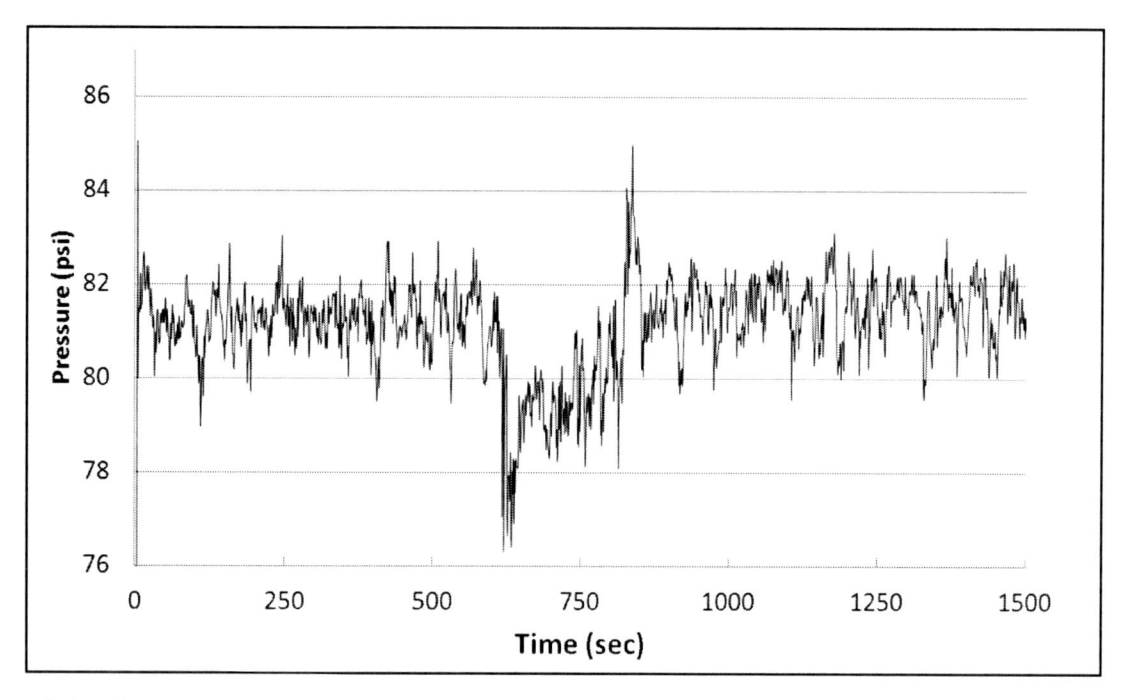

Figure 3-1 Chart of pressure logger system pressures

Typically, pressure measurements are taken at hydrants. However, pressures can also be taken from other locations within the system. These can include pump stations, tanks, meter vaults, air release valves, or blow offs. Regardless of where measurements are taken, care must be taken to ensure that the location is not affected by local conditions. For example, highly variable flow conditions may result in severe pressure fluctuations.

Taking readings at a customer meter location where usage is sporadic can also result in pressure fluctuations due to head loss or turbulence in the service line as flows change. If other options are available, taking pressure readings at meters should be avoided. However, in the absence of other alternatives, the user must make sure there are no PRVs on the service line upstream of the gauge. Gathering pressures at such a location will almost certainly introduce error in the results. If obtaining pressures from such locations, it is important to make sure water is not being used by the customer.

3.3.2. Flow Measurements

Flow measurements are an integral part of multiple water distribution system tests that can be performed, including C-factor tests, diurnal demand measurements, pump tests, and fire flow tests. Flow-measuring equipment in water supply includes devices such as hydrant Pitot gauges, Pitot tubes, strap-on meters (those that do not require a tap into the flow stream of the pipe), and master meters (such as Venturi meters).

3.3.2.1. Hydrant Pitot Gauges. Hydrant Pitot gauges (Figure 3-2) measure the discharge indirectly by measuring the velocity head with a pressure gauge. Figure 3-3 shows a hand-held Pitot gauge in use. The gauge measures pressure at the hydrant discharge point where the water is no longer under pressure. The flow equation for a Pitot gauge is

$$Q = 29.83 \times p^{0.5} \times d^2 \, C_H \tag{3-1}$$

Where:

Q = rate of flow in gpm

p = gauge pressure of the flow stream in psi

d = diameter of hydrant discharge port in in.

C_H = hydrant discharge coefficient

$$Q_{,gpm} \times (6.3 \times 10^{-5}) = Q_{,m^3/sec}$$

The magnitude of a hydrant discharge coefficient, C_H, depends on the physical transition from the hydrant barrel to the discharge port, which is typically (A) 0.9 for round and smooth, (B) 0.8 for sharp and square, and (C) 0.7 for ports protruding into the hydrant barrel as shown in Figure 3-4. See AWWA Manual M17, *Installation, Field Testing, and Maintenance of Fire Hydrants,* for a more detailed discussion on field procedures for flow testing.

Hydrant Pitot gauges should be calibrated using a dead-weight tester. The gauge resolution must be within 0.5 psi (3.45 kPa) to accurately measure pressure heads below 10 psi (68.95 kPa). Flowing hydrant pressure is used to measure and calculate the flow rate as previously described. However, a flowing hydrant cannot be used for a system pressure measurement because of friction loss in the hydrant valve and barrel. Therefore, pressure measurements should be taken from the hydrant upstream from the flowing hydrant as described in more detail in the following sections.

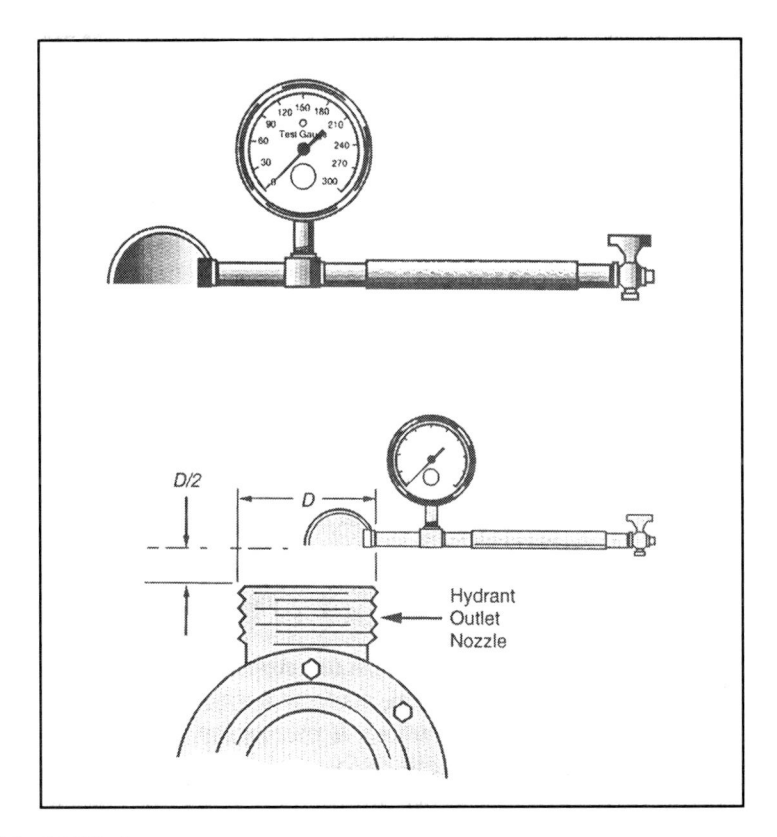

Figure 3-2 Hand-held Pitot gauge

Figure 3-3 Hand-held Pitot gauge in use (photo by Tom Walski, Bentley)

Other Pitot gauges available include diffusers and/or dechlorinators. These have been developed to minimize erosion, discharge of chlorinated water to the environment, and reduce other damage that can result from flow discharging directly from a hydrant nozzle. If diffusers are used (Figure 3-5), it is recommended that care be taken when using the discharge measurement data because the Pitot could be measuring discharge flow that is still under pressure. This may result in inaccurate discharge measurements if traditional discharge coefficients are directly applied to a diffuser (Walski and Lutes 1990). Engineers are encouraged to have the device manufacturer provide the necessary information such that discharge measurements can be adjusted accordingly.

3.3.2.2. Pitot Tubes. Pitot tubes measure velocity heads within a pipe. Velocity heads are typically small for nominal flows in water mains. For a flow velocity of 0.5 ft/sec (0.15 m/sec), the equivalent velocity head is only a 0.05 in. (0.0013 m) column of water, which is difficult to measure. To get an accurate test of head loss, generally higher than normal flow should be induced. Theoretical velocities are adjusted with an instrument coefficient from laboratory calibrations. For meter and pump tests, a variety of flows should be planned and the flow measurement device chosen appropriately for the task.

Flow rate in a pressurized pipe is computed by multiplying the cross-sectional area of the pipe by the average velocity of the flowing water. Inside diameter is used to compute the cross-sectional area, the value of which is adjusted for the obstruction of the Pitot tube, as well as any corrosion or tuberculation inside the pipe. The average velocity may be determined from a velocity profile.

Figure 3-6 illustrates a method for determining velocity profiles in which a pipe is divided into 5 rings of equal area. Average velocity is the mathematical average of 10 measurements, 5 on each side of the pipe center. The ratio of the average velocity to the center velocity is called the *velocity factor* (VF), which is unique for each gauging point. Figure 3-7 shows typical profiles at similar flow rates for two different gauging points.

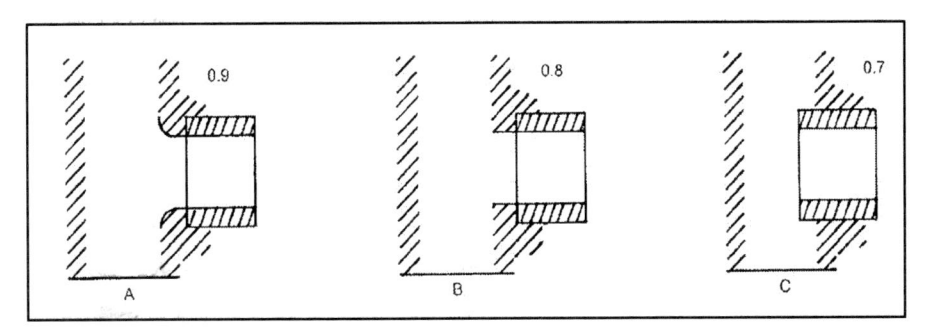

Figure 3-4   Three general types of hydrant outlets

Figure 3-5   Diffuser with pressure logger (photo by Gregory Brazeau, Draper Aden Associates)

Insertion points for Pitot tubes are typically at air release valves and 1-in. (25.4-mm) corporations. However, larger corporations can be used with proper fittings depending on the testing equipment available. The best locations for stable velocity profiles are along pipes having at least 10, preferably 20, diameters of straight pipe upstream of the gauging point.

3.3.2.3. Strap-on Meters. Strap-on, or clamp-on, portable meters are becoming more popular as their technology advances, allowing for more accurate readings of flow within potable water systems (Figure 3-8). Such meters are advantageous as they do not require insertion of a device into the pipe for measurement. Flow can be obtained by simply strapping the meter onto the outside of the pipe. Using ultrasonic or Doppler technology, the meter detects the velocity of fluid within the pipe and determines the amount of flow passing by the meter. When using these meters, their limitations must be considered, such as having the necessary length of pipe upstream and downstream to obtain a uniform flow for accurate readings. The user must also know the type and thickness of pipe, which are critical to the meter's accuracy. Very low flows can also have an effect on meter accuracy. Generally, the higher the velocity, the higher the meter accuracy.

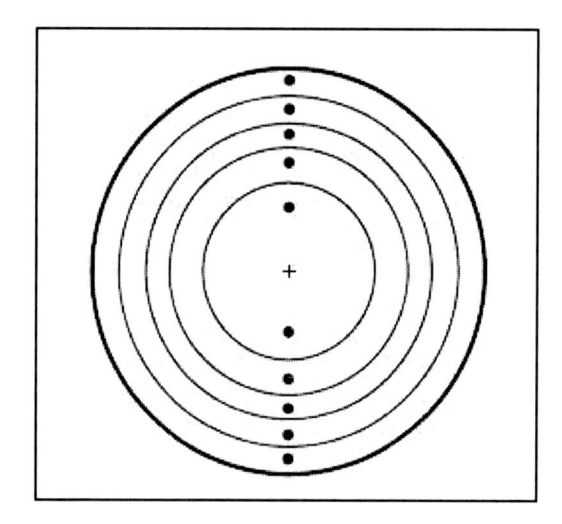

72 COMPUTER MODELING OF WATER DISTRIBUTION SYSTEMS

Figure 3-6   Traverse positions within a pipe

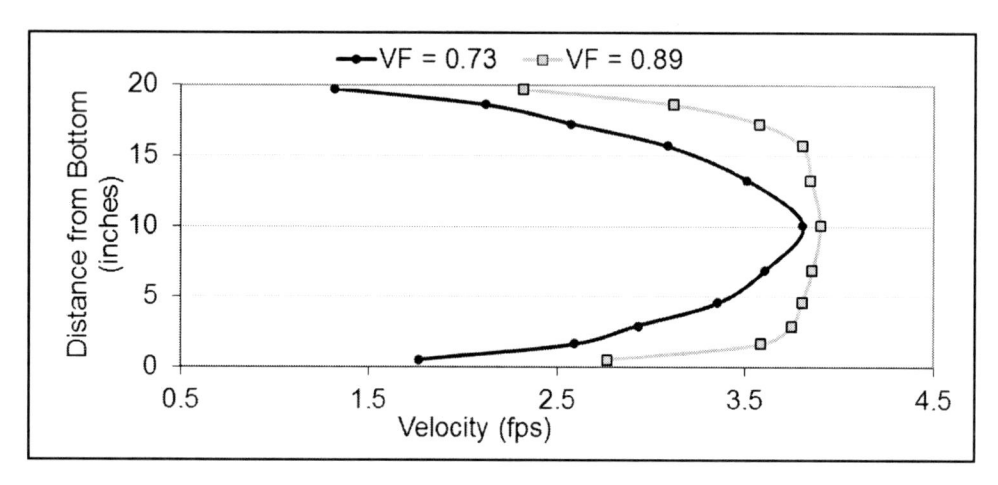

Figure 3-7   Typical velocity profiles at two different gauging points (VF = velocity factor)

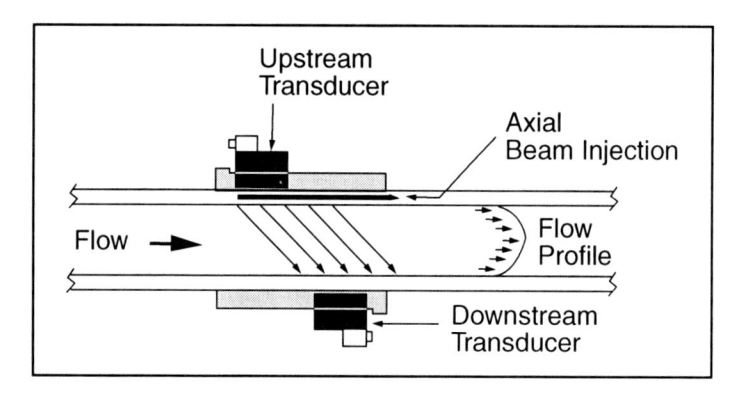

Figure 3-8   Schematic of a strap-on flowmeter

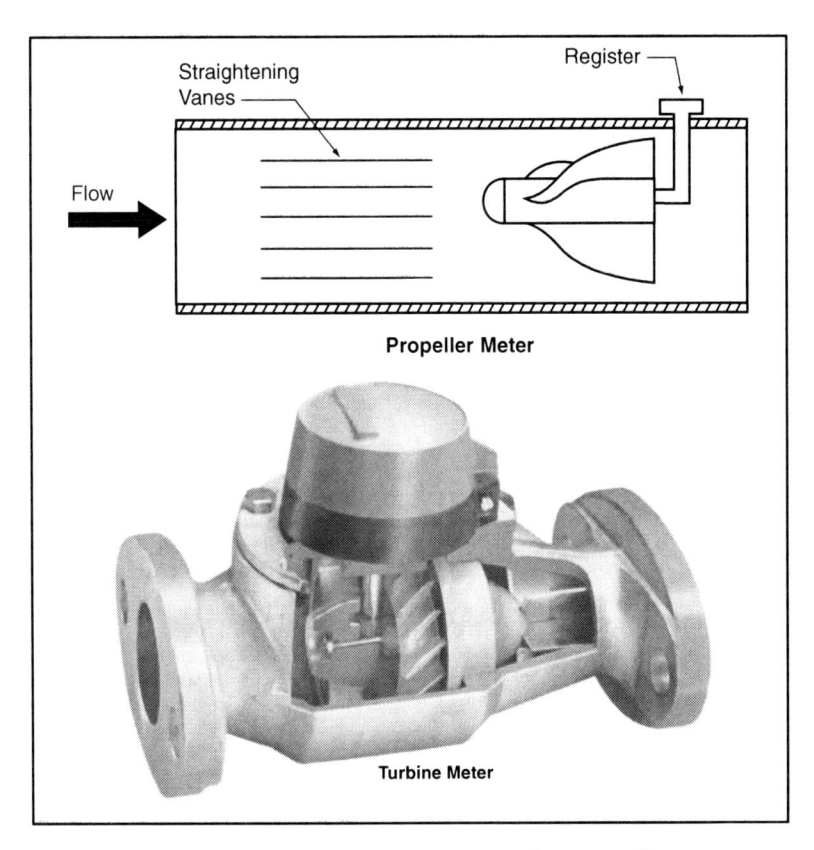

Figure 3-9 Schematic of propeller flowmeter and picture of turbine flowmeter

3.3.2.4. Master Meters. Master meters are flowmeters that are usually installed permanently in a pipeline to measure relatively large flows of water supplied or transferred. Master meters are typically present at water treatment plant discharges, larger transfer pump stations, purchased supply points, emergency supply locations, or at district metering points. District metering is typically used to measure flow within a smaller area of the system and often to identify water loss within the system. See AWWA's Manual of Water Supply Practices, M36, *Water Audits and Loss Control Program,* for further discussion concerning district metering. Meters typically used for master metering include propeller/turbine meters, Venturi meters, and magnetic meters, all of which are discussed further in the following sections. More details about these and other types of meters can be found in AWWA Manuals M6, *Water Meters—Selection, Installation, Testing, and Maintenance,* and M33, *Flowmeters in Water Supply.*

3.3.2.4.1. Propeller/Turbine Meters. Propeller meters have been commonly used as master meters in the past but are less common today (Figure 3-9). They consist of a propeller mounted axially in the line of flow within the pipe. As water passes through the meter, the flow induces a rotation of the propeller that is calibrated to measure the flow rate.

Turbine meters are typically not used for very large flows at water treatment plants but are often used for metering at pump stations and for large users or wholesale customers. Turbine meters translate the mechanical motion of a rotating turbine in water flow around an axis into a volumetric rate of flow. The turbine wheel is set in the pipeline. The flowing water pushes the turbine blades, imparting a force to the blade surface and setting the rotor in motion. When a steady rotation speed has been reached, the speed is proportional to fluid velocity.

3.3.2.4.2. Venturi Meters. The Venturi meter is widely used as a master meter as shown in Figure 3-10. It consists of a Venturi tube (also called a converging-diverging nozzle) inserted directly into a pipeline. Because the diameter changes are gradual, there is very little friction loss. Static pressure measurements are taken at the throat and upstream of the diameter change. These measurements are traditionally made by manometer. A typical Venturi tube with manometer is shown in Figure 3-11.

The calculation of flow using a Venturi meter assumes a horizontal orientation and frictionless, incompressible, and turbulent flow. Bernoulli's equation is applied for points 1 and 2.

Because:

$$Q = V_1 A_1 = V_2 A_2$$
$$P_1 - P_2 = (\rho/2) \times (V_2^2 - V_1^2)$$

Then:

$$Q = A_1 \sqrt{\frac{2(P_1 - P_2)}{\rho[(\frac{A_1}{A_2})^2 - 1]}} = A_2 \sqrt{\frac{2(P_1 - P_2)}{\rho[1 - (\frac{A_2}{A_1})^2]}} \qquad (3\text{-}2)$$

Where:

Q = flow

A = area

P = pressure

ρ = density of the fluid

V_1 = (slower) fluid velocity where the pipe is wider (A_1 area)

V_2 = (faster) fluid velocity where the pipe is narrower (A_2 area) as seen in the Figure 3-11.

h = differential head ($P_1 - P_2$)

It should be noted that the equation shown assumes no friction loss through the meter. Typically a Cv factor is applied to account for friction losses.

3.3.2.4.3. Magnetic Meters. Magnetic meters are often used as master meters and consist of a coil of wire around the pipe through which a current is induced as ions in the water pass through the pipe (Figure 3-12). These meters are calibrated so that the induced electrical current correlates to the flow rate in the pipeline. Magnetic meters are usually installed permanently, although temporary installations are also possible, and some magnetic meters are available as strap-on meters.

Figure 3-10 Existing Venturi tube (photo by Jerry Higgins, Blacksburg Christiansburg VPI Water Authority)

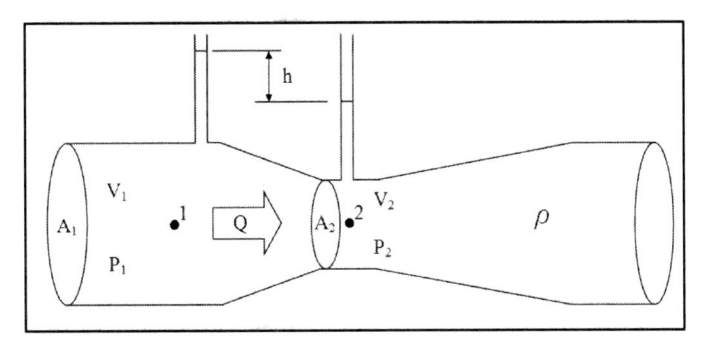

Figure 3-11 Typical Venturi tube with manometer

Figure 3-12 Magnetic meter (photo by Robbie Cornett, Washington County Service Authority)

3.3.2.5. Meter Calibration Test. A meter calibration test compares master meter registration to flow rates measured by another instrument, such as a test meter or an inserted probe. Meter calibration tests are used in verifying production records, conducting pump tests, and calibrating computer models.

Meter calibration tests compare meter registration to the measured flow in a pipe, rather than simply checking electronic calibration of a transducer. Many meters include a primary device, such as a Venturi tube, and a secondary element, such as a transducer to convert physical output to an electronic signal. A malfunctioning primary device causes meter error even if the secondary element is calibrated correctly. For example, a Venturi meter may provide inaccurate readings because of leakage between the Venturi tube and transducer, even though the transducer is functioning correctly.

Meter calibration tests are conducted for a range of flow rates. The range should cover what is expected to flow through the meter, including high and low flows. At high flow rates, a meter may provide readings that are sufficiently accurate, but the same meter may provide inaccurate readings at lower flows that are more typical of normal operating conditions. Conditions in a shop test may not duplicate influences of upstream piping, turbulence, or other field conditions.

3.3.3. Supervisory Control and Data Acquisition (SCADA)

Operational SCADA information should also be collected during testing. If available, information to collect should include flows, pressures, tank levels pump, and valve status information. Such information will be useful during the model calibration process.

3.4. WATER DISTRIBUTION SYSTEM TESTING

3.4.1. Fire Flow Tests

Fire flow tests can be used to determine friction factors in pipes and are generally used for model calibration or localized system investigation, such as a suspected closed valve. These tests consist of simultaneously taking flow and pressure measurements at selected locations.

First, static pressure measurements on the test hydrant are taken under normal conditions. Next, the flow is measured on a nearby hydrant (flow hydrant), while dynamic pressure (residual pressure) measurements are taken on the test hydrant under the condition of steady flow (Figure 3-13). In larger networks or where larger mains serve the hydrant, two or more hydrants are opened to provide the necessary flows and resulting head losses. The flow rate is calculated from hydrant Pitot gauge readings and the hydrant discharge port size. A flowing hydrant (flow hydrant) should never be used for a residual pressure measurement due to the friction loss in the hydrant valve and barrel. Therefore, pressure measurements should only be taken from a second hydrant (test hydrant) upstream from the flowing hydrant.

To achieve sufficiently accurate test results, there should be a minimum 10 psi (68.95 kPa) pressure drop (static pressure minus residual pressure). Ideally, the pressure gauge or pressure logger should have a minimum 0.5 psi (3.5 kPa) resolution to provide good accuracy. Gauges should be calibrated with a dead-weight tester and pressure loggers should be calibrated before every set of tests.

In addition to helping model calibration by comparing measured flows and pressures from the fire flow test to modeled flow and pressures, fire flow tests can also be used to determine the available fire flow at a certain pressure. Derived from the Hazen-Williams formula, the results are often expressed as flow rates available at 20 psi (137.9 kPa):

$$Q_A = Q_M \times \{(S - 20)/(S - R)\}^{0.54} \tag{3-3}$$

Where:

Q_A = the rate of flow available at 20 psi

Q_M = the rate of flow measured during the test

S = the static pressure in psi

R = the residual pressure in psi

gpm × (6.3 × 10^{-5}) = m^3/sec

The minimum pressure during fire flows is 20 psi (137.9 kPa), as recommended by the National Fire Protection Association (NFPA). A minimum pressure of 20 psi (137.9 kPa) prevents backflow or groundwater contamination and provides head for overcoming friction losses in the hydrant branch, hydrant, and suction hoses to pumper trucks during a fire.

```
┌─────────────────────────────────────────────────────────────┐
│     ↖ Flow Hydrant                          Test Hydrant    │
│    ─────────◆────────────◀──────────────────────◆────────   │
│                             Pipe Flow                       │
└─────────────────────────────────────────────────────────────┘
```

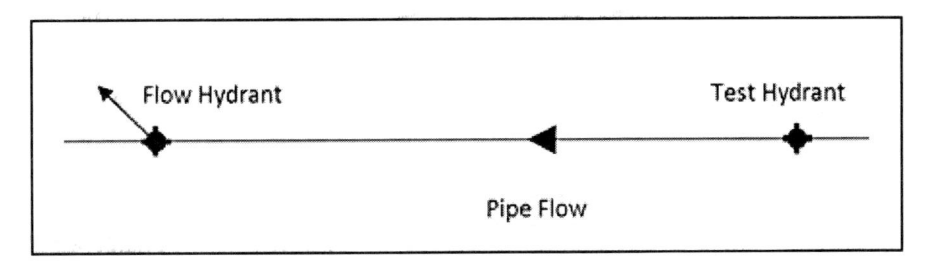

Figure 3-13 Fire flow test configuration

Models can also calculate available fire flows. Comparing the calculated available fire flow at 20 psi (137.9 kPa) from field tests can also help calibrate the model. However, comparing model results to a measured flow and pressure is preferable to using the calculated available flow at 20 psi (137.9 kPa) because of possible complications from operational changes temporarily used during a fire, such as check valves opening, flow reversals, or pump changes. The model would need to reflect these temporary changes when simulating available fire flows if they are used for calibration purposes.

For fire flow tests to be useful for calibration, critical boundary conditions, pump status, and reservoir levels should be recorded during the time of the test so that the model can be run under the same conditions for comparison. Critical boundary conditions may include PRV settings. PRVs are sometimes used on pressure zone boundaries, instead of closed valves, so that low pressures will trigger the valves to open under fire flow conditions (low pressures). Under normal conditions, the PRVs are closed to separate pressure zones. The date and time should also be recorded as system demand conditions will change over the course of a day and seasonally.

Fire flow tests in large systems have only limited value in calibrating models. Because most of the head loss occurs in the area near the flowing hydrant, measurements from fire flow tests allow only limited calibration in the region of the hydrant, and this does not provide a general indication of model accuracy.

Another enhanced approach to fire flow testing, at least with respect to time and labor efficiency, is the use of continuous-recording pressure loggers while simultaneously releasing water from multiple hydrants (sometimes called *system stress testing*). When compared to normal protocols for conducting flow tests, this procedure can be accomplished by smaller crews and performed faster and can result in more data that could be used for model calibration (Grayman et al. 2006). However, care must be taken in the placement of pressure loggers and selection of hydrants to flow so that an area can be properly calibrated in the model without creating a surplus of useless data.

3.4.2. Roughness Testing

Roughness testing can be performed to determine the C-factor for the Hazen-Williams equation or roughness factor for Darcy-Weisbach equation. Discussion here will be limited to the Hazen-Williams equation. C-factors are friction coefficients used in the Hazen-Williams formula, an approximation of pipe roughness. Hydraulic models use the Hazen-Williams formula or other head loss equations, relating flow to head loss in each pipe element within the model. Like Moody diagrams or other equations, the Hazen-Williams formula, although empirical, provides sufficient accuracy within the normal range of velocities and temperatures in a water distribution system. As mentioned in chapter 2, if C-factor testing is done at very high velocities (velocities > 10 ft/sec [3 m/sec]), a significant error can result when the resulting C-factors are used to predict head loss at low velocities. (Walski et al. 2007) Therefore, velocities should be kept under this threshold when testing.

Measuring C-factors for each and every pipe is not practical; therefore, assumptions are made based on a sample of C-factor measurements. The sampling should include all combinations of pipe sizes and materials, as well as old and new pipes. Larger pipes tend to have higher coefficients (less roughness). Careful consideration should be given to pipe installed prior to 1950. Unlined cast-iron pipes should be included in the sample, as C-factors vary widely from less than 25 to over 100, depending on the pipe age and water quality. If it is unknown whether waterlines are lined or unlined, maps from the 1940s can help to identify unlined pipes because the use of unlined cast-iron pipes mostly ceased in the 1950s. It is also important to measure C-factors near sources where treatment processes may have coated the inside of pipes.

Usually, the C-factors used in hydraulic models account for losses in bends, valves, and other fittings. Losses at these fittings are typically small at normal velocities in water distribution systems.

The procedure for conducting field tests involves obtaining nominal diameter, flow, head loss, and length of test section and then solving the Hazen-Williams formula for the C-factor coefficient. The formula for computing the C-factor of a pressurized circular pipe is

$$C = 3.551 \times Q \times D^{-2.63} \times L^{0.54} \times H^{-0.54} \tag{3-4}$$

Where:

C = C-factor coefficient

Q = rate of flow in gpm

D = diameter of pipe in in.

L = length of pipe in ft

H = head loss in ft

Field results are verified by measuring head losses at two different flow rates and comparing resulting C-factors. C-factors should agree within 10 percent. Another check consists of running two test trials on the same pipe but using different equipment for each trial.

Usually, flow must be artificially induced to produce velocities and head losses high enough for achieving accurate measurements. In pipes 12 in. (305 mm) and smaller, high flow rates are induced by opening one or more hydrants. In larger pipes, it is necessary to open multiple hydrants or use other methods, such as opening blowoffs, throttling or closing off parallel mains or other pipes, operating pumps, or filling tanks to increase pipe flows.

Pipe diameters used in C-factor calculations are the same diameters used in the model, which are typically nominal pipe diameters. The length of pipe used in calculating a C-factor is the distance along the pipe between the pressure measurement points. Tests that include more than one pipe diameter between pressure sensors should be avoided. It is also preferable to select a location with minimal demand to ensure consistent flow through the pipe section.

Head loss measurements are equivalent to the drop in the hydraulic grade line (HGL) between the inlet and outlet of a test section. HGL are taken at hydrants, air valves, corporations, or other taps. Measurement techniques depend on field conditions and the amount of head loss measured.

3.4.2.1. The Parallel Hose Method. The parallel hose method measures head loss using a hose or pipe connecting the ends of a test section to a differential pressure transducer or manometer, as shown in Figure 3-14. This method measures head loss directly without knowledge of elevation. Portable test equipment includes reels of ⅛-in. (3.2-mm) or ¼-in. (6.4-mm) elastic tubing or rubber hose.

The parallel hose method generally requires at least 2 ft (0.6 m) of head loss to achieve accurate test results, provided differential pressure measurements are accurate within ±0.1 ft (0.29 kPa). Pipe lengths and flow rates that produce measurable velocities and losses without exceeding the range of the transducer should be selected.

The parallel hose method is not suitable for testing some pipes, as the practical limit for the length of hose is approximately 3,000 ft (914 m). Some short test sections require extremely high flow rates to produce measurable head loss. Having to deal with such long hose lengths can make this method impractical.

Sources of error include leaks and air in the hose. If possible, a valve should be closed immediately downstream of the test section, and the section should be checked for zero head loss at zero flow.

3.4.2.2. The Two-Gauge Method. The gauge method uses pressure gauges at the inlet and outlet of a test section. Pressure measurements along with known elevations define the HGL. Head loss is the HGL at the outlet subtracted from the HGL at the inlet, as shown in Figure 3-15. An accurate test requires a loss of at least 10 ft (3.1 m) using pressure gauges and elevations accurate to within ±0.5 ft (0.2 psi) (1.4 kPa) of head. This level of accuracy requires gauge calibration with a dead-weight tester and accounting for the heights of instruments from benchmarks.

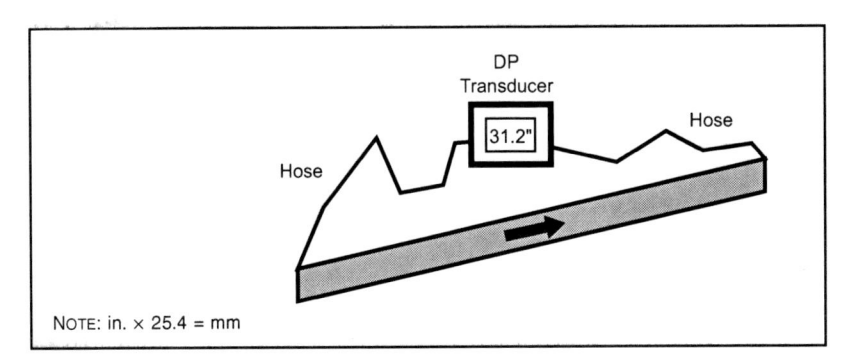

Figure 3-14 Parallel hose method for head loss

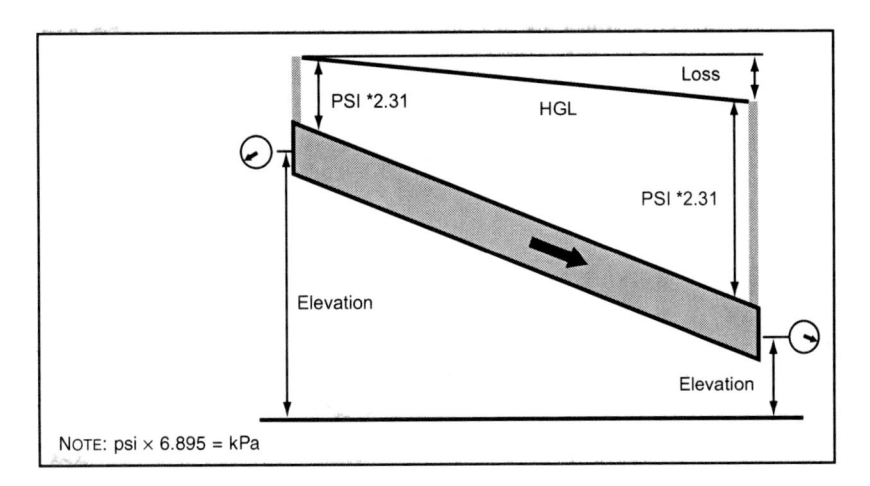

Figure 3-15 Gauge method for head loss

An advantage of the gauge method is that test lengths are increased to achieve additional head loss. However, isolating long test sections requires closing many valves. Another disadvantage is the need to know the elevations at which the pressure gauge measurements were taken. If elevation data are not readily available, it is possible to close a valve downstream of the test section and record pressures at zero flow. The difference in static pressure is converted to feet of water and used as the elevation of the higher gauge, assuming the elevation of the lower gauge is zero or datum.

3.4.2.3. Other Methods. Other methods of measuring head loss or pressure are possible under certain field conditions. For instance, reservoir levels define HGLs in some cases. The height to which water rises in clear hoses connected to pipe taps indicates the hydraulic grade line directly; however, this method is only practical when pressures are relatively low. Parallel pipes with no flow are used to transmit the HGL from one end of a test section to the other. Some methods of calculation use pressure measurements at several flow rates to eliminate the need to know elevations.

3.4.3. Pump Tests

Flow and head data collected from pump tests are entered into the model for simulating hydraulic performance and for calculating energy costs. Pump tests are used to determine the actual pump curve of a pump in the field. Assuming an existing pump's design curve, or shop tested pump curve from the manufacturer, is accurate has the potential of introducing significant error into the model, as true performance is sometimes quite different from design parameters. This difference in performance could be caused by worn impellers or other equipment deterioration, undocumented equipment changes, trimming of impellers, or other field conditions.

As shown in Figure 3-16, pump tests can reveal how total dynamic head (TDH) and efficiency truly vary with flow. Most models generate pump curves based on at least three operating points, each associated with a unique flow rate, TDH, and efficiency. Tests usually include normal conditions, design conditions, induced flow, throttled flow, and shutoff head.

During pump tests, master meters are often used for measuring flow. The throttling of flow and shutoff conditions is often achieved by using the downstream valve. Throttling of flow will provide data to the left of the operating point on the pump curve. Induced flow is achieved by opening hydrants or blowoffs, filling a tank, or adjusting operations at other pump stations. Inducing flow will provide data to the right of the operating point on the pump curve. If the pumps are equipped with variable speed drives, the speed of the pumps can be changed to determine points on the system head curve.

TDH is defined as the energy added by a pump per pound of water flowing through the pump. TDH measured in foot-pounds per pound is dimensionally equivalent to feet of water. The equation for the TDH added by a pump is

$$\text{TDH} = (P + V + Z)_{OUT} - (P + V + Z)_{IN} \qquad (3\text{-}5)$$

Where:

TDH = total dynamic head in ft ($\times$ 2.989 = kPa)

P = pressure head in ft

V = velocity head in ft

Z = elevation of pressure sensor in ft

$_{OUT}$ refers to the pump discharge

$_{IN}$ refers to the pump suction

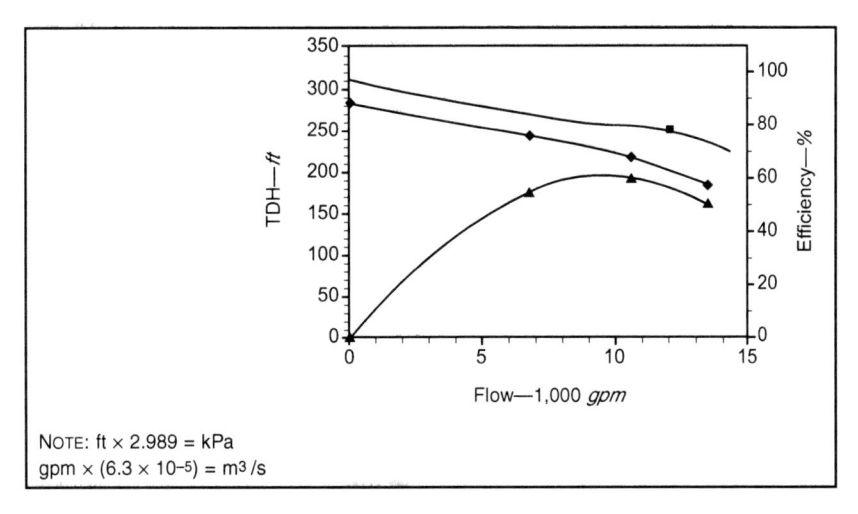

Figure 3-16 Pump tests (TDH = total dynamic head)

Velocity head is computed by squaring velocity (in feet per second) and dividing by 64.4 ft/sec² (two times the acceleration of gravity). Velocities are calculated by dividing measured flow rate by the cross-sectional area of the pipe. Velocity heads can be significant, especially for induced flows.

Efficiency is calculated by dividing the pump's output hydraulic power by the input electrical power to the drive motor. The equation is

$$E = 0.01885 \times Q \times \text{TDH}/KW \tag{3-6}$$

Where:

E = efficiency in percent
Q = rate of flow in gpm
TDH = total dynamic head in ft
KW = electrical power in kW

Electrical power is measured directly using portable test equipment. Electrical power is calculated from measurements of voltage, amperage, and power factor. Power factors must be measured to verify accuracy. Electrical power may also be calculated from the rotational speed of power meter disks using the watt-hour per revolution factor (K_H) and multipliers, which are stamped on most meters.

3.4.4. Hydraulic Gradient Tests

Hydraulic gradient tests reveal how HGLs vary along the length of a pipeline. Data points along the HGL are computed by adding elevation to each measurement of pressure head in feet. Hydraulic gradient tests provide data for calibrating computer models. Often, in combination with a computer model, the information can be used to locate closed or partially closed valves, under- or overloaded mains, and can help determine where pipeline improvements may be warranted.

Hydraulic gradient tests consist of taking simultaneous flow and pressure measurements at intervals along a pipe path under steady flow conditions. As shown in Figure 3-17, pressure measurements are taken along large trunk mains located between a plant and a tank. Pressure measurements are taken at the water treatment plant or supply point, key facilities (major water users), on the inlet and outlet of pump stations, at tanks and major pipe intersections, and where pipe diameters transition. At key pipe sections, the flow rate is measured.

Pressure measurements in units of psi are converted to feet of water and added to pipe elevations to obtain HGLs. HGLs are then plotted against distance, as shown in Figure 3-18. The plots show the accumulation of head loss along pipe sections. The slope of the HGL will steepen in sections with significant hydraulic restrictions where head losses are greatest.

To be truly useful, hydraulic gradient tests should be conducted under conditions of high flow when head loss is significant or when high flow can be induced. In small systems, a simple way to induce high flow is by opening hydrants. In large systems, it may not be sufficient to open hydrants. In such cases, the test can be conducted when demand is high, when a tank is being filled, or by turning on pumps at appropriate locations.

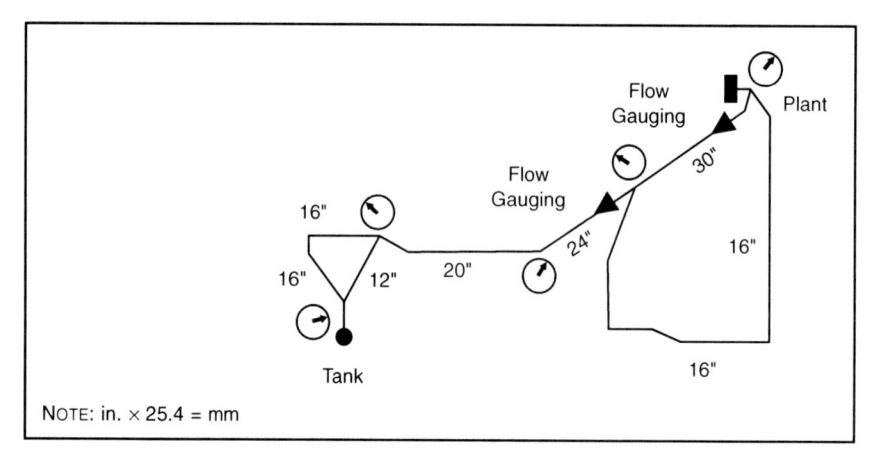

Figure 3-17 Hydraulic gradient layout

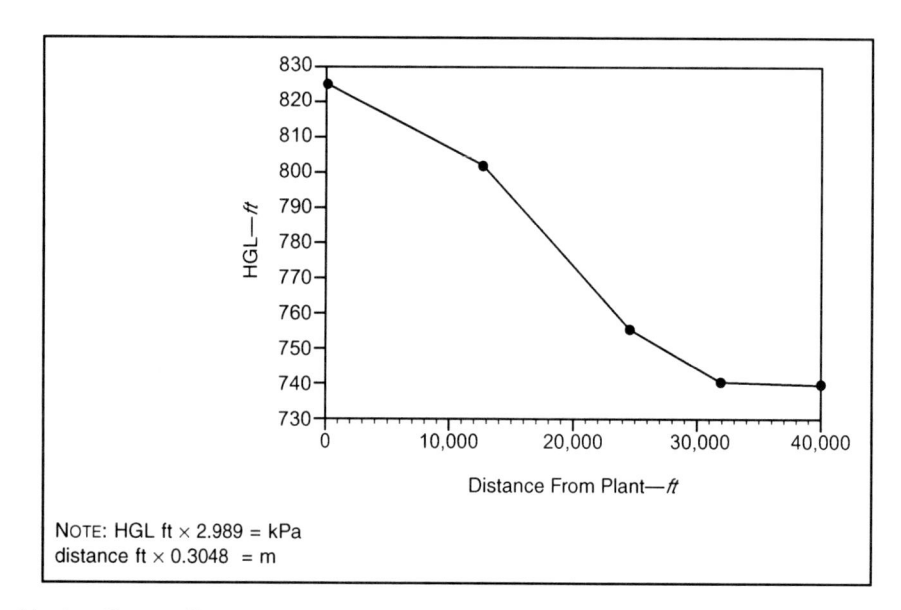

Figure 3-18 Hydraulic gradient test

3.4.5. Other Tests

Some models require conducting other types of tests. Modeling PRVs requires accurate downstream pressure measurements at known elevations. Pressure and flow measurements are useful in modeling other types of control valves. Pressure and tank level measurements are necessary for checking the accuracy of data from water plant charts and SCADA systems.

3.4.6. Testing Considerations

Any time a test is performed that stresses the water system, the overall effect on the water system must be taken into consideration and customers that will be affected should be notified. When performing such tests, the system will see decreases in water pressure, and, due to higher velocities, sediment can be stirred up into the system causing discoloration of water. This can be alarming to customers if they are not notified to expect these changes in the system during the testing process. When hydrants and valves are operated during the testing process, the utility should anticipate problems and be prepared to perform maintenance where needed. The utility should also be prepared to repair damaged landscapes caused by the high volume of water hydrants can produce. During the planning process, all these issues need to be taken into consideration.

3.5. DATA QUALITY

Data collected for the purpose of model calibration must be more precise than data collected for routine operation and control. Calibration data will be used to adjust pipe roughness and demands. However, much of the time the head loss in water distribution systems is very small, and errors in the measurement of head loss are of the same order of magnitude as the head loss itself (Walski 2000; Walski et al. 2007). Such data are helpful in checking boundary heads and ground elevations but are useless for adjusting pipe roughness because at such low flow rates, head losses cannot be measured accurately.

The key to data collection for calibration is to ensure that head loss is much greater than the error in measuring head loss. This is achieved either by minimizing errors in head loss measurements or increasing head loss. Head loss is maximized by taking measurements during peak demand periods or by conducting fire hydrant flow tests. If valves are known to be throttled, they should be opened prior to testing. Throttled valves can create high head losses but also introduce unknown minor losses.

Head measurements consist of both pressure and elevation measurements. Pressures are read with a gauge calibrated to 0.5 psi (3.5 kPa) or better. Elevation data should be at least as good (Walski 2007). While reading elevations from a USGS topographic map with 20 ft (4 m) contour intervals are acceptable for model building, the elevation of calibration data points should be significantly more accurate. Surveying the elevation of the gauge (rather than the ground elevation) is usually the most accurate way to determine if maps with 2 or 3 ft (1 m) accuracy are not available. A global positioning system (GPS) can be used to determine the elevation of pressure sensors but the GPS will need to have submeter or better vertical accuracy. The GPS should also be calibrated to nearby known benchmarks.

Adjusting demands and pipe roughness in models adjusts the slope of the HGL. Therefore, it is essential to know the boundary heads when calibration data are collected. For example, knowing that the pressure at a point in the system is 40 psi (275.8 kPa) without knowing the water level in nearby tanks, the operation and discharge head of nearby pumps, or the setting and status of nearby pressure-reducing valves, renders such data useless for making calibration model adjustments. Relying on SCADA systems for this information helps if changes are gradual, such as changes in tank water level, but is not useful when pumps can turn on and off, and PRVs can open and close during a flow test because the polling interval of a SCADA system may miss a hydrant flow test. This requires stationing personnel at key points in the system during tests or installing data loggers to capture test results.

Readings through all flowmeters should be taken at a time corresponding to pressure measurements. It is very helpful to know the flow rate through meters during a hydrant flow test if it is suspected that flow rates change during the test.

3.6. REFERENCES

Alden Research. 1995. *Calibration of an Insert Traversing Pitot Tube and Pitometer.* Computer Recorder. Holden, Mass.: Alden Research Laboratory.

American Water Works Association (AWWA). 2008. Manual M31, *Distribution System Requirements for Fire Protection.* Denver, Colo.: AWWA.

AWWA. 1999. Manual M6, *Water Meters—Selection, Installation, Testing, and Maintenance.* Denver, Colo.: AWWA.

AWWA. 2011. Manual M58, *Internal Corrosion Control in Water Distribution Systems.* Denver, Colo.: AWWA.

AWWA. 2006. Manual M33, *Flowmeters in Water Supply.* Denver, Colo.: AWWA.

AWWA. 2009. Manual M36, *Water Audits and Loss Control Programs.* Denver, Colo.: AWWA.

Bush, C.A., and Uber, J.G. 1998. Sampling Design Methods for Water Distribution Model Calibration. *Journal of Water Resources Planning and Management,* ASCE, 124:6:334–344.

Grayman, W.M., Maslia, M.L., and Sautner, J.B. 2006. Calibrating Distribution System Models With Fire-Flow Tests. *Opflow,* Vol. 32, No. 5.

Strasser, A., Diallo, N., and Koval, E.J. 2000. Development and Calibration of Denver Water's Hydraulic Models for a Treated Water Study. In *Proceedings of the Annual Conference.* Denver, Colo.: AWWA.

Walski, T.M. 2000. Model Calibration Data—The Good, The Bad, The Useless. *Jour. AWWA,* 92(1):94–99.

Walski, T.M., and O'Farrell, S.J. 1994. Head Loss Testing in Transmission Mains. *Jour. AWWA,* 86(7):62–65.

Walski, T.M., and Lutes, T.L. 1990 Accuracy of Hydrant Flow Tests Using a Pitot Diffuser. *Jour. AWWA,* 90(7):58–61.

Walski, T.M., Chase, D.V, Savic, D.A., Grayman, W.M., Beckwith, S., and Koelle, E. 2007. *Advanced Water Distribution Modeling and Management.* Exton, Pa.: Bentley Systems.

AWWA MANUAL M32

Chapter 4

Hydraulic Calibration

4.1. INTRODUCTION

Hydraulic models are only as good as the accuracy of their predictions of actual water system performance. The best way to verify that a hydraulic model provides results that closely match actual field conditions over a wide range of operating conditions is to perform model calibration. This chapter will provide guidance on the steps required to calibrate a hydraulic model, describe the various types of calibration, and describe the issues and pitfalls to keep in mind when conducting a model calibration.

Hydraulic models are essentially mathematical representations of the water system. They must contain accurate data defining the water usage in the system, the physical characteristics of the piping and facilities in the system, and the control settings of equipment and devices needed to replicate accurate system performance. A model may complete a simulation based on the information provided, but, if this information does not accurately reflect the system being analyzed, the result will have limited benefit and may actually lead to erroneous conclusions. Calibration can be tedious and time consuming. However, the potential benefits of a well-calibrated model for making accurate engineering decisions far outweigh the potential cost of mistakes caused by incorrect model results.

4.2. WHAT IS CALIBRATION?

Calibration is the process of comparing the results of model simulations to actual field data and making corrections and adjustments to the model to achieve close agreement between computer-predicted values and field measurements. Typical comparison values include pressures, flow rates, and reservoir water levels. A well-calibrated model will stay in close agreement with field measurement over a wide range of operating conditions.

Discrepancies between model results and field data can be due to many factors including errors in model construction, errors in reported field data, unknown constraints or controls in the water distribution system, and inaccurate assignment of model variables. Some of the parameters related to physical facilities that may require correction during calibration include system connectivity, pipeline diameters, node elevations, control valve settings, check valve direction, and pump characteristics. Some of the more variable or estimated data that may require adjustment include assignment of minor losses, pipe roughness coefficients, and demand peaking factors.

It is important to remember that field data include a certain level of uncertainty and is too often inaccurate or simply wrong. It should not be immediately assumed that there is something wrong with the model if it is not able to match the observed data. Inaccuracies should be investigated carefully.

4.2.1. Calibration, Reconciliation, and Validation

The term *model calibration* is deemed by some to be a misnomer because it implies that the model will be adjusted to match the actual system. However, in many cases, there are situations in the system that cannot be simulated in the model, such as unknown closed valves. For this reason, some suggest that the term *model reconciliation* is a better term than model calibration. For purposes of this manual, the generally accepted historical term *calibration* will be used. It is acknowledged that the degree of calibration is highly dependent on the accuracy of available data and potential unknowns in the distribution system. The term *reconciliation* acknowledges that when model results do not match recorded data, the recorded field data may be incorrect or there may be unknown anomalies in the distribution system, such as closed valves.

Understanding that a distribution system model has inherent limitations can prevent inappropriate use of the model. Adjusting input data to achieve a higher degree of calibration or *best-fit* between modeled and measured results, without sufficient justification, will not negate the inherent limitations in the data. The frequent use of a model, along with a high degree of familiarity with the system, is the best method for establishing model validity and will help the user to understand the limitations of the calibrated model.

Validation is a term often used to refer to the process of checking the results of a hydraulic model following an update process. Validation may also be used to refer to comparing the model to a different set of field data than that for which the model was calibrated. For example, *calibrating* on August 12 and then *validating* against conditions on August 14. Validation is considered by some to be not as formal as the calibration process and may involve a check of results to verify that pressures and flows are within expected and accepted ranges. The term *validation* is used in many ways by different modelers and is not as well defined a term as calibration.

4.2.2. Reconciling Discrepancies

Model results and measured values should be compiled into tables and graphs. The difference between measured and predicted values should be analyzed to see how closely the model reflects the actual system. If many of the differences are great, something may be wrong with the model setup or there may be errors with connectivity. When only a few differences are significant, the modeler should concentrate on the areas where large differences exist. Reconciling differences at this time requires patience and ingenuity. Sometimes tweaking the model will bring simulated and measured results closer; sometimes talking with operators will reveal additional information that was not included in the original model. Calibration is an iterative process throughout which the modeler must carefully make assumptions, because calibration is not just about minimizing differences but more about getting the model to correctly represent the system.

Unknown conditions in the distribution system can often result in calibration inaccuracies. Any model assumptions made in the adjustment process should be the result of clear and logical findings in the system. For example, a large leak or a mistakenly closed valve may cause pressures discrepancies, which should be corrected in the field and replicated in the adjusted model.

A key point of model calibration is a full understanding why the model and field data do not agree. Once these differences are understood, adjustments are easier to identify and apply.

4.2.3. Types of Calibration

Model calibration can be categorized several ways as summarized below:
- Hydraulic versus water quality simulation
- Steady-state versus extended-period simulation
- Macro- versus microcalibration

Currently, most models are calibrated only to hydraulic parameters to verify that pressures, flows, and water levels can be accurately simulated. However, calibration to water quality parameters is becoming more common. Two common water quality calibration parameters include chlorine residuals and disinfection by-products (DBPs). Water quality calibration requires additional data to be input into the model; for example, chlorine calibration requires input of initial chlorine concentrations, chlorine bulk decay curves, and wall decay coefficients. Further discussion of water quality calibration can be found in chapter 7.

Steady-state calibrations are by their nature purely hydraulic calibrations. They simulate distribution system hydraulic performance at a given moment in time, usually a high demand condition such as maximum hour or hydrant flow. Extended-period simulation (EPS) calibrations can be either hydraulic or water quality calibrations. EPS analyses are a series of steady-state simulations linked together to approximate the behavior of a system over a period of time. EPS analyses model changes in demands, operational controls for pumps and valves, and storage facility water levels. EPS analyses are required for water quality simulations and should be conducted using a model that has been hydraulically calibrated under EPS conditions.

Macrocalibration generally refers to calibrating the entire system or an entire pressure zone to recorded data captured by recording charts and/or a SCADA system. Macrocalibration focuses on large discrepancies between observed and predicted values and helps identify gross model errors, such as erroneous model connectivity, nodal elevations, or operational settings.

Microcalibration generally refers to analyses that verify localized conditions in a specific area are being accurately simulated. Microcalibration can be equated to fine-tuning the model in a specific area. A typical microcalibration is the use of simulating recorded fire hydrant flows to verify the correctness of localized piping and valve closures. In smaller systems and in systems with minimal historical data capture, the use of calibration to fire hydrant test results may be used as the basis for macrocalibration of the model.

There is usually an overall macrocalibration performed when a model is first constructed or updated. However, if a model is used for a specific design study, it is often desirable to investigate the area in question and conduct a microcalibration with a fire hydrant flow test as a check of model accuracies in the study area.

4.2.4. Calibration Goals

The goal of calibration is guided by the model end use. A model calibrated for master planning, for instance, may not be calibrated sufficiently for water quality analysis. The real guideline for determining if a model is calibrated appropriately is whether the end result is capable of supporting the decisions to be made. Calibrating should be stopped when the cost of additional calibration exceeds the benefits for supporting decision making.

The required level and focus of calibration depends on the intended use of the model. If its end use is for pipe sizing, fire flow comparisons are critical. If it is used for water quality studies, tank level fluctuations and chlorine decay are much more important.

There are no established standards for hydraulic calibration. Calibration criteria have, however, been published by several entities as described in the following sections. It is important to note that these sources present only suggested guidelines and not regulated standards. It is also important to recognize that these goals were initially developed for steady-state hydraulic calibrations.

All models are approximations of actual represented systems. In a network model, both the mathematical equations used in the model and the specific model parameters are only numerical approximations. The goal in calibration is to reduce uncertainty in model parameters to a level such that the accuracy of the model is commensurate with the decisions to be made based on the model results.

4.2.4.1. ECAC Draft Calibration Guidelines. In 1999, the AWWA Engineering and Computer Applications Committee (now the Engineering Modeling Applications Committee) released the draft report *Calibration Guidelines for Water Distribution System Modeling* (*ECAC Calibration Guidelines Report*). This report presents a set of model calibration criteria. These criteria have been reproduced in other publications, but it is important to recognize that these criteria were never approved by the committee because of their controversial nature. For that reason, they are not reintroduced here but are mentioned only because they may be encountered in other publications.

4.2.4.2. General Guidelines. The following are a commonly referenced set of guidelines:

- Hydraulic grade lines (HGL) predicted by the model should be within 5 to 10 ft (2.2 to 4.3 psi) of those recorded in the field.

- Water level fluctuations predicted by the model should be within 3 to 6 ft (0.9 to 1.8 m) of those recorded in the field.

The following applications have calibration guidelines based on these two criteria in some form:

- Master planning for smaller systems
- Master planning for larger systems
- Pipeline sizing
- Fire flow analysis
- Subdivision design
- Distribution system rehabilitation study
- Rural water system with no fire protection
- Emergency planning

The last two applications in this list have a suggested accuracy guideline of 10 to 20 ft (3 to 6 m) for HGL predicted by the model, or twice the allowable range for the others.

The lower accuracy guideline for HGL of 5 ft (1.5 m) (pressure of 2.2 psi [15 kPa]) would typically be applied to models used for design and operations evaluations. The higher accuracy guideline of 10 ft (3 m) (pressure of 4.3 psi [29.6 kPa]) would typically be applied to models used for long range planning, or to what is referred to in this manual as *master planning*.

4.2.4.3. United Kingdom (WRc) Guidelines. Steady-state calibration criteria were established in 1989 in the United Kingdom by the Water Authorities Association and independent research company, WRc (formerly World Research Centre). Steady-state performance criteria in the United Kingdom are summarized in the following list:

- Modeled flows agree to within:
  - 5 percent of recorded flow when flows are greater than 10 percent of total demand (for transmission mains generally 16-in. (4.06-mm) diameter and larger with no service connections)
  - 10 percent of recorded flow when flows are less than 10 percent of total demand (for distribution mains generally 12-in. (3.05-mm) diameter and smaller with service connections)

- Modeled field test pressures agree to within:
  - The greater of 1.6 ft (0.5 m) or 5 percent of recorded head loss for 85 percent of test measurements
  - The greater of 2.31 ft (0.75 m) or 7.5 percent of recorded head loss for 95 percent of test measurements
  - The greater of 6.2 ft (2 m) or 15 percent of recorded head loss for 100 percent of test measurements

4.2.4.4. Automated Calibration. The process of adjusting model parameters so that the model reproduces results as measured in the field is usually an iterative, or trial-and-error, process that involves a significant amount of time. Automated calibration methods are becoming more available in commercial modeling packages, and their use may increase in the future. However, automated procedures may mask inaccuracies that might be uncovered with a thorough trial-and-error method. Care should be taken if automated procedures are used that such a situation does not occur. Automated calibration can work well once the source of the error is known. If it is known that errors in assigned pipe roughness are causing calibration inaccuracies, automated calibration procedures can greatly help refine the assigned roughness values. However, if the error is actually due to a closed valve, automated calibration procedures will not find it. The results must be interpreted critically to avoid assignment of unreasonable values.

4.2.4.5. Calibration Accuracy Reporting. Increasing regulatory focus on water quality in the distribution system has increased the value of a model that has the ability to run water age and source trace analyses. By allowing modeling results to be used for the Initial Distribution System Evaluation (IDSE) required by the Stage 2 DBP Rule, the rule acknowledges that an extended-period simulation (EPS) model is an effective tool for evaluating distribution system water quality. The absence of quantitative calibration criteria in the Stage 2 DBP Rule is significant because, by the absence of specific criteria, the guidance manual acknowledges the difficulties in defining acceptable level of calibration for quantitative purposes.

Calibration accuracy data can be presented in several ways that can hinder direct comparison of models. All statistical analysis results should be documented sufficiently to identify the statistical methods and assumptions made and should include a discussion of deviations. Forced data (using flow control valves, wells modeled as negative demands, etc.) should not be included in the evaluation of calibration accuracies.

Calibration criteria as presented in the ECAC calibration guidelines report can best be used to establish parameters for use in automated calibration procedures.

The calibration accuracies of an iteratively calibrated model can be improved with the application of automated calibration procedures. Considering the fact that today's modeling software has this built-in capability, it should be considered for use as a part of a careful iterative calibration process. However, care should be taken to verify that system anomalies are adequately identified prior to using automated calibration procedures.

It is equally important that modeling engineers and utilities recognize the limitations of available data. Calibration accuracies can be improved with additional field testing and the identification of anomalies.

The costs for obtaining additional and more precise field data and the costs for additional efforts to improve model calibration accuracies must be considered and weighed against the benefit of improved calibration accuracies.

4.2.5. Data Sources and Errors

There are three general categories of data required to build and hydraulically calibrate a distribution system model. These are (1) physical facilities data, (2) demand data, and (3) operational data. Physical facilities data generally includes system maps and pipeline construction drawings, topographical (elevation) data, drawings of pumping, control valve, and storage facilities, and pump curve information. Demand information generally includes metered sales data, nonrevenue water data, and calculated demands based on recorded flows and storage facility water levels. Operational data generally includes information on pump on/off times and controls, storage facility fill rates as appropriate, and control valve settings. Supplemental data may be available that could include field tests to evaluate pipe roughness factors (physical data) or recorded demands for large uses (demand data).

The sources of data for model calibration vary widely from one utility to another. Some utilities have well-organized electronic data sources and sophisticated SCADA or other real-time monitoring equipment while others may have little information and less reliable equipment. Utilities will find future calibration efforts easier by developing strategies to enhance their ability to procure accurate calibration data. Potential data sources are presented in the following sections with a discussion of associated best practices and cautions.

4.2.5.1. System Maps. It is important to have accurate maps that correctly identify the pipeline sizes and connectivity. In the past, most system maps were in a simple drawing format with no underlying database. In some locations, especially in older systems, system mapping may be unreliable and incomplete. Many utilities now incorporate system mapping within a geographic information system (GIS), which may include information such as pipeline diameter, installation year, material, and valve locations to name just a few. Most commercially available hydraulic modeling software is designed to make model construction from GIS mapping an easier process than in the past. It is important to recognize, however, that electronic information stored in GIS is often based on older mapping and may be unreliable and incomplete in certain areas.

4.2.5.2. Elevation Data. To obtain accurate pressures, elevations must be assigned to each node in the distribution system. Elevation data in electronic format are readily available and can be used within the modeling software to assign elevations to model nodes. However, the accuracies and limitations of available elevation data must be known. Elevations at pumping stations, storage facilities, and all locations where pressure is recorded for comparison should be based on more detailed information such as construction drawings or actual surveys.

It is important to understand the accuracy of the elevation data source. Topographic maps and digital elevation models based on those maps can have significant inaccuracies. Global positioning systems (GPSs) can be used to obtain data, but it is necessary to understand their levels of accuracy because those values can be misleading. Elevation data based on light detection and ranging (LIDAR) imaging are generally very accurate. LIDAR is a method of aerial surveying used to develop topographic maps.

4.2.5.3. Supervisory Control and Data Acquisition (SCADA). If the utility has a SCADA system, this is the preferred source of operational data. SCADA systems can usually provide data on pump flows, pump discharge pressures, reservoir elevations, and, in some cases, even pressure-reducing-valve (PRV) flows. These data are very useful for setting up the model for the specific calibration scenario because the data for a snapshot in time are usually available. Data from SCADA systems can also help develop diurnal demand factor by conducting water balance analyses based on inflow into the system and changes in storage. It is important to recognize that SCADA systems can report data to higher precision than justified. The accuracy of the sensor and any analog to digital conversion must be understood as well. When SCADA data are not available, operating information may be available from operators' hourly logs or on ink pen recorders.

4.2.5.4. Pressure Recording Devices. Pressure data may be available from chart recorders or may be electronically collected and stored through a SCADA system. For model calibration purposes, it is important to have pressure measurement devices that are accurately calibrated. Electronic pressure gauges can be out of calibration due to inaccuracies in the recording device and due to improper setup of the electronic signal that converts the mechanical pressure to an electronic signal that can be interpreted by the SCADA system. It is important to recognize that the SCADA system can be programmed to report pressure at any given elevation other than the actual pressure recorded. For example, the pressure measured in the field is relayed to the SCADA system, where a constant value can be added to the measured pressure before the pressure is written to a data record. This is sometimes done to record the approximate pressure at a different elevation than the recorded location.

The elevation at any pressure recording device must be known, and its level of accuracy should be commensurate with the accuracy of the pressure reading itself. For example, a pressure recorded to an accuracy of 1 psi (6.9 kPa) (2.3 ft [0.7 m]) combined with an elevation recorded to an accuracy of only 20 ft (6 m) results in hydraulic grade accurate to no greater than 20 ft (6 m).

Temporary pressure recording devices have advanced from the old pressure ink chart recorders. Electronic pressure loggers (recorders) are available that can be attached to hydrants, or to other accessible points in the distribution system such as valve manholes or vaults, and will collect data at any interval defined by the user. These devices usually sample pressure every few seconds and record maximum, minimum, and average pressures for the selected interval. These are useful for EPS calibration, but due to costs, they may need to be supplemented by less expensive pressure gauges. It is preferable to convert all measured pressures to HGL before making comparisons, rather than comparing pressures. See chapter 3 for additional information on system pressure testing.

4.2.5.5. Flow Recording Devices. Flow data may also be available from chart recorders or may be electronically collected and stored through a SCADA system. Many pumping stations incorporate flow recording devices. It should be noted, however, that flow recording devices may be out of calibration due to inaccuracies in the recording device and/or improper setup of electronic signaling interpreted by the SCADA system.

Flow recording devices are not as versatile as pressure recording devices for temporary monitoring needed for calibration purposes. In many cases, flow rates can only be estimated by using insertion instruments, which are not always easy to install. Ultrasonic flowmeters can be more accurate for large diameter pipes, but installation requires a certain amount of space and the equipment can be expensive. The accuracy of ultrasonic flowmeters can be questionable for smaller diameter pipes, especially when used to measure lower flows. They are more suited for wastewater flow measurements. See chapter 3 for additional information on flow testing and measurement.

4.2.5.6. Pump Curves. Levels of confidence in modeling results can be significantly impacted by the source of pump curve data. Depending on the age of a facility, pump curves may or may not be available and can be difficult to obtain. The best pump curve information is recent data taken from certified results of pump tests conducted after installation of the pumps or from pump tests for older facilities. Although older facilities may have certified curves that were accurate at the time of installation, they may not represent the current performance of the pump. Operators sometimes trim or replace impellers, which will affect performance of the pumps. In these cases, a pump test may be needed to evaluate current pump performance. In the absence of pump test data, manufacturer's standard curves are the next best source of initial pump curve data.

Manufacturer's standard curves are often provided in lieu of curves from actual pump tests for modeling purposes. When no pump curve can be obtained, usually an operating point (design flow and design head) can be utilized. Most modeling software packages have routines to estimate pump curves based on input of design operating points. However, the estimated curve generated by the model may not accurately represent the actual pump characteristics.

SCADA data can also be used to develop pump curves using recorded flow, suction, and discharge pressures at a pump. When only one pump is operating, it is relatively easy to determine actual points on a pump curve and compare them with the original curve. This can readily identify whether the original curve is correct or if additional testing is required. See chapter 3 for additional information on pump testing procedures.

4.2.5.7. Control Valves. Control valves maintain a setting that keeps operating parameters such as flow, upstream pressure, or downstream pressure constant for the duration of the setting. Some utilities receive water from a wholesale water provider through a turnout, which has settings for both the pressure at which the water is provided and the flow setting. Other types of control valves include PRVs and pressure-sustaining valves that allow water transfer from one pressure area to another and storage facility fill valves that control the rate of flow into a reservoir to refill it after being depleted. Throttling valves may be used to induce head losses within a system for various reasons such as filling a ground-storage reservoir without dropping the pressure in the distribution system too low. In this case, the operator may set the percent open on a throttling valve based on maintaining an acceptable upstream pressure in the system. It is important to gather as much information as possible on control valve settings during calibration in order to accurately simulate actual performance.

Many control valves have smaller diameter piping than the adjacent piping and minor losses may need to be incorporated into the model for the model to accurately simulate actual conditions when the valve is fully open. Some PRV stations have no method for capturing and reporting actual pressures and flows. The actual set point of a PRV may not be accurately recorded and/or may have been adjusted in the field without documenting such a change. It is relatively easy to verify control valve settings in the field, and this should be done as part of the model calibration whenever possible.

It is important to know what type of control is provided by the control valves included in the model. A PRV will keep a constant pressure downstream of the valve, as long as the upstream pressure is higher than the PRV setting and the downstream pressure is lower than the PRV setting. A pressure-sustaining valve will prevent upstream pressure from falling below its setting, protecting upstream customers from low pressures. A relatively easy way to calibrate these devices within the model is to compare pressures and head both upstream and downstream of the valve. Common errors in modeling PRV stations include setting the wrong elevation for the valve and/or using an incorrect pressure setting. If flow is measured at or near the PRV, flow can be logged and compared with model predictions. Though rarely used, PRV manufacturers are now offering equipment that will measure flow through these devices. Having this additional information can be very helpful during model calibration.

4.2.5.8. Unmetered Flows. Unmetered flows are difficult to estimate. One place where unmetered flows are prevalent is at PRV stations. Many utilities provide PRV stations that allow water from higher pressure zones to be delivered to lower zones. In many cases, there are no meters at these PRV stations. These flows can be estimated using billing records, although adjustments and assumptions are required to convert billing records from monthly time frames to average daily flows. Inaccuracies in estimated unmetered flows should be considered when evaluating model calibration.

4.2.5.9. Metered Sales Data. The basis of allocation of demand to the model is generally on an average annual day, with spatial distribution determined from an analysis of meter sales data. Many analytical methods can be used to determine the spatial distribution of average day demand using metered sales data. GIS tools are increasingly used to accurately locate each meter in the system, associate sales from that meter to the assigned location, and to then use this information to determine the allocated sales. Errors can arise from incorrectly located meters.

4.2.5.10. Nonrevenue Water. Nonrevenue water is the difference between metered usage and the water delivered to the system. Nonrevenue water generally includes lost water (leakage and unauthorized use), under-registration of meters, and water use for such things as firefighting, flushing of mains and sewers, and street cleaning. It can usually be estimated fairly accurately for the entire system but may be difficult to estimate for smaller areas or individual pressures zones. Assumptions are usually necessary to allocate nonrevenue water to the system. These assumptions may not accurately account for higher loss in older areas or for any unknown but significant leaks in the system. If possible, nonrevenue water should be determined for discreet areas such as pressure zones, but this is often not feasible. Typical methods of allocating nonrevenue water in the system are to assign it uniformly to all nodes or to allocate it proportionally to all nodes based on demand. Other distributions may account for pipe age and material. The assignment of nonrevenue water can be considered an adjustment parameter during calibration if such adjustments can be justified. For more information on nonrevenue water see AWWA's Manual M36, *Water Audits and Loss Control Programs*.

4.2.5.11. Demand Peaking Factors. Allocated annual average day sales and nonrevenue water are factored, or peaked, in the model to simulate demands that have occurred during the calibration period. The actual peaking of demands by user class and location is unknown. Assumptions must be made to peak the allocated demands that may impact the calibration accuracies of the model. For large users such as industries, hospitals, and commercial institutions, water usage data can be collected to determine a typical water usage pattern.

Most utilities have very good water production data, along with facility operational reports. This data can be used to create historical demand summaries that can be used to set up hydraulic models. Production data have to be adjusted to take into account leaks and other nonrevenue water. Once summaries have been aggregated by year, month and day, it is possible to create peaking factors, which are ratios between high demand periods and the average daily demand. The most common peaking factors are maximum day to average day, which reflects seasonal demands, and peak hour demand to maximum day demand, which reflects diurnal variations. These values are very system specific and depend on the amount of industrial customer usage and the amount and timing of irrigation, among other factors.

4.2.5.12. Pipe Roughness Coefficients (C-factors). The roughness coefficient (C-factor) used in the Hazen-Williams equation is not known throughout the entire system with total certainty because it is not feasible to examine and test every pipe. When the model is first constructed, an initial set of C-factors is established for all pipelines. The initial C-factors used may be arbitrary values based on pipe age, material, and diameter. Sometimes, utilities have historical C-factor measurements, which can be used in regression analyses to determine the estimated C-factor for each pipe in the model based on age and material. Additional information on testing to determine C-factors in presented in chapter 3.

Regardless of how the initial C-factor was established, as part of the calibration process, the C-factor can be adjusted to improve the agreement between model predictions and actual measurements. This can be done manually for all pipes, for certain areas of the system, or automatically through software calibration tools.

4.2.5.13. Minor Losses. Minor losses are frequently neglected in distribution system models. However, when a specific area is experiencing a significant head loss, as when a valve is known to be partially closed or certain known fittings are present, which cause the system to exhibit a significant head loss, it is important to calibrate this minor loss by assigning an appropriate minor loss coefficient (K value) to the pipeline. This can be especially important at pump stations. When comparing suction and discharge pressures at a pump station, it is important to know where the pressure gauge is located. Ideally, the pressure gauge will be located adjacent to the pump. There are many fittings and valves inside a pump station that will create additional minor losses. It is important that these be taken into account during microcalibration.

4.2.5.14. Connectivity. Whether planning to conduct a steady-state or an EPS calibration, it is important to check system connectivity. This step is very important, particularly when models are constructed using data imported from GIS. In many instances, GIS will represent pipelines in a way that is not hydraulically correct. For example, nodes that are very close to one another may appear to be connected when viewing them in the GIS; however, they may not be connected. In the model, pipes must connect to the same node in order to be hydraulically connected. Current modeling software includes tools that allow the modeler to perform connectivity checks, ensuring the model has the correct connectivity. See chapter 2 for more information concerning connectivity.

4.2.5.15. Mass Balance. Mass balance is another important error check. The modeler should verify that total node demands for a specific run match the intended scenario. When all pumps are off, the total demand should be equal to the outflow from storage facilities. Conducting mass balance checks on demands is a good way to perform quality assurance in the model.

4.3. STEADY-STATE CALIBRATION

A steady-state simulation represents a snapshot of the system under one set of conditions. A common approach has been to perform a steady-state calibration first followed by an EPS calibration. However, if an EPS model is developed, the model may be directly calibrated for EPS.

Steady-state calibrations, based on simulation of historical conditions using SCADA and other operational data, are typically conducted during the peak demand hour on a maximum demand day because the demands are highest, resulting in greater energy (head) loss throughout the water system. Another approach to steady-state calibration is to calibrate to a set of induced high flows generated from open and flowing hydrants. Both types of these steady-state calibrations are discussed in this section.

Without significant head loss, the HGL can be close to the static grade line, which minimizes potential differences between predicted and measured values. To effectively assess the level of accuracy of model predictions, it is important to stress the system and verify that the model will correctly represent the changes in HGL. A stressed condition with high head losses produces more meaningful comparisons between field measurements and model predictions.

Artificial demands can be placed on the system to induce high head losses and calibrate under lower demand conditions. Stressing smaller pipes is relatively easy to do using fire hydrants. Flow from hydrants causes high flow rates and measurable head loss. Opening hydrants also provides an opportunity to conduct field flow measurements. Stressing larger pipes requires more effort but can be done by filling tanks and operating large pumps.

During a steady-state calibration process, the most significant comparison values are HGL (calculated from pressures and node elevations) at selected nodes. The nodes are preselected and measurement is conducted for the specified time period. It is better to select locations in various pressure zones and, preferably, farther away from reservoirs to allow for head loss accumulation. It is best to make comparisons in terms of hydraulic grade rather than pressure because hydraulic grade accounts for both pressure and elevation. It is relatively easy to spot anomalous data when hydraulic grade is used.

4.3.1. Peak Hour Calibration

Using inflow into the system from production data and changes in storage, it is possible to verify hourly demand if sufficient operational data are available. Historical demand information also provides ratios between maximum day demand (MDD) and peak hour demand (PHD). Based on this information, the model can be set up to simulate a maximum hour scenario, which can be used for calibration.

Appropriate calibration hours are selected to verify system performance under various operating conditions. As an example, the peak hour is useful for looking at the system for high head loss and for checking whether appropriate C-factors are used in the model. The lower demand period can be used during reservoir refill to look at inflow into reservoirs and verify total dynamic head and pump station flows.

The model needs to match the state of the system, such as boundary conditions, at the time of calibration. If pumping stations are off, the model will have pumps set to off. Whatever is occurring in the water system during the time of calibration should be replicated in the model. PRVs should be set at the same setting as in the field. Utilities usually have lists of facilities and their settings, including valves and PRV stations. The starting water levels in reservoirs should match the levels at the time of calibration.

4.3.2. Fire Flow Calibration

Fire flow calibration involves conducting fire hydrant flow tests. These tests can be important for calibration because they simulate high demand and create significant head loss, which provides a great opportunity to test how close the model mimics the performance of the water system. This type of test not only helps identify incorrect roughness but also indicates closed valves and/or incorrect connectivity. These tests can also help differentiate between errors caused by hydraulics versus those caused by demands (Walski 1983). Additional information on hydrant flow testing can be found in chapter 3.

Hydrant flow testing provides both static and residual pressures at the area where the testing is conducted. Replication of this test in the model can be done in many ways. First, the flows measured during the flow test are applied to corresponding nodes in the model. Then, the model is run to verify that the residuals predicted by the model match the residuals obtained from the hydrant flow test. It is a good practice to convert all measured pressures to head and make comparisons in terms of HGL rather than comparing pressures. If discrepancies are found, the modeler will need to troubleshoot the cause of the differences and make the necessary adjustments to bring observed and calculated values closer. It is important to identify and duplicate in the model the boundary conditions (tank levels, pump operation, etc.) to ensure the usefulness of this test method.

4.3.2.1. Identifying Closed Valves. Fire flow calibration can be a useful tool for identifying inadvertently closed valves. In one utility system's example, initial fire flow calibrations did not match well in one area of the system. An additional series of focused fire flow analyses were further conducted in the area where an improperly closed valve was suspected to be the culprit. Model analyses were conducted using the focused fire flows; and after a series of iterative analyses, the closing of one valve in the model resulted in accurate simulation of all the fire flow tests. Operators later verified that there was a closed valve at that location that should have been open.

4.3.2.2. All-Pipes Models. Many utilities have implemented all-pipes models in recent years as new water quality regulations have created the need for more detailed models. While some may argue that all-pipes model calibration requires more measurement points to cover more areas of the system and ensure that calibration is as thorough as possible, others maintain that it may be easier to calibrate an all-pipes model because there is no need to adjust roughness to account for missing pipes.

Fire flow calibration in all-pipes models can be taken a step further by modeling fire flow out of tested hydrants as emitters. However, caution should be exercised in relying too heavily on this method because most of the head loss is in the hydrant lateral and barrel, which results in calibration of the hydrant rather than the model. If this method is used, both the flow from the tested hydrant and the residual pressure should be compared with the predicted values from the model.

4.3.3. Presenting Calibration Results

Steady-state model calibration results are typically presented in tables in which predicted and observed values are shown for each measurement point. For HGLs, the nodes may be listed along with a description of the point. The observed and measured values, differences, and percentages of these differences can also be presented. Pump station flows and PRV flows may also be presented along with upstream and downstream PRV pressures or HGLs.

The goal in presenting calibration results is to provide assurance to model users that the model represents as accurately as possible the real water distribution system. When discrepancies remain after model calibration, an acceptable explanation for the differences and the resulting limitations of the hydraulic model should be provided. Figures 4-1 and 4-2 present example steady-state calibration results for flows and hydraulic gradients.

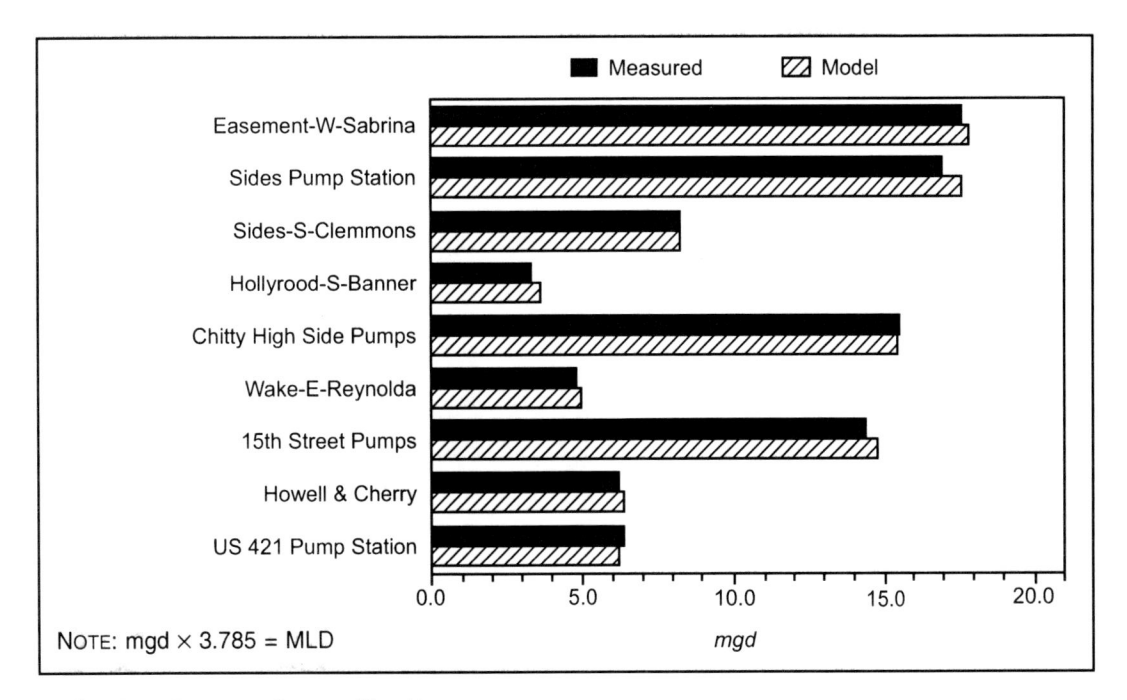

Figure 4-1 Steady-state flow calibration

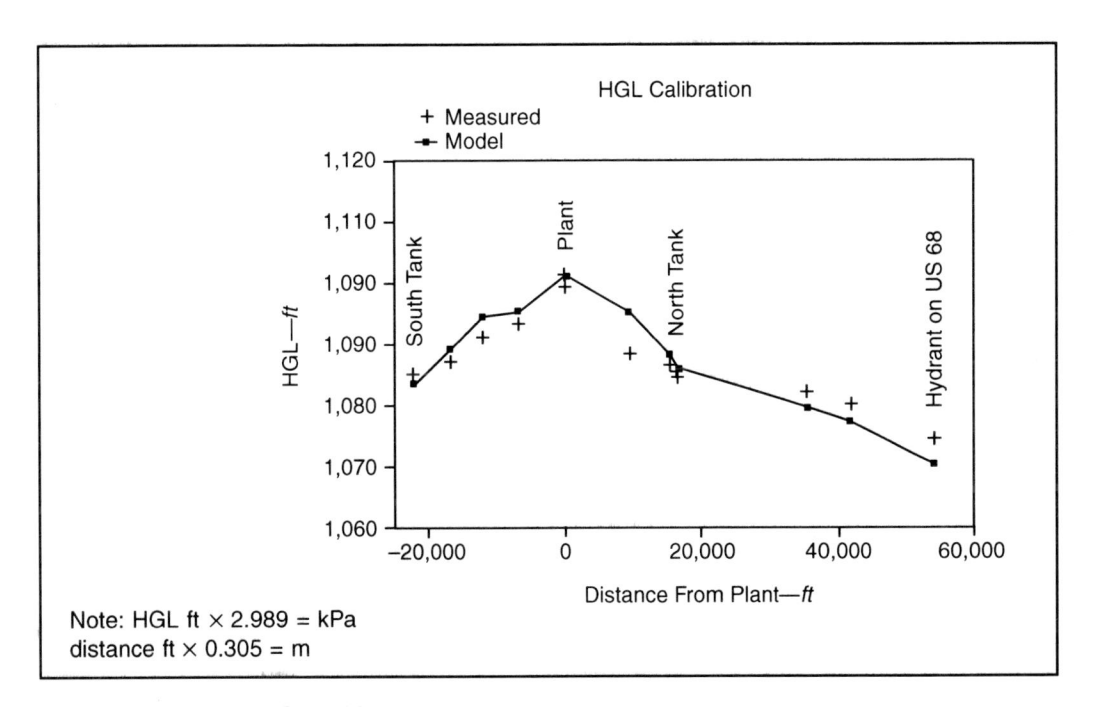

Figure 4-2 Steady-state HGL calibration

4.4. EPS CALIBRATION

EPS calibration methods include comparisons of model results to measurements made in the field over time and to tracer test simulations. Tracer tests simulations are covered in more detail in chapter 7 in discussions on water quality calibration. Hydraulic calibration under maximum day conditions is described in the following section. Other demand conditions may be calibrated to ensure that a model simulates system performance under other conditions, such as average day demands for evaluation of energy costs and minimum day demands for evaluation of water quality evaluations.

4.4.1. Maximum Day Calibration

A typical EPS hydraulic calibration simulates the distribution system for a 24-hour period of high demand. Other periods of time have been used such as 48-hour or 7-day periods. A 7-day EPS calibration is rarely conducted but may be done to simulate changing diurnal usage patterns and operations that may occur on a weekend. A 24-hour EPS is often used because it simulates the diurnal supply and demand conditions that are typically repeated daily. Longer calibration periods may be justified where typical operating conditions might repeat at a period longer than 24-hours. The benefit of conducting longer calibrations must be balanced against the effort to calculate and input data for demands and operational controls for longer periods.

Similar to the previous peak hour steady-state discussion, high demands are required to stress the system and provide a better comparison of modeled results to recorded data. An EPS calibration will normally start with the same modeled facilities and allocated annual average day demands as a steady-state calibration. Additional data are then required to simulate the system over the designated calibration period.

4.4.1.1. Model Controls. For an EPS calibration, it is necessary to input model controls to tell the model when to activate certain facilities to simulate actual conditions. The two most common operational controls required for EPS calibration are pumps and control valves. Storage facility water levels must also be entered accurately.

Model controls for pumps and valves are often initially input using time-of-day controls that reflect actual times of operation as recorded on the day of calibration. While logical controls that regulate operation of one facility based on specific conditions at some other facility or location in the system are available in the model, e.g., pump on/off based on water level in a storage facility, these may make calibration more difficult. It is common practice for some operators to initiate operation of facilities without a strict set of operational parameters. Using time-of-day controls for calibration removes an unknown parameter and allows the modeler to focus on flows, pressures, and water levels, rather than on whether or not the operational controls are correct. That is, it ensures boundary conditions are accurately modeled during the calibration period. It is important to recognize that calibration using time-of-day controls makes it difficult to extrapolate the model to new situations. Other logical controls should be considered for addition following the initial time-of-day calibration to increase the ability of the model to simulate differing situations.

Pump controls may include on/off status and pump speed. Care should be taken to identify all pumping units that are equipped with variable frequency drives (VFDs) and to obtain pump speed as a percent of full speed during calibration, whenever possible. However, pump speed is often not captured in historical operational data reports. Where VFDs were operating during the calibration, it may be necessary to estimate the speed if not recorded. It may be necessary to estimate speed based on flow measurement data from the pump station.

Valves that require control generally consist of reservoir fill valves, throttling devices that may be modulated, and motor operated valves that may be moved from open to closed position during routine system operation. PRVs, pressure-sustaining valves, and altitude valves are not modulated on a routine basis and generally should not have variable set points during a calibration period. For valves that are modulated, it is possible to model the valve as a flow control valve or as a throttling valve with a specified position (percent open/closed). Many times the position of a modulating valve is not captured in the historical operational data reports. EPS calibrations can be simplified by simulating valves that modulate as flow control valves, and inputting recorded or calculated flow rates.

Once the model is calibrated sufficiently with the valve modeled as a flow control device, the modeler may elect to then calibrate the throttling characteristics of the valve so that modeled flows match recorded flows at given, or recorded, valve positions. This is only possible if both the valve position and the flow through the valve are recorded. Flow rates through reservoir fill valves may not be recorded but can be calculated based on water level in the reservoir and any recorded pumping or outflow from the reservoir.

Storage facility water levels required for EPS calibration can be categorized as (1) those that must be set at each time interval or (2) those that are allowed to fluctuate based on other parameters within the model. Water levels that must be set at each time interval are those that establish the hydraulic gradients in the system at each time interval. One example is the level of a finished water storage facility that provides suction to the high service pumps that deliver water from the treatment plant to the distribution system. In this case, a finished water storage facility, or clearwell, may be modeled as a variable head reservoir and a pattern input to the model to vary the water level in the reservoir to mimic what was recorded during the calibration period. If the water level is not recorded, it may be acceptable to model this situation with a fixed head reservoir that reflects typical or average levels.

Another example of a water level requiring a known level at each hour is a well supply where groundwater level is simulated by a reservoir. In both of these cases, water levels are necessary to establish the starting supply conditions for the model analyses.

Except for facilities required to establish starting supply conditions, all other storage facilities are generally allowed to fluctuate. This includes all floating storage facilities and other facilities where water is delivered via throttling valves and then repumped as needed. The fluctuation in these storage facilities is generally a calibration parameter where the modeled water level is compared to the recorded water level.

4.4.1.2. Demand Peaking Factors. The second significant piece of additional data required for EPS calibration is the addition of data to allow the model to simulate changing demand conditions. Patterns are applied to the allocated demands for each calibration increment. The patterns represent a variable peaking factor that is applied to the allocated demands to vary the demand conditions in the model. Differing demand peaking factors may be developed for each class of demand allocated to the model. However, it is important to recognize that information on variations in peaking by user class is usually not well known. However, in some cases the input of variable peaking factors by class may be required to achieve successful calibration. Examples of this might include a golf course known to take the majority of its water at certain times of day or a wholesale customer known to use water according to specific characteristics or whose hourly use may be recorded in operational data reports. An example of typical hourly demand factors that might be applied during an EPS calibration is shown on Figure 4-3.

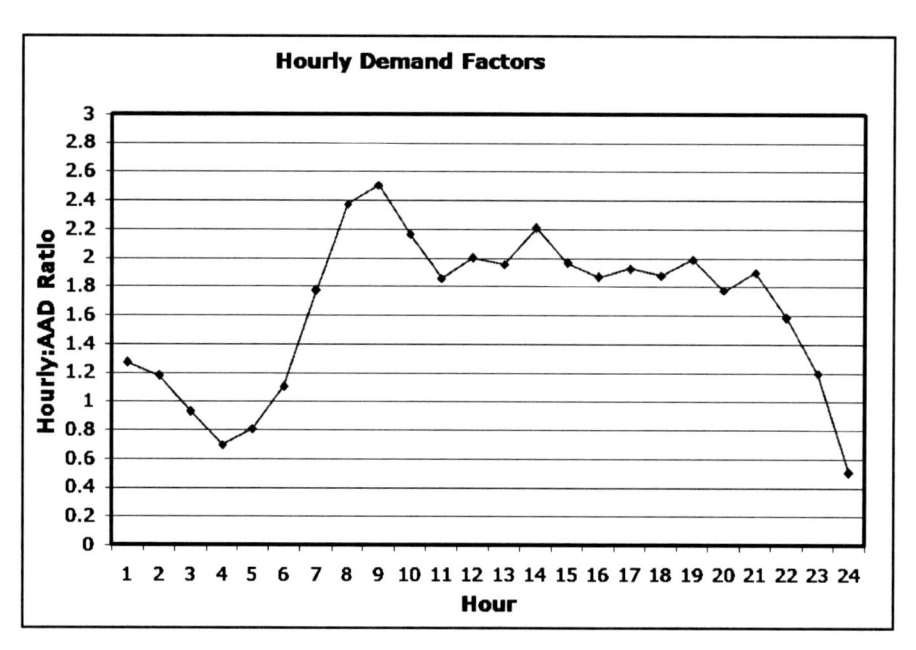

100 COMPUTER MODELING OF WATER DISTRIBUTION SYSTEMS

Figure 4-3   EPS hourly peaking factors

When the system serves large water-using industries, it may be advisable to install data loggers on their water meters to better understand and simulate their use patterns, thereby removing significant unknowns and potential error sources.

4.4.1.3. Fire Flows. An EPS calibration can be conducted with a simulated fire flow superimposed on the system. This may be done for the same reasons as explained for a steady-state calibration—to supplement existing operational data or to conduct microcalibration. Fire flows for EPS calibrations are generally simulated using demand patterns. One method to accomplish this is to assign a demand unit of one to a fire flow node. The demand pattern would contain values of zero when the hydrant is not flowing, and an appropriate peaking factor when the hydrant is flowing such that the factor demand matches recorded flow at the hydrant.

4.4.2. Adjustments to Model for Extended-Period Calibration

Making adjustments for EPS calibration can be difficult. There are many types of model adjustments that can be made. The following are some considerations for calibrating extended period models.

4.4.2.1. Pump Curves. There are many possible reasons that the flow output of a pump station indicated by SCADA could be considerably different than the model results. Differences in pressure losses on the suction side or discharge side can affect the output of the pumps. This is especially true with head losses within station piping. These are often underestimated, and high velocities in the pump discharge piping can lead to much higher head losses. If system head losses are correct, pump curves could be incorrectly characterized. Pump impellers wear over time potentially reducing pump output. Adjusting the pump curve can help match model results to SCADA data.

4.4.2.2. Valve Positions. It is important to confirm the minor loss characteristics of valves. A flow control valve works by imparting head loss to match upstream and downstream pressures for the flow set point. Throttle control valve positions are not normally set with precision. When the valve is throttled to a significant amount, a small difference in position can make a large difference in the head loss through that valve.

4.4.2.3. Control Logic. It is possible that the control logic programmed into the model does not match that programmed into the SCADA system. The response time of some instruments may be long enough that a pump or valve will change its actual status at a different time than indicated by the model. More common is a change in status that occurs between model time steps.

4.4.2.4. Diurnal Demand Patterns. Adjustments to demand patterns for specific customers or localized areas may be required to achieve acceptable calibration. As previously mentioned, a large irrigation customer or wholesale customer, or a large industrial customer may have a significant impact on actual system performance. Specific diurnal patterns may need to be estimated and input into the model.

4.4.2.5. Pipe Condition. If other adjustments do not seem to be effective or if local anomalies prove difficult to calibrate, adjustments to pipe C-factors may be in order. Adjustments to C-factors made during steady-state calibration may have overlooked some areas not covered by hydrant testing, which could affect how different parts of the system are served.

4.4.2.6. Closed Valves. Improperly closed valves may also be a cause of poor calibration results. It may be necessary to assume closed valves at certain locations to achieve acceptable calibration results. If closed valves are assumed in the model, they should be field verified. Additional field testing may be warranted to identify and verify assumed and potential improperly closed valve locations.

4.4.2.7. SCADA Errors. The possibility of errors in the SCADA system should not be discounted as a reason model results do not agree. Older instruments can show drift in their readings, so the model results may actually be more accurate than the SCADA readings. Some SCADA values are calculated within the SCADA system based on the summation of values from several instruments. These types of instruments are more prone to calculation errors than other instruments.

4.4.3. Presenting Calibration Results

EPS model calibration results are typically presented in graphs in which model-predicted and field-observed values are shown for each measurement point. Typical values that are compared include flows, pressures, and water levels. The guidelines shown in this chapter can be used to assess the quality of calibration results. Figure 4-4 provides an example of an EPS calibration graph that compares the modeled water level to the recorded water level.

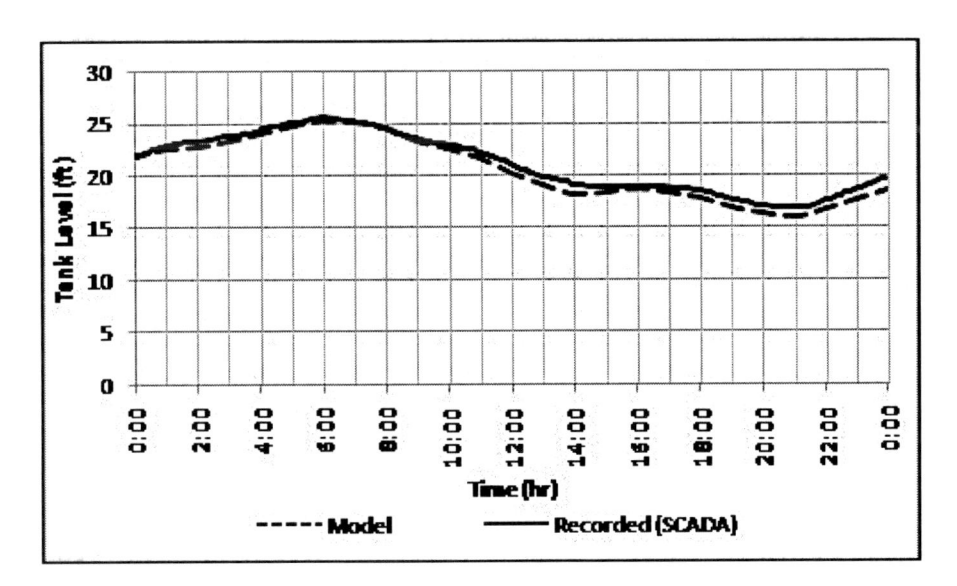

Figure 4-4　EPS water level calibration

Care must be taken when evaluating statistical calibration accuracies. In many cases, a pump station only pumps for several hours during the day. When the pumps are not operating, modeled flows will by default match recorded flows exactly when using time-of-day controls. Flows at hydro-pneumatic pump stations should likewise not be included in the calibration accuracies. Hydro-pneumatic pump stations are regularly used to increase pressures to smaller areas of the distribution system.

Calibration accuracy can be calculated in many ways. All statistical analyses results should be documented sufficiently to identify the statistical methods used and the assumptions made and should include a discussion of deviations. Forced data, such as using flow control valves or wells modeled as negative demands, should not be included or should be clearly identified in the evaluation of calibration accuracies, because they will necessarily improve the reported accuracy of the model.

4.5. REFERENCES

American Water Works Association Engineering and Computer Applications Committee. 1999. Calibration Guidelines for Water Distribution System Modeling. In *Proceedings of AWWA Information Management and Technology Conference.* Denver, Colo.: AWWA.

Cesario, A.L. 1995. *Modeling, Analysis and Design of Water Distribution Systems.* Denver, Colo.: AWWA.

Chase, D. 1999. Calibration Guidelines for Water Distribution System Modeling. In *1999 Proceedings of the AWWA Information Management Technology Conference.* Denver, Colo.: AWWA.

Cruickshank, J.R., and Long, S.J. 1992. Calibrating Computer Models of Distribution Systems. In *1992 AWWA Computer Conference Proceedings.* Denver, Colo.: AWWA.

Edwards, J.A., Cole, S., Brandt, M. 2006. Quantitative Results of EPS Model Calibrations With a Comparison to Industry Guidelines. *Jour. AWWA,* 98:11:72–83.

Grayman, W.M., Rosssman, L.A, Deininger, R.A., Smith, C.D. Arnod, C.N., and Smith, J.F. 2004. Mixing and Aging of Water in Distribution System Storage Facilities. *Jour. AWWA,* 96:9:70–80.

Hirrel, T.D. 2008. How Not to Calibrate a Hydraulic Network Model. *Jour. AWWA,* 100:8:70–81.

Ormsbee, L.E., and Lingireddy, S. 1997. Calibrating Hydraulic Network Models. *Jour. AWWA,* 89:2:42–50.

Speight, V., and Khanal, N. 2009. Model Calibration and Current Usage in Practice. *Urban Water Journal,* 6:1:23–28.

Speight, V., Khanal, N., Savic, D., Kapelan, Z., Jonkergouw, P., and Agbodo, M. 2010. *Guidelines for Developing, Calibrating, and Hydraulic Models.* Denver Colo.: Water Research Foundation.

US Environmental Protection Agency (USEPA). 2003. *The Stage 2 DBPR Initial Distribution System Evaluation Guidance Manual, EPA-815-D-03-002 (Draft).* Washington, D.C.: USEPA.

USEPA. 2005. *Water Distribution System Analysis: Field Studies, Modeling and Management, A Reference Guide for Utilities, EPA-600-R-06-028.* Cincinnati, Ohio: Office of Research and Development.

Water Authorities Association and WRc. 1989. *Network Analysis—A Code of Practice.* Swindon, UK: WRc.

Walski, T.M. 1983. Technique for Calibrating Network Models. *Journal of Water Resources Planning and Management,* 109:4:360–372.

Walski, T.M. 1990. Sherlock Holmes Meets Hardy-Cross or Model Calibration in Austin, Texas. *Jour. AWWA,* 82:3:34–38.

Walski, T.M., and O'Farrell, S.J. 1994. Head Loss Testing in Transmission Mains. *Jour. AWWA,* 86:7:62–65.

Walski, T.M. 1995. Standards for Model Calibration. In *AWWA Computer Conference Proceedings.* Denver, Colo.: AWWA.

Walski, T.M. 2000. Model Calibration Data: The Good, the Bad, and the Useless. *Jour. AWWA,* 92:1:94–99.

Walski, T., Chase, D.V., Savic, D.A., Grayman, W.M., Beckwith, S., and Koelle, E. 2003. *Advanced Water Distribution Modeling and Management,* 1st Ed. Waterbury, Conn.: Haestad Press.

AWWA MANUAL M32

Chapter 5

Steady-State Simulation

5.1. INTRODUCTION

Hydraulic model simulations are divided into two main categories: steady-state and extended-period simulations. A steady-state model simulation computes the state of the system (flows, pressures, pump operating attributes, valve position, and so on) at a given point in time. Steady-state analysis is sometimes compared with taking a snapshot picture of the distribution system. Typically these snapshots are taken during worst-case conditions such as peak demand times, fire protection usage, and system component failures. Steady-state simulations are often times used for basic infrastructure-related design problems.

Extended-period simulation is a technique for modeling a distribution system using a series of steady-state analyses that are performed over a specified time period at selected intervals. Extended-period simulations are useful in seeing the effects of water usage over time in order to observe system behavior such as tank level fluctuations and pump and valve operations. Extended-period simulations are often used for master planning and operational studies and are required for water quality simulations. More discussion on this topic can be found in chapter 6.

Many hydraulic problems in a water distribution system can be evaluated with steady-state simulations. The types of problems that can be analyzed include:

- Sizing mains, pumps, and reservoirs
- Performing fire flow analysis
- Assessing the impact of system changes such as new large customers or housing developments
- Planning water main replacement or rehabilitation projects
- Planning the layout of a water distribution system
- Establishing set points for control valves and variable speed pumps
- Locating hydraulic restrictions such as undersized pipes or closed valves

This chapter details the following general topics:

- *Selecting limiting conditions for design scenarios:* The modeler determines the design conditions under which the model predicts distribution system performance.

- *Establishing design criteria:* A model compares distribution system performance against a set of performance criteria to determine the adequacy of the system.

- *Developing system improvements:* The processes for several categories of model simulations are described to explain the process of steady-state hydraulic modeling.

A calibrated model is required to conduct the steady-state analyses described in this charter. Additional detail on model calibration is provided in chapter 4.

Hydraulic modeling is partly science and partly interactive trial and error that require the modeler to be knowledgeable in water distribution system hydraulics and be able to apply sound engineering judgment to the different situations presented to them during the modeling process. This chapter is intended as a general guide, pointing out important considerations and providing guidance on the scientific principles used in system analysis. The modeler should use this information as a guide to help tailor specific requirements to best fit their system conditions.

5.2. SYSTEM PERFORMANCE ANALYSES

If a distribution system operates satisfactorily under the most severe demand conditions, it should operate satisfactorily under all conditions. For this reason, the most limiting demand conditions should be established and simulated using the model. These are points at which system demands and pressures are at their highest and lowest points.

Under normal operations, the maximum hour demand is the most extreme condition experienced by a water system on a single day and the maximum day demand is the most extreme condition that can be experienced during the year. The diurnal demand curve under maximum day demand conditions can be used to assist with the selection of these limiting demand conditions as illustrated in Figure 5-1. Additional detail on diurnal demand curve development is provided in chapter 2. Points A and B in Figure 5-1 represent the two times during the day when the instantaneous demand equals the average production rate from the supply sources, assuming the supply source production rate is constant throughout the duration of the day. Point D represents the peak hour demand while Point C represents the minimum hour demand. At any point in time, the difference between the hourly demand and the system production represents the flow into or out of the storage facilities. This assumes that production and pumping facilities are provided at a constant rate all day equal to the maximum day demand. If the production and supply facilities cannot deliver at a rate equal to or greater than the maximum day demand, there is a deficiency that needs to be resolved.

At the minimum hour demand rate (Point C), the demand for storage replenishment is at its maximum. This can be a limiting demand condition to be analyzed to determine whether the distribution system can adequately fill the storage tanks or to assess maximum pressures in the system.

At the peak hour demand rate (Point D), the flow out of the storage reservoirs is at its maximum. The system must be assessed under this condition to ensure that the customers have sufficient pressure and to identify the most serious bottlenecks in the system.

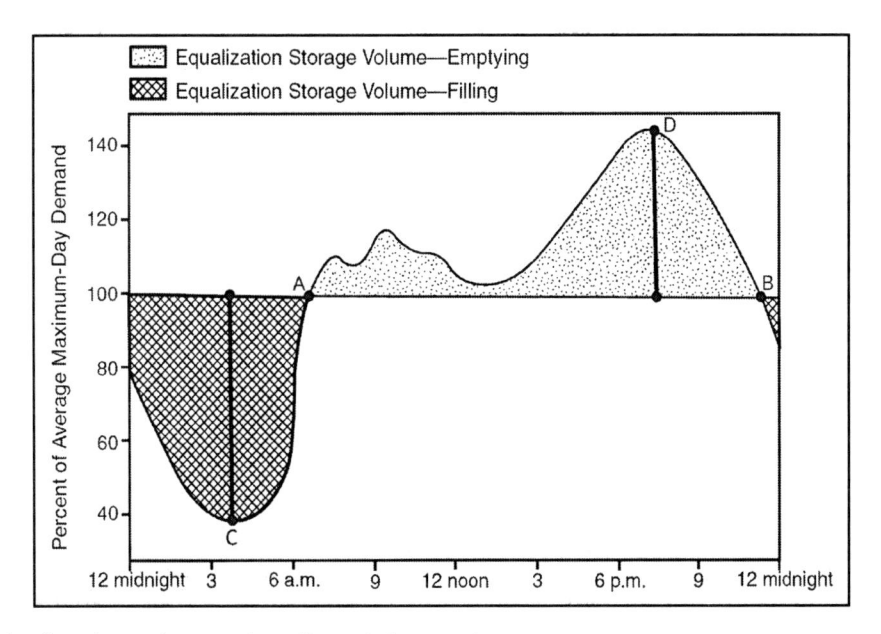

Figure 5-1   Idealized maximum day diurnal demand curve

Another important limiting demand condition not reflected on the diurnal demand curve is the required fire flow demand. These demands are community specific and should be assessed as an additional flow superimposed on the maximum day demand. For steady-state analyses, this condition could correspond to a fire at Point B on the curve on Figure 5-1 where storage volume is at its lowest. Fire flow conditions are often studied using a combination of maximum day demands or peak hour demand and reduced storage availability.

The most common steady-state scenarios are average day, maximum day, maximum hour, the minimum hour of the average day, and maximum day plus fire flow. These demand conditions are coupled with minimum and maximum tank levels, pumps being turned on and off, and/or controls valves being opened and closed to assure that the system can handle worse-case conditions. The choice of which condition to model depends on which question the modeler is trying to answer. A common approach is to use steady-state simulations initially to evaluate the system and develop the conceptual design of improvements, and then use extended-period simulations to verify the final design. Table 5-1 summarizes some typical analysis types and the corresponding condition used to simulate the network.

Building individual models for each of these scenarios is not necessary. Most modeling software programs allow storage of multiple scenarios within the same model file. By doing so, changes in the model configuration do not need to be transferred between multiple model files. The modeler should instead develop a base model containing the information required to construct the model and use peaking factors and model scenarios to change the demands to reflect the different conditions. Typically, the base model is set up to reflect either the average day conditions or the maximum day conditions. Average day has the advantage of being able to correlate total system demands to the information from the utility billing systems. Maximum day is advantageous because it is easily adjusted to the most common design scenarios, either maximum hour or fire simulations. Maximum day demands may also exhibit different spatial allocation of demands.

Table 5-1 Typical model scenarios

Purpose of the Analysis	Recommended Steady-State Demand Scenario
Studies of normal operation	Average day, maximum hour
Production and pumping requirements	Maximum day
Sizing of pipelines—small systems	Maximum day plus fire flow*
Sizing of pipelines—large systems	Maximum hour and maximum day*
Sizing of storage facilities	Maximum hour†
Tank filling capabilities	Minimum hour of average day†
System reliability during emergency or planned shutdown	Condition when the emergency or shutdown is likely to occur
Model calibration	Condition during time when measurements were collected

*Both maximum day plus fire flow and maximum hour conditions are generally studied for sizing of pipelines regardless of system size.

†These analyses are best studied using extended-period analyses as discussed in chapter 6.

The following sections discuss the major demand scenarios that are normally studied under steady-state conditions in more detail to assist the modeler in analyzing the system under varying conditions. As previously indicated, it is critical to have a calibrated water model prior to running any of the following scenarios. A noncalibrated model could provide erroneous results providing useless or, worse, dangerous information.

5.2.1. Maximum Day Demand Conditions Analysis

Maximum day demand is the largest quantity of water consumed on any one day of the year, assuming normal system operations. Maximum day model runs assess the ability of the system to meet demands with adequate residual pressures and maintain levels in elevated storage or high-level ground storage facilities. At a minimum, a system should be capable of providing adequate flow and pressure under maximum day demand conditions. In a maximum day model run, system pump capacity (or inflow conditions for gravity feed systems) should be able to meet or exceed the demand of the system while tanks are filling. Pumps that are not able to meet or exceed the maximum day demands should be of concern. Tanks that are not filling could be of concern depending on the situation. If a single tank provides storage to a single pressure zone, not being able to fill the tank under a maximum day condition would be a concern. However, if multiple tanks are within the same pressure zone, one tank could be filling while the other empties. This may not be as much of a concern. However, with this condition it is recommended that an extended-period simulation be performed to assure all system tanks can be refilled by the end of the day. Maximum day is a critical condition because it is the largest demand to be supplied completely from production, without using system storage.

Any water supplied by tanks must be replenished within a 24-hour period. Otherwise, tank levels would drop from one day to the next. Eventually, there would be inadequate storage available for daily operations or emergencies. Some systems may allow storage to deplete slightly during a maximum day event with the understanding that the tanks would be refilled on a subsequent date when demands are less than the system's maximum.

To execute a maximum day demand simulation the following model settings are necessary:

1. Set the model demand conditions to simulate a maximum day condition.

2. Set the initial or operational conditions for the pumping facility serving the pressure zone of interest with the proper number of pumps operational as desired by the modeler while allowing for some spare pumping capacity. If not assessing existing pumps, the modeler can introduce an inflow rate at the source point instead.

3. Set tank elevations at a point near their overflow elevation or the point where the pumps are turned off and the tanks are considered full.

The results of the existing system maximum day model analysis can be used to identify deficiencies in the existing system and plan short-term improvements to address these deficiencies. The goal for future year maximum day model runs is to identify water lines and facility improvements that enable the system to provide an adequate water supply to customers under future demand conditions as well as to address potential future deficiencies and corrective measures.

5.2.2. Maximum Hour Demand Conditions Analysis

Maximum hour (or peak hour) demand represents the single hour of the day with the highest system demand. Maximum hour model runs are used to assess the ability of the distribution system to maintain sufficient residual pressures under worst-case conditions and maintain reasonable head loss within piping systems. Maximum hour model runs can also be used to examine whether the storage tank draining rates are excessive. However, because tank draining rates are a function of time, it is best to make this determination using EPS. Maximum hour is most commonly the limiting design condition for large water systems. Maximum hour usually exceeds maximum day plus fire in large systems with a maximum day greater than 7 to 10 mgd (30 MLD to 38 MLD).

To execute a maximum hour demand simulation the following model settings are necessary:

1. Set model demand conditions to simulate the maximum hour demand condition. Maximum hour demand is modeled using peaking factors from maximum day demands or input directly in the demand data. Peaking factors are obtained from measurements or SCADA records.

2. Set the initial or operational conditions of the pumping facility serving the pressure zone of interest to the desired setting. If a worse-case scenario is being sought, the modeler could choose to turn all pumps off. However, if this is not a standard system condition (such as a system with little or no system storage), the modeler could choose to leave the pumps on.

3. Set tank elevations at a point near the normal low level. If this information is unknown, it is likely reasonable to assume the tank elevation to be half-full as a worst-case scenario.

The results of the maximum hour demand simulation are reviewed to assess whether system pressures remain adequate under worst-case conditions and that head losses within the network remain reasonable. If low pressure exists at a location near a water storage tank, high ground elevation relative to the tank overflow elevation is likely the cause of the low pressure. In this case, the opportunity to increase the low pressure may not be available and booster pumping may be indicated. However, if low pressure is at a location a distance away from the nearest tank, the problem is likely inadequate system capacity (pipe friction losses) to serve the location.

The maximum hour demand simulation allows comparison of pressures under worst-case conditions for various alternatives and helps simplify decisions for proposed improvements. If the model predicts pressures that are not within design criteria, additional improvements can be analyzed. Additional discussion on pressure and piping design criteria can be found later in this chapter.

Results from a maximum hour demand simulation can also help to examine whether storage tank draining rates are excessive. Tank depletion rates during the maximum hour should be checked against tank capacities. A rule of thumb is that a tank should not exceed a drain rate in million gallons per day equal to three times its volume in million gallons, which assumes using half the volume in four hours. A maximum hour analysis would likely provide the worst-case conditions for drain rate. However, it is recommended that an EPS simulation also be performed to provide confirmation of these results. Providing adequate storage is key to avoiding excessive tank drain rates. Further discussion on sizing of water storage tanks is provided later in this chapter and in chapter 6.

5.2.3. Average Day Conditions Analysis

Average day demand is defined as the total water used in a year divided by the number of days in the year. This assumes the system is not affected by seasonal or intermittent demand changes. In these instances, the modeler must determine the appropriate average day demand that applies. Average day simulations model the normal operation of a system under typical conditions. Calculated flows are compared to annual average production and billing records. Steady-state simulations of average day are often used as the basis for extended-period simulations that study power costs or water quality.

Pump conditions can be verified and new pump selections can be optimized using average day model runs. The pumps should be operating at or near their selected design points at high efficiency while supplying demands.

5.2.4. Fire Flow Analysis

Fire flow simulations are a common application of steady-state modeling. Fire simulations can also be used for model calibration. After simulating fire flows and identifying deficiencies, models are used to design improvements. For accurate results it is not only necessary to begin with a calibrated model but one that has accurate ground elevations within 2 ft (0.6 m) to calculate residual pressures within 1 psi (7 kPa).

The sophistication of current modeling programs allows the user to set the desired fire flow rate as well as an upper limit to determine available fire flows. Some programs can determine the available fire flow at all applicable points in the system with one model execution. At the location where the fire flow occurs, the flow can be added to the system demand or can replace the system demand as the situation dictates. Some modeling programs allow the use of pressure-dependent emitter coefficients that can represent flow from a hydrant. Depending on the situation, several limiting parameters such as residual pressure or pipe velocity should be set to determine fire flow availability.

Minimum residual pressure should be set such that the fire flow will be limited based on the point at which pressures drop below a predefined level within the water system. The limiting residual pressure can be set for the whole system, the same pressure zone as the fire flow, a different pressure zone, or the point of the fire flow depending on the situation. As a general rule, the minimum pressure is set to 20 psi (138 kPa) within the pressure zone while the fire flow is occurring. However, residual pressures should be set higher in instances where higher residual pressures are required, such as for sprinkler systems. When running such scenarios, the modeler needs to be aware of high points that may already be below 20 psi (138 kPa), typically near water storage tanks or on the suction side of booster pumping stations. If these nodes are not removed from consideration as limiting conditions, results will indicate that no fire flow is available. The modeler also needs to make sure high points are appropriately identified in the model as well and included as limiting locations. If these points are missing, the results could be incorrect. These situations need to be taken into consideration when performing a fire flow simulation.

Maximum pipe velocities can also be a major concern in a water distribution system. High pipe velocities can lead to high head loss conditions and possibly transient flow (water hammer) concerns, potentially causing damage to the water system. Because limiting fire flow due to pipe velocities is an available option in most of the current models, consideration should be given to its use to avoid excessive pipe velocities. For more information on how pipe velocities can lead to transient flow see chapter 8.

To execute a fire flow simulation the following model settings are necessary:

1. Set model demand conditions to simulate either maximum day or maximum hour demand condition. Sometimes regulatory requirements will dictate what demand conditions are required.

2. Set the initial or operational conditions of the pumping facility serving the pressure zone of interest to the desired setting. Typically pumps are turned off to ensure fire flows can be achieved if there is a pump failure. However, the modeler must determine the proper condition based on system conditions.

3. Set tank elevations at a point near the normal low level. If this information is unknown, it is likely reasonable to assume the tank elevation to be half-full as a worst-case scenario. The minimum tank level can also be set to a level that provides the needed fire flows without causing pressures to drop below minimum thresholds.

4. Set the desired fire flow rate for individual nodes in the model as well as the limiting parameters such as pressure and velocity. For skeletonized models, localized piping may need to be added to allow for more accurate modeling.

The results from a fire flow simulation indicate whether the system is capable of providing adequate fire flow protection under the worst system conditions. Key information provided from a fire flow simulation is generally as follows:

- Available fire flow
- Residual pressure at the fire flow location
- Minimum residual pressure with the zone the fire flow takes place
- Minimum overall system pressure
- Maximum pipeline velocities

These results help determine if adequate fire flow is available or if fire flow is inadequate due to inadequate pipe sizes or high pipe velocity, or if fire flow is limited due to loss of pressure at a localized high point. System improvements can then be analyzed to assure adequate fire flow is provided.

To determine if tank storage is adequate for a specific fire flow condition, it is recommended that an extended-period simulation be performed over the duration of the fire flow. See AWWA Manual M31, *Distribution System Requirements for Fire Protection,* for further discussion concerning required fire flow rates and durations.

5.2.5. Minimum Hour Demand Conditions Analysis

Minimum hour demand is the point at which the smallest amount of water is used during the day. This typically occurs in the early hours of the morning. Minimum hour runs assess the maximum pressures the system will have to accommodate during the day. Excessively high pressures can cause damage to the system components. Coupled with transient flow conditions that could occur as pumps come on or off, high pressure can be a major concern. Therefore, this analysis can help the modeler to determine key areas of vulnerability and provide solutions to mitigate high pressure concerns.

To execute a minimum hour demand simulation, the following model setting are necessary:

1. Set model demand conditions to simulate the minimum hour demand condition.

2. Set the initial or operational conditions for the pumping facility serving the pressure zone of interest with the proper number of pumps operational as desired by the modeler. If not assessing existing pumps, the modeler can introduce an inflow rate at the source point instead.

3. Set tank elevations at a point near their overflow elevation or the point where the pumps are turned off and the tanks are considered full.

The results of the minimum hour demand simulation will indicate where pressures in the system are high or above desired levels under a worst-case condition for high pressures. These results may be used to indicate where pressure-reducing valves may be needed.

5.2.6. Emergency Conditions and Reliability

The limiting conditions previously discussed assume that all components of the system are operable. Water systems are also subject to emergency situations and planned maintenance activities when certain components are no longer available to meet the water demands. Examples of emergency situations include earthquakes, well contamination, major power failures, or transmission main failures. Planned maintenance activities include water treatment plant shutdowns, reservoir cleaning or repairs, and transmission main valve repairs. System analyses assess the degree to which the system is relied on to function adequately when system components are inoperable. The reliability designed into a water distribution system is a policy decision that is based on a number of factors, including the frequency of the condition occurring, the contingency plans to mitigate the effects to the customers, and the cost to the utility.

Power failures and critical water main failures are situations that sometimes may occur, so the distribution system should be designed to maintain a minimum level of service to customers. The minimum level of service is defined in terms of a particular water demand scenario, such as "the average day demand should be met during a line break." Currently, some modeling programs allow the user to identify the criticality of selected mains by determining the number of customers that would be without water if the main was shut or the portion of the system that would experience compromised levels of service such as reduced pressures. Steady-state analyses can give the modeler an idea of the level of service that could be expected during a power failure. Extended-period analyses are better for assessing how long a system can reasonably function with the loss of critical elements.

It is important to note that even though a system may be capable of meeting the maximum hour criteria, a critical line break or a 6-hour power outage may debilitate the system to a greater degree. Analyzing a distribution system under emergency conditions requires engineering judgment. The important objectives of such analyses are to identify potentially vulnerable portions of the system and identify cost-effective solutions to improve the reliability. These alternatives are assessed against other utility policies to determine which approach best meets the needs of the utility in maintaining supply to the customers. Some utilities plan for different levels of service during emergency conditions as part of their overall system emergency response plan. In emergency conditions, some limiting criteria, such as minimum pressures, may be relaxed.

5.3. SYSTEM DESIGN CRITERIA

Design criteria define system capabilities by specifying the performance requirements of the system components. Thus, whether the objective of the analysis is the design of a new system or improvements to an existing system, the design criteria define the potential solutions and are the standard against which system performance, both observed and predicted, is compared to determine sufficiency of service.

The computer model predicts the performance of the distribution system under various demand conditions. To identify the deficiencies, the model-predicted performance is compared to the established design and operational standards. Inadequate system pressures generally indicate deficiencies in a system. These deficiencies are caused by any of the model components including piping, pumping, and storage or caused by inaccuracies in the assumed system operating conditions.

5.3.1. Pressure Design Criteria

When assessing the adequacy of a system, the first parameter to check is the predicted pressure. There are generally three design pressures that are defined for each community: maximum pressure, minimum pressure during peak hour, and minimum pressure during a fire flow. The range of pressure fluctuations at a single point experienced over a 24-hour period should also be kept to less than 20 or 30 psi (138 or 207 kPa).

The maximum pressure refers to the maximum pressure that customers experience. It is often in the range of 90–110 psi (620–759 kPa). In mountainous regions, pressures ranging from 100–130 psi (689–896 kPa) are not uncommon due to varying topography. It is important, however, to conform to any limits imposed by local building codes or regulatory agencies. In some communities, the maximum allowable pressure is 80 psi (551 kPa) for internal plumbing. In these instances, the distribution system is sized for the higher pressure and individual pressure-reducing valves can be installed on the service lines to customers.

The minimum pressure during peak hour refers to minimum pressure at customers' taps during normal system operation. This value is typically in the range of 35–50 psi (241–345 kPa) and ensures that there is adequate pressure to the second story fixtures within a property if internal plumbing is configured correctly. Historically, minimum pressures as low as 25 psi (172 kPa) were acceptable in the design of water distribution systems; however, when using this pressure, the utility can expect customer complaints because of the lower pressures. When determining minimum pressures, modelers must be aware of the elevations of the facilities being served in relation to node elevations in the model. Providing minimum pressure at the model node may not provide adequate pressure at the facility being served. The minimum pressure also affects the design of lawn irrigation systems. The lower the pressure, the more piping and sprinkler heads required to ensure adequate coverage. It is important to note that in some communities where residential fire sprinkler systems are required by legislation, the minimum acceptable pressure is 50 psi (345 kPa) for them to operate properly. Backflow preventers are required in many commercial, offices, industrial, and retail buildings. These devices result in a 5–15 psi (34–103 kPa) head loss, so minimum system pressures should be set accordingly. It is important to check the regulatory requirements prior to determining the minimum acceptable pressure during peak hour demand conditions.

The recommended minimum pressure during fire flows is 20 psi (138 kPa). This value has been recommended by the National Fire Protection Association (NFPA) in the United States and by the Fire Underwriters Survey (FUS) in Canada. The pressure is the minimum desired pressure in the distribution main closest to the flowing hydrant. The value of 20 psi (138 kPa) is used as it allows for adequate supply of water to the pumper vehicles while overcoming the friction losses in the hydrant branch, hydrant, and suction hoses to the pumper vehicle. This standard assumes typical hydrant branch configurations; if the hydrant branches are very short or very long, the standard is adjusted to reflect the decreased or increased losses because of friction in the hydrant branch. Some modeling programs allow for input of data related to hydrant branches to account for such situations. In addition to confirming the minimum 20 psi (138 kPa) at the hydrant location, the modeler should also be concerned about maintaining 20 psi (138 kPa) in other parts of the network to ensure that the system will remain sufficiently pressurized to prevent water main damage caused by collapse. Another consideration is the potential for backsiphonage of nonpotable water from customer properties or infiltration of groundwater around the pipes when very low pressures are seen in the water lines.

Depending on the extent of distribution system deficiencies, the distribution system model may predict negative pressures. Predicted negative pressures would result in creating a vacuum in the system and potentially an occurrence of backflow contamination or cavitation. Though negative pressures can only drop to the value of vapor pressure (−14.7 psi or −101.3 kPa) at the ambient temperature, the model could predict greater negative pressures (which are not physically possible) because the mathematical model is forced to supply the full amount of all nodal demands. The significance of these excessively low predicted negative pressures is that the system demands cannot be met and that there are serious deficiencies either in the system or in the model. In some cases, negative pressures could actually occur in limited areas of the system, for example in high elevation areas. However, in general, a deficient distribution system simply fails to meet demands. Most modeling programs allow the user to input pressure dependent demands to better represent conditions under low pressures.

5.3.2. Piping System Design Criteria

In general, a distribution system is considered to have deficient pipe looping or sizing if the following conditions are seen under normal operating conditions:

- Velocities greater than 4–6 ft/sec (1.22–1.83 m/sec)

- Small pipe diameters (less than 16 in. [406 mm]) having head losses greater than 5–7 ft/1,000 ft (5–7 m/km)

- Large pipe diameters (16 in. [406 mm] or greater) having head losses greater than 2–3 ft/1,000 ft (2–3 m/km)

It should be noted that these conditions are only general rules-of-thumb and are provided as a general guide for identifying areas where improvements are most needed. The head loss criteria are dependent on the pipe roughness and minor losses in the pipe. Pipes with a high head loss gradient may be excessively rough, have partially closed valves or some other obstruction, or simply be undersized.

Although none of these conditions are without exception, they are a concern to the modeler as they can indicate inefficient system design or operations due to aged pipes, partially or fully closed valves, and improperly sized tanks all of which require additional pumping and which result in wasted energy. The modeled solution may have the least capital cost, but the operating costs are substantial over the life of the pipe. It is important to note that as velocity within a pipe doubles, head losses increase by a factor of four. The key point to understand is that when pressures are adequate during low-flow conditions but drop below standards during high demand periods or fires, the pipes may be undersized. To find the bottlenecks in the system, look for pipes with the highest velocities and head loss conditions. It should also be recognized that higher velocities within pipe make the pipe more vulnerable to transient flow problems (see chapter 8). Therefore, it is important for the modeler to consider these conditions when designing piping systems.

The design of new pipelines should address two basic features: sizing and routing. The size of the pipe is determined by the maximum flow rate carried, typically either the maximum day plus fire flow condition or the maximum hour condition for distribution mains. Either the maximum hour flow rate or the maximum storage replenishment rate is typically the maximum flow for the large transmission mains. The pipe size is selected to ensure that velocity and head loss limits previously described are not exceeded.

The second pipeline design feature to establish is routing. In general, looping of mains should be incorporated wherever possible. A looped system provides supply from two or more directions for large demands, such as fire flows, which keeps velocities lower and provides backup and redundant supply paths in the event that a section of pipe is removed from service. The location of the new pipe is determined by the availability of right of way, construction easements, and common sense. When locating new pipelines, consideration should be given to the projected growth characteristics of the service area and where major demand areas are expected to be located in the future. Routing studies consider the impact of development staging. Development restrictions may be required to limit properties needing large fire flows until a certain portion of the network and looping is completed.

5.3.3. Pumping Systems Design Criteria

Distribution systems have both high-service pumps and booster pumps. High-service pumps are generally defined as pumps that obtain water from a reservoir open to the atmosphere (usually the system source water) while booster pumps are generally defined as pumps that obtain their water from within the pressurized distribution system. Regardless of the pump type, the pumps must be sized to meet the full range of system demands, and all of the limiting demand conditions must be considered. These include all of the extreme high flow conditions, average demands, and minimum demand conditions. Consideration of the duration of each condition should be made when selecting the appropriate pump size. The best choice may be a series of parallel pumps of varying sizes that can be operated to most efficiently match the pumping requirements.

In systems with elevated or high-level ground storage, constant speed pumps are typically used for high service and booster pumping, and storage is used to equalize the pumping rate over the range of water demands. For systems without such storage tanks or with insufficient equalization storage, it is generally desirable to use variable speed pumps to efficiently change the pumping rates to meet the system demands.

As indicated previously, high-service and booster pumps are installed as multiple pump installations. The pumping installation should be designed to deliver the maximum day demand at a minimum. As a precaution, an additional allowance ranging from 10 to 15 percent for future pump deterioration and wear with any one of the pumps, preferably the largest, out of service could also be provided. Generally, this is not necessary because pumps are typically designed to meet a future projected demand condition. Individual pumps and combinations of pumps are sized to meet the range of demands from average day demand to maximum day demand plus fire flow. When variable speed pumps are used, it is not necessary to have all pumps as variable speed. However, for redundancy, the design should allow for maintenance of each pump individually while still maintaining the ability to vary pumping rates with system demands. This is accomplished through the use of two variable-speed pumps per pumping station or, if multiple pumping stations supply the same pressure zone, one variable-speed pump per location. It should be noted in multiple pump installations where not all pumps are variable speed, the variable speed pumps should be sized larger than the constant speed pumps to allow for proper operation of the variable speed drives.

A family of system head curves is developed for each pump station. These curves are combined to develop characteristic curves for the pump station based on the different pumping configurations and storage tank levels. If possible, pump rating curves (pump head-capacity curves) provided by the pump manufacturer should be field verified on a regular basis to ensure the pumps are performing as expected. This type of pump verification is especially important prior to model calibration. A typical pump rating curve and a system head curve used to select the pump size are shown in Figure 5-2. The pump will operate where the two curves intersect, and the pump should be sized to operate at its highest efficiency as indicated by the pump's efficiency curve.

The system head curve represents the required head for a range of flow rates. It is a relatively easy curve to develop for a single pipe. The system head curve is calculated by adding the static head, the head loss through the pipe, and the pressure required at the end of the pipe. However, the situation is more complex in a pipeline grid where water is conveyed to many points and via various routes. The hydraulic model is used to simulate the varying demand conditions, and the required system head curve is developed based on the pressure standards. This curve is plotted and the pump selection is made based on the percentage of time each condition is expected to occur. Figure 5-3 illustrates this selection process. Some models can provide the modeler with the system head curve as an output under specific system conditions.

Nationwide, about 4 percent of US power generation is used for water supply and treatment. Electricity represents approximately 75 percent of the cost of municipal water processing and distribution according to the Department of Energy (DOE). Of this, pumping typically accounts for 75 to 80 percent of the power consumed by water utilities. Consequently, selecting the most efficient pumps for the range of demands must be evaluated. Figure 5-4 shows the typical efficiency range for a pump rating curve. The specific information on pump efficiency is available from the pump manufacturer. The modeler selects pumps that show the highest efficiencies over the expected ranges of demands. For multiple pump applications establishing the optimum sequential ordering of pumps has a substantial impact on the power consumption and cost to the utility. Pump optimization strategies are best evaluated using extended-period simulations. Additional discussion can be found in chapter 6.

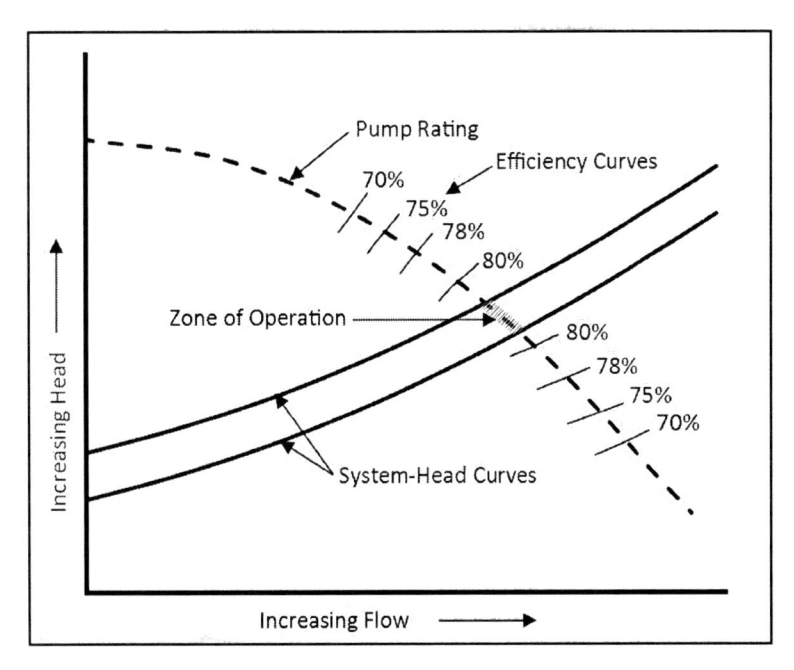

Figure 5-2   Pump rating curve versus system head curve

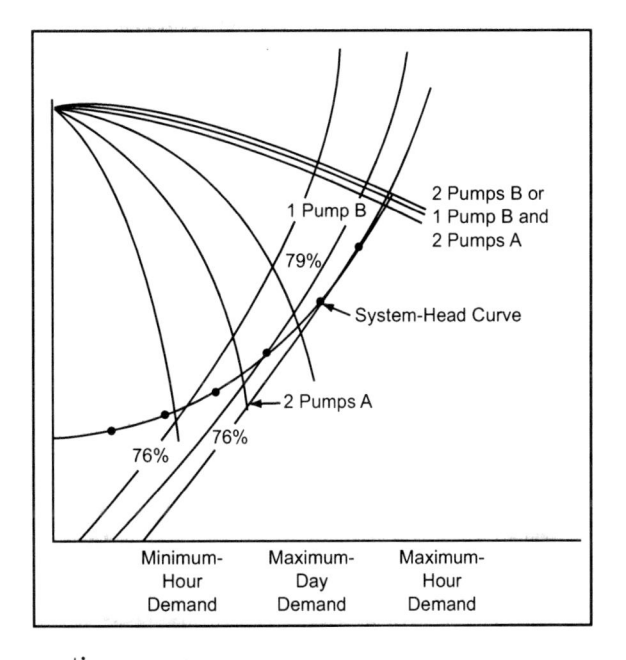

Figure 5-3   Multiple pump rating curves

5.3.4. Storage Facilities Design Criteria

Storage facility design criteria must address many interrelated factors, including system storage requirements, elevated or high-level ground storage versus ground storage, and the number and location of the storage facilities. The volume of storage required is classified into three primary components: equalization storage, fire storage, and emergency storage.

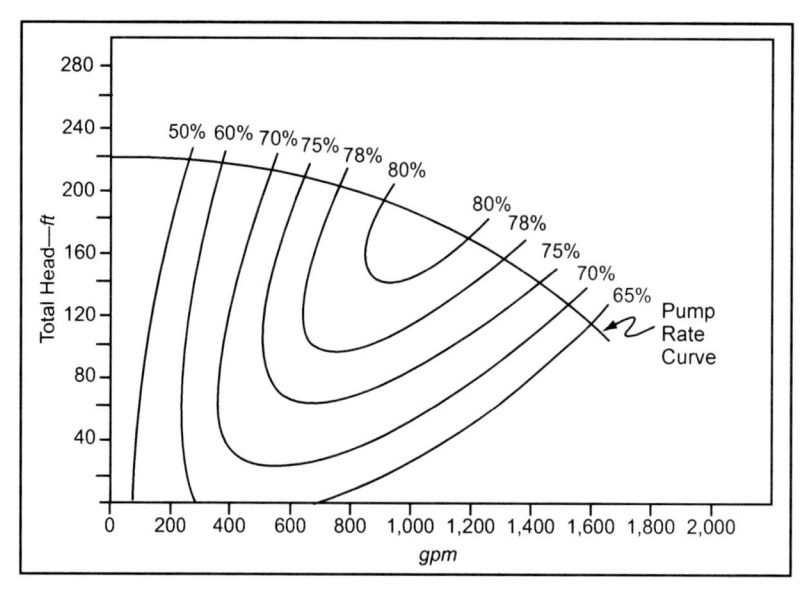

Figure 5-4 Pump efficiency curve

5.3.4.1. Equalization Storage. Equalization storage is the amount of water required to meet demands in excess of the production and delivery capabilities. This storage is generally less expensive to provide than the production facilities, pumping, and piping that are required to meet all instantaneous demands. The amount of equalization storage maintained by a community should be determined based on a comparison of the production capabilities versus the demands expected on the system. In most communities, this is based on the maximum day condition and ensuring that the equalization storage is sufficient to meet the demands that exceed the production capabilities. A typical way to calculate equalization storage volume is to plot the hourly hydrograph for maximum day demands as shown in Figure 5-5. The area under the diurnal curve but above the average hourly demand line is the equalization volume required. For large systems, the equalizing storage requirement is typically 15 to 20 percent of the average demand over a 24-hour period, but equalizing storage could exceed 30 percent for small service areas or arid climates.

5.3.4.2. Fire Storage. The fire protection needs of a community typically are determined by the Insurance Service Office (ISO) in the United States and by the Insurance Advisory Organization (IAO) in Canada. The requirements calculated by these organizations are based on the building types, land use, water supply facilities, and the response capabilities of the local fire department. The fire storage volume is determined by multiplying the required flow duration by the maximum fire flow in each service area of the distribution system. AWWA Manual M31, *Distribution System Requirements for Fire Protection,* includes tables showing the recommended duration for various fire flows. Factors, such as redundant piping and alternate supplies, are considered in determining the final requirements for the community.

Large fire flows may require more than one storage facility and a secondary piping route to the site. Typically, fire storage is obtained from reservoirs located within the same pressure zone as the fire though some smaller pressure zones must rely on pumping to supply needed fire flows.

5.3.4.3. Emergency Storage. Emergency storage provides water during events, such as pipeline failures, equipment failures, power outages, pumping system failures, water treatment plant failures, raw water contamination events, or natural disasters. The amount of emergency storage is a policy decision based on an assessment of the risk of failures and the desired degree of system dependability. As an example, a system operator may desire a full day of emergency storage in order to provide enough storage in the event of a systemwide power failure. An assessment must be made of the type and nature of the emergency condition, including the frequency, intensity, and duration. In general, a vulnerability analysis, such as described in AWWA Manual M19, *Emergency Planning for Water Utility Management,* should be used to determine emergency storage requirements. Some state or provincial regulations indicate the minimum emergency storage required for the community based on the average or maximum daily demands. In addition, water quality should be a part of the equation when determining the amount of emergency storage to provide. The utility may relax its minimum pressure requirements under emergency conditions as long as negative pressures are avoided.

5.3.4.4. Effective Storage. *Effective storage* is defined as the volume of water available in a tank above a level that will provide pressures above the minimum allowable pressure under peak hour or maximum day demand conditions during normal operating conditions. The level at which this occurs is known as the *minimum effective elevation*. The minimum allowable pressure and demand conditions for identifying the minimum effective elevation may be based on regulatory requirements or determined by the utility. Storage above the minimum elevation can provide the minimum allowable pressures when a tank may be near depletion. Any storage within a tank below the minimum effective elevation is considered ineffective storage but may be available for emergency purposes. Providing all storage above the minimum effective elevation may not always be practical but should be considered when placing a tank.

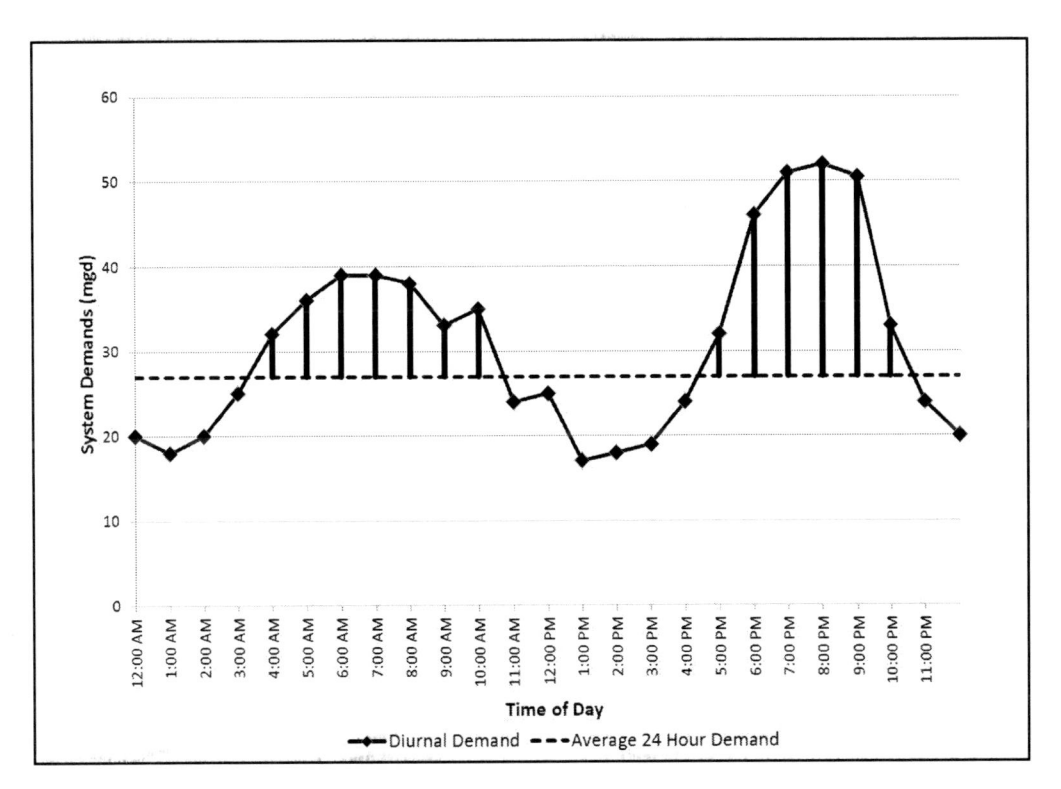

Figure 5-5 Equalization storage requirements for maximum day conditions

5.3.4.5. Storage Allocation. In evaluating system storage and developing operational plans for its use, the allocation of storage for equalization, fire, and emergency must be addressed. It is useful to consider storage schematically. Figure 5-6 shows how the three storage components are positioned in the reservoir. Storage for equalization occupies the top portion of the reservoir. During the day, this volume increases and decreases as demands are pulled from the reservoir when demand exceeds production rates. Fire storage is positioned in the reservoir, under the equalization storage. Emergency storage occupies the bottom portion of the reservoir. The importance of designating levels to the different storage components is to ensure that the volume has the correct hydraulic grade line for its intended purpose when dealing with elevated or high-ground level storage reservoirs. A fire may occur when the equalization storage has been completely consumed for the day, and the remaining hydraulic grade line must be sufficient to deliver the required flows at the proper pressures.

5.3.4.6. Types of Storage. The evaluation of the type of storage to construct includes high-level ground storage, elevated storage, and ground storage with pumping (see Figure 5-7). If there is high ground in or near the service area, high-level ground storage connected directly to the system is used. This type tank is generally found in mountainous or hilly areas where high ground is not difficult to find. However, in areas where the topography is much flatter, the use of elevated storage, ground storage with pumps, or a combination of both is more common. If a ground storage tank's height is greater than its diameter, it is commonly known as a *standpipe*.

There are advantages to elevated storage, high-level ground storage, and ground storage tanks. Both elevated and high-ground storage tanks can maintain the system hydraulic grade line without the need for continuous pumping, which results in a simpler operation. Both are considered more reliable because water is delivered to the system from the tank by gravity and can provide water during a power failure. Finally, both also provide more stable system pressures.

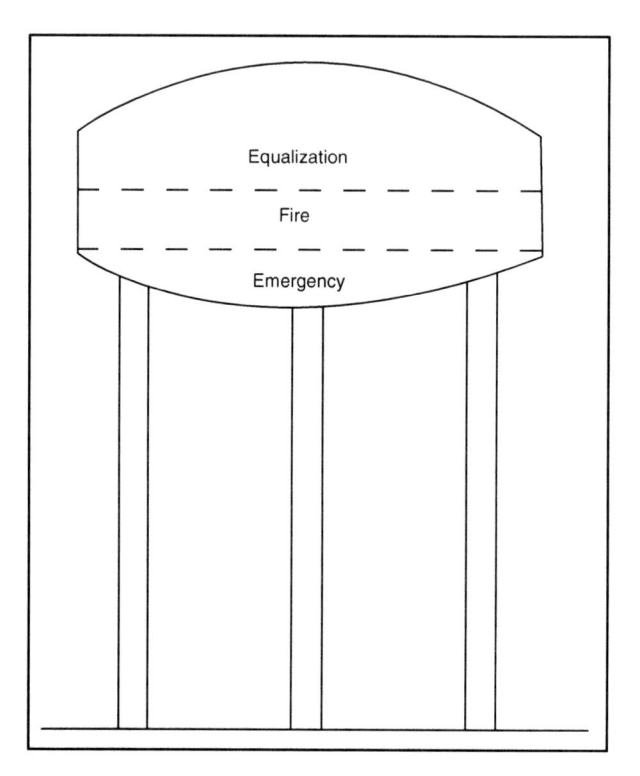

Figure 5-6 Storage allocation

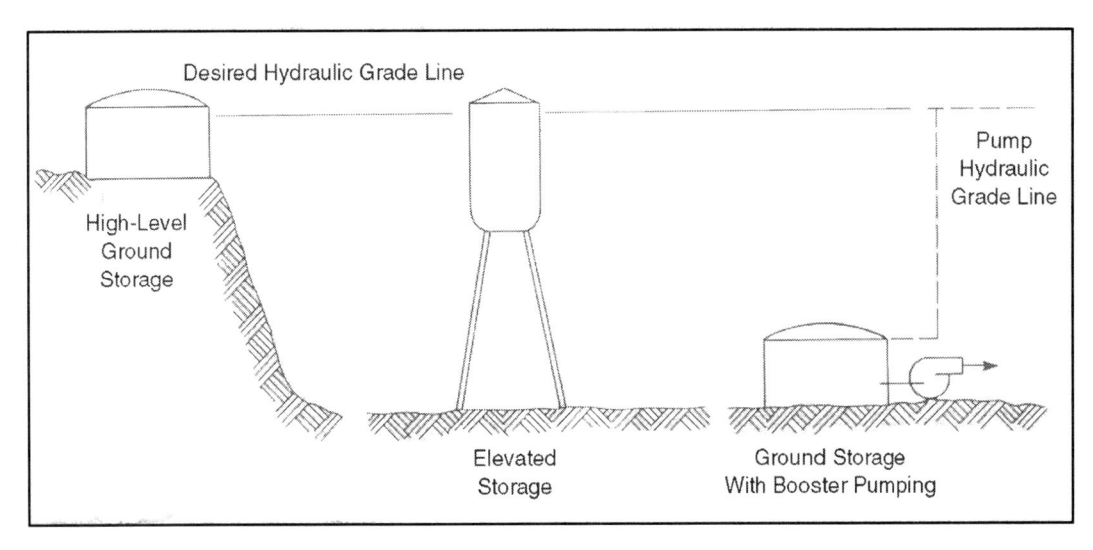

Figure 5-7 Types of storage and elevation

The major advantage for both high-level ground storage and ground storage is the reduced construction costs per unit of volume. Both types of ground storage can be built to hold large volumes of water, whereas the size limit of elevated storage is about 4 MG (15 ML). Ground storage facilities with pumps can be adapted to changing system hydraulics by changing pumps and controls. In contrast, elevated or high-level ground storage tanks must operate at the specific hydraulic grade line defined by their geometry. Ground-level tanks can have a disadvantage of a large percent of the tank's contents falling below the minimum level for effective storage. This can lead to low turnover and stagnation unless means are provided to enhance turnover. Water quality issues in tanks are described in more detail in chapter 9. Frequently, as water systems expand, the required hydraulic grade line and water volumes change. This sometimes results in situations where either there is little turnover in the elevated reservoir because the system hydraulic grade line is higher than the tank, or conversely, the tanks deplete too quickly during high-demand periods. These impacts must be considered in storage analyses of future conditions.

When comparing types of storage to provide, there are several other considerations beyond hydraulics. The operation and maintenance costs for steel elevated tanks are generally higher for painting and corrosion protection than ground storage. Power costs vary according to the conditions under which the facility is operated. Ground storage may have higher power costs because of the associated pumping requirements. Elevated storage can reduce power costs by allowing lower pumping rates during peak hours when power costs are highest.

5.3.4.7. Number and Location of Storage Facilities. The factors to consider when determining the number and location of storage facilities include the size of the reservoir, the location of other system components, hydraulic elevation, and fire flows.

The required number of reservoirs is determined using a trial-and-error analysis. The analysis determines the cost-effective number of storage reservoirs, taking into consideration the impact of cost, piping, pumping, and storage and water quality. A key factor to establish is the limiting size of each reservoir. The reservoir type (elevated versus high-level ground versus ground storage) determines the limiting factor. In general, the reservoir should be as large as practical but small enough to be filled within a reasonable amount of time. The tank's operation and design should be able to maintain water quality through the promotion of mixing and turnover while avoiding stagnation within the tank. The use of individual reservoirs for each region to be supplied is compared against the cost of regional reservoirs supplying multiple areas. The cost comparison includes reservoir construction and operating costs, as well as the associated pumping and piping required for each alternative.

Reservoir location is also an important design consideration. Reservoirs should be connected to major transmission mains so that high flow rates are adequately transferred to and from them without excessive head losses. Typically, reservoir sites should be located at the extremities of the system opposite from the production source, allowing the reservoir to serve new growth occurring beyond the existing service area. Depending on the situation, this may not always be the case. Some larger systems may have centrally located storage to better serve a larger proportion of the system from the tank, especially if the sources are located at the system extremities.

For effective fire flows, the storage facility should be located to provide coverage over the area served by the tank, while avoiding unnecessary overlap with other storage sites. It is beneficial to locate the site as close as possible to the areas requiring the largest fire flows.

Water quality is another consideration. Large tanks with little turnover produce stale water and poor water quality. Reservoirs should be located and sized in a way that allows water to be circulated regularly. See chapter 9 for further discussion on tank mixing.

5.4. DEVELOPING SYSTEM IMPROVEMENTS

There are many uses for a distribution system model, but the primary uses of models are master planning design, operational studies, and water quality studies. The primary objective of most steady-state models when they are produced is for master planning, or to develop and evaluate potential system improvements for the future. Other uses of steady-state models include assessment of proposed local system changes (such as subdivisions or new large customers), determination of control valve settings, and emergency planning.

Comprehensive plans should be developed for system improvements that address the needs and deficiencies identified in the system analysis. The plans developed should include both an assessment of infrastructure requirements and an assessment of changing consumption patterns to identify potential optimum solutions. Some typical plans that utilities develop are listed in the following subsections.

In system modeling, a field-calibrated model is used to evaluate a distribution system for both existing and projected design scenarios. The design criteria that have been established before any scenarios are run are used to ensure that the proposed system meets the customers' needs.

Before running each analysis, the modeler prepares an outline or index of scenarios that describe each situation to be simulated (e.g., 2025 peak hour, 2050 peak hour). It is suggested that all of these runs are executed within the same model and that these different runs are broken out into different scenarios using the modeling software's scenario manager. Each scenario is either named or numbered, and a printout is maintained of the results of each analysis. Notes are taken on each scenario and its results so that the evaluations are readily understood by anyone referring to them in the future.

5.4.1. Master Planning

Master planning is defined as the development of long-range plans for the growth of the water distribution system. Typically master plans are developed for the 5-, 10-, and 25-year projections. They include the assessment of the most cost-effective solutions for the location of new production or storage facilities and the expansion of the large transmission main network to support growth through the expansion of the water system and from increase in water demands within the core interior portion of the community. Master plans should also analyze how to address existing system deficiencies, optimize system performance, and interconnect separate systems for emergencies.

Master planning not only looks at the construction and expansion of water infrastructure but also assesses the impact of changes in customer demands caused by water conservation or shifting demands to other times of the day through customer education programs. The ability to increase or decrease demands in the model enables the utility to assess the changing effects of demands.

With the technology advancement of computers and modeling software, the ease to produce an all-pipes model for master planning purpose is easier than ever before. The use of GIS is also increasing the efficiency in developing all-pipes models. GIS has also improved the method of distribution of the existing demands and also future demands, which makes the master planning process easier than ever.

5.4.2. Subdivision Planning

Subdivision planning is a localized analysis that considers the staging and growth patterns of new areas in the distribution system. The master planning model identifies the transmission main that supplies the volume of water required to the area from the production source. The subdivision model outlines the details of how the area should grow in a sequential manner.

The modeler considers the effects of staging from the perspective of system reliability by assessing the number of lots on a single feed. Fire flows are confirmed at each stage and, if insufficient flows are available, contingency plans developed to either limit high-density development or provide additional looping to safeguard property.

In some communities, adjacent developments can affect the ability of a neighborhood to grow in a cost-effective manner. Subdivision modeling is used by utilities to identify these problems and address them before they occur.

5.4.3. Rehabilitating Neighborhood Distribution Mains

Rehabilitating neighborhood distribution mains is similar to subdivision planning, as it deals with the design of water systems to individual customer lots. The primary difference is that because the rehabilitation is occurring in existing areas, additional costs are incurred because of increased pavement and landscape reconstruction and the requirement to maintain customer supply during construction.

In some communities, this type of analysis is done in conjunction with the wastewater and transportation departments to ensure efficient use of resources if all three infrastructure components need to be upgraded in the same year.

Existing neighborhoods should also be periodically assessed to determine whether the utility is providing the necessary fire flows. Over time, the type of development in a neighborhood changes as single-family homes are replaced with multifamily buildings. The carrying capacity of a pipe may degrade with age, depending on the corrosive nature of the water and soils. These two factors sometimes result in a water system that does not deliver the appropriate fire flows for the development. A detailed model of an existing neighborhood is used to identify the most cost-effective solution to upgrade the firefighting capabilities.

5.4.4. Outage Planning

If a community maintains a calibrated model of its transmission and distribution system, the model can be used to assess the impact of an upcoming outage on the system. The modeler should assess three main issues: reservoir volumes, minimum system pressures, and firefighting capabilities.

Reservoir volumes are reduced either because of a reservoir shutdown for cleaning or maintenance or by the shutdown of a transmission main that fills the reservoir or supplies water to a neighborhood. In these cases, the duration of the shutdown is assessed against production capabilities and the network analyzed to ensure the needed water is moved to the other locations to maintain customer supply at adequate pressures.

The minimum pressures in the water distribution network are analyzed during the planned shutdown. If there is a major change and the shutdown is expected to cover an extended period, residents in the affected area should be notified in advance. It is especially important to identify critical customers, such as hospitals, nursing homes, or large industrial or commercial customers, that should not lose water service. The utility call center should also be notified so qualified personnel can respond effectively to any customer complaints caused by the reduced level of service.

Finally, the firefighting capabilities in the immediate area of the shutdown are assessed. Contingency plans are developed with the local fire department to identify how to respond to a fire in the affected area during the shutdown.

5.4.5. Local Impact Analysis

If a utility has a model that is well calibrated for current conditions, the model can be used to assess the potential impact of localized changes in the system. New customers may want to know what level of service could be expected if they locate to a certain area. The model can tell them the range of pressures that can be expected as well as the available fire flows. If the new customer would create a potential deficiency in the system, the utility can use the model to assess what improvements are needed to provide adequate service to that customer while maintaining equal or better service to existing customers.

5.4.6. Control Valve Set Points

Modeling analyses can identify smaller areas within the system that may require adjustment of pressures in that area or flows to that area. For example, customers in an area of lower elevation may experience higher pressures (typically greater than 100–110 psi [690–760 kPa]) than customers at higher elevations. Generally, such small areas could be served by one or more pressure-reducing valves that lower pressures in that area to acceptable levels. The model is used under a range of demand conditions to determine the pressure settings for the valves as well as the size. At a particular location, two or more pressure-reducing valves may be used to cover the expected range of flows to the area. Doing so affords the utility more control and helps avoid cavitation in the valves. In other cases, areas with higher elevations could see excessive drops in pressure as flows move through the area to serve adjacent lower elevation neighborhood. In these cases, a pressure-sustaining valve could be used to restrict the amount of flow passing through the higher elevation neighborhood to prevent excessive reductions in pressure.

5.5. CONTINUING USE OF THE MODEL

A utility that invests the time, effort, and money in performing a computer-assisted analysis of its water distribution system should consider maintaining the model for continued use. A hydraulic model is a corporate asset and should be treated as such. This requires the support of a modeling team that ensures that the model remains current and sufficiently calibrated for its intended use. The utility should develop a protocol or work flow that defines the process and frequency for model updates from GIS files, as-built drawings, or other data sources. The model should be recalibrated on a regular basis or after a major capital project or operational change to ensure that the model represents real world conditions. Steps should also be taken to ensure that the model can communicate with other data sources such as GIS, CMMS (computerized maintenance management system), CIS (customer information system), asset management system (AMS), and SCADA to enable utilities to evaluate a variety of data sources simultaneously when making business decisions, and to maximize the leverage of a utility's modeling investment.

The integrity of the base model that represents the physical and operational elements of the existing system is paramount. One member of the modeling team should be responsible for updating the base model and maintaining backup copies. Documentation of the model is critical to ensure that any new users are aware of the assumptions and limitations that were involved in the development of the model. Periodic checks against actual operating conditions help identify any changes in the distribution system that affect the accuracy of the model. If internal resources are not sufficient, the utility should consider contracting all or part of the model activities to outside consultants. It is essential that the modeler be properly trained in hydraulic analysis, computer modeling, and long-term model maintenance.

5.6. REFERENCES

AWWA. 2008. Manual M31—*Distribution System Requirements for Fire Protection*. Denver, Colo.: AWWA.

AWWA. 2001. Manual M19—*Emergency Planning for Water Utility Management*. Denver, Colo.: AWWA.

Boulos, P., Lansey K., and Karney, B. 2006. *Comprehensive Water Distribution Systems Analysis Handbook for Engineers and Planners*. Broomfield, Colo.: MWH Soft.

Cesario, A.L. 1995. *Modeling, Analysis and Design of Water Distribution Systems*. Denver, Colo.: AWWA.

Chase, D. 1999. Calibration Guidelines for Water Distribution System Modeling. In *Proceedings of the AWWA Information Management Technology Conference*. Denver, Colo.: AWWA.

Cruickshank, J.R., and Long, S.R. 1992. Calibrating Computer Models of Distribution Systems. In *1992 AWWA Computer Conference Proceedings*. Denver, Colo.: AWWA.

Doe, S.R.K., and Duncan, C.T. 2008. Naming Conventions and Documentation for Distribution System Modeling. *Jour. AWWA*, 100(9):132–138.

Electric Power Research Institute (EPRI). 1999. *A Total Energy and Water Quality Management System*. Palo Alto, Calif.: EPRI.

Fire Underwriters Survey. 1981. *Water Supply for Public Fire Protection—A Guide to Recommended Practice*. Toronto, Ont.: Insurance Bureau of Canada.

National Fire Protection Association (NFPA). 1985. NFPA Standard 291-1985, *Recommended Practice for Fire Flow Testing and Marking of Hydrants*. Quincy, Mass.: NFPA.

US Department of Energy (DOE). 2006. Energy Demands on Water Resources—Report to Congress on the Interdependencies of Energy and Water. Washington, D.C.: DOE.

Walski, T.M. 1995. Standards for Model Calibration. In *AWWA Computer Conference Proceedings*. Denver, Colo.: AWWA.

Walski, T.M., and O'Farrell, S.J. 1994. Head Loss Testing in Transmission Mains. *Jour. AWWA*, 86:7:62–65.

Walski, T., Chase, D.V., Savic, D.A., Grayman, W.M., Beckwith, S., and Koelle, E. 2003. *Advanced Water Distribution Modeling and Management*, 1st Ed. Waterbury, Conn.: Haestad Press.

This page intentionally blank.

AWWA MANUAL M32

Chapter 6

Extended-Period Simulation

6.1. INTRODUCTION

Extended-period simulation (EPS) is a technique for modeling a distribution system where a series of steady-state simulations at specified intervals are performed over a specified time period. This technique simulates the way a system changes in response to changing demands and operational conditions. While steady-state models provide results on flows and pressures in the system, extended-period models go further to also provide results on system storage fluctuations and operational controls. Extended-period simulations greatly improve water system planning and operation at a relatively low cost. Extended-period simulation distribution system models are also used to refine water supply and distribution system improvements. Furthermore, simulation of water quality with distribution system models normally uses extended-period simulations.

Typically, extended-period analyses strictly for hydraulics are simulated over several hours or days, such as a 24-hour period, during average and maximum demand days while water quality analyses are simulated over several days, weeks, or even months. These extended-period simulations are analyzed at user-selected time intervals, from minutes to hours.

In addition to steady-state analysis performed on water supply and distribution systems, an extended-period simulation can further analyze both existing and future water systems to test system-operating procedures. While a steady-state analysis is a useful tool to size pipelines and supply facilities, an extended-period simulation analysis provides a great deal more information about system operating characteristics and how the water system responds to changing demands or emergency situations. This includes information such as proper booster pump sizing, determining fluctuating reservoir levels and sizing the reservoirs, and more efficient system control valve settings. Extended-period analyses can also be used for optimizing energy use and for operator training.

Under continually changing consumption rates that water systems experience, storage facilities within a water system permit more uniform pumping rates over time, and hence, more efficient system operation. Reservoirs also provide reserves for firefighting and other water supply emergencies. For simple systems containing only one floating storage facility, water level fluctuations are easily calculated from incremental net changes in system inflow and outflow. However, in complex systems, reservoir level prediction and booster pump interaction become more complicated because of hourly variations and pressure zone demands on each reservoir.

For such complicated water systems, extended-period simulations are extremely useful in determining how a system behaves under changing demands, how a stressed system reacts in emergencies, or in determining the best locations and sizes of future storage and pumping facilities. For each reservoir, the volume used as floating storage, the minimum emergency reserve, and the speed of refilling are easily calculated. An extended-period simulation uses a steady-state analysis subroutine to determine pressures and flows in the distribution system and a subroutine performing a time integration of reservoir node flow in or out of the system to track reservoir volume changes.

When modeling individual pressure zones separately, it is impossible to accurately predict how boundary conditions of one zone impacts an adjacent zone under varying supply and demand conditions. Experience shows that entire water systems, with multiple pressure zones, must be modeled together in order to accurately simulate real water system operation when pressure zones are interconnected. For these complex interconnected water system pressure zones, extended-period simulations model how an entire water system interacts dynamically via varying reservoir water levels, booster pumps turning on and off, and pressure and flow control valve operation between pressure zones.

A prerequisite for modeling water quality in distribution systems is to have a hydraulic model that has been fully calibrated for extended-period hydraulic conditions. Water quality parameters that can be simulated include the age of water in the system, the percent of water derived from a source, and the growth or decay of chemical constituents.

6.2. INPUT DATA FOR HYDRAULIC EPS MODELING

All of the data in a steady-state simulation are also used in an extended-period simulation. Along with this data, additional data are required to describe how demands, pumps, valves, reservoir levels, and water quality parameters change over time. For water quality modeling, input data requirements include identification of water sources, reservoir mixing characteristics, and the rates of growth/decay of water quality constituent. This section describes the additional information useful for an extended-period hydraulic model. Additional information useful for an extended-period water quality model is described in chapter 7.

6.2.1. Verifying System Operation

Results of steady-state computer analyses on an existing system are first verified to ensure that they agree closely with actual system operations. Only then should extended-period simulations be conducted. Information on how booster pumps are controlled in an extended-period computer run is input to turn pumps on and off in similar sequence as in the real system. Other system equipment, such as isolation or control valves, must have information related to how they are controlled. Reservoir water levels and pumped flows during an extended-period analysis should agree with historical field observations of water level fluctuations and flow rates. Regulating valves and general distribution system pressures contained in computer output are also similar to conditions experienced in the field.

6.2.2. Establishing the Extended-Period Simulation Time Interval

The extended-period simulation time interval is set based on two factors: the interval necessary to achieve the objectives of the simulation and the interval that is appropriate for equipment controls to function properly. A reservoir storage analysis, for example, might run for 24 hours and have a computation time interval of 30 minutes or 1 hour. A water quality simulation could run for several days with computation time intervals of usually less than 5 minutes to accurately capture the changes in water quality that could be missed with a longer time interval. Changes in equipment controls are often simulated with patterns, curves, or profiles that describe the changes that are to take place over time. Nowadays, simple or complex control statements can be used to simulate operation of valves and pumps. The computational time interval should not be so long that significant changes in the patterns are missed. On complex models, the time interval also should not be so short that the time to complete an extended-period simulation run is unnecessarily long. However, increases in computational speeds makes this problem less of an issue. In systems with hydropneumatic tanks, the time interval must be short (1–10 minutes) to capture the full fill-and-draw cycle of these tanks.

6.2.3. Diurnal Demand Variations

Often, the development of the diurnal demand curve that defines the variations in demand in the system over time can be the most time-consuming portion of extended-period modeling. It is also one of the areas of modeling requiring the most use of the modeler's judgment. This is because steady-state demands, which are generally based on meter readings over the period of several weeks or months, are associated with an hourly pattern for water use based on system production rather than water use. The modeler must decide whether to use a single diurnal curve for all demands or to associate different curves with different categories of customers and then make sure the combination of different customers and diurnal curves match to the systemwide patterns. Development of diurnal curves is discussed in chapters 2 and 5 as well as later in this chapter.

6.2.4. Booster Stations and Wells

As with steady-state simulations, extended-period simulation data include actual pump performance curves used in the input data. Booster stations having multiple pumps operating on system pressure also contain control information to allow the computer program to determine when each pump turns on or off. A simple method of modeling multiple pumps is to model each pump individually and establish settings to simulate actual pump settings in the real system. As system pressure or reservoir level drops, additional pumps "turn on" in the model. Whenever a modeled pump turns on or off, basic system boundary conditions will change. Therefore, an extended-period calculation and output are generated at that specific time interval. Care should be taken to not set pump on/off pressure settings too close, to avoid several short-duration time interval calculations. Also, modelers should check that pump starts and stops do not adversely affect other modeled components, such as other nearby system booster pumps and control valves and that system pressure data accurately reflect real system reactions. In the cases of well pumps, the modeler must decide how to represent the drawdown in water levels within a well as the drawdown may not occur instantaneously. It may be prudent to assume that it occurs as soon as the pumps are activated.

6.2.5. Other Supply Sources

Connections to regional water supply systems, major transmission pipelines, and interconnections to other water distribution systems can be modeled as a negative demand at each node corresponding to these sources. These demands can be constant or vary following a predefined pattern. These connections can also be modeled as constant-head or pattern-defined varying-head reservoirs. The extended-period results should be checked to ensure that flow rates represent typical average flow rates for the demand condition specified.

6.2.6. Tanks and Reservoirs

To model reservoir water level fluctuations, specific tank input data are required. These data include maximum and minimum hydraulic grade line elevations, tank diameter, and external sources of inflow or outflow, such as any wells and booster stations that are connected directly to a reservoir. For noncircular reservoirs, a depth-to-volume relationship is calculated based on reservoir capacity and water surface height. The modeler is cautioned to confirm how the modeling software handles tanks when they reach their maximum level. Some software may allow flow to continue to the tank without an increase in level (i.e., overflow), while others may stop flow altogether to the tank.

By calculating flow rates into and out of reservoirs at each time interval during a simulation, a volumetric change in storage is computed, and a new water surface elevation calculated. Simulations begin at any time interval. However, it is often useful to begin simulations at midnight or some other low-demand period and allow the water system to cycle through an entire 24-hour day, or other "typical" operating cycle. Reservoirs are typically at or near full late at night. Normal reservoir water level fluctuations during daily demand periods can range from 5 to 15 ft (0.6 m to 4.6 m) of water depth. Starting reservoir water levels are set based on actual reservoir operating records for typical average and maximum day demands. In the absence of actual records, reservoirs are assumed to be at or near full at midnight.

6.2.7. Control Valves

Proper control valve operation is critical to performing accurate extended-period simulations. Typical control valves include pressure-reducing or sustaining valves, flow control valves, and altitude valves. Proper pressure or flow settings are important if these elements are to accurately represent the real system. As with pumps, whenever these elements open, close, or modulate, extended-period simulation time interval calculation occurs. Altitude valves are complex control valves that can be installed as two-way valves or one-way valves with a check valve in parallel. Pressure-reducing and/or sustaining valves can be installed in parallel with different flow settings to accommodate a wide range of flow rates.

The modeler should carefully check how the other system elements respond to these changes. For example, an altitude valve closing when a reservoir reaches its maximum water level results in instantaneous system pressure increases of up to 10 psi (69 kPa) or more, thus causing system pumps or control valves to change operation. While this may be a normal response in the real system, such model results should always be verified by actual operating data or discussions with operations personnel.

In other cases, when several parallel pressure-reducing valves are feeding a pressure zone, instability in the model's calculation engine can occur.

6.3. EXTENDED-PERIOD SIMULATION SETUP

6.3.1. Objective

One of the most important uses of modeling is to identify and remedy major system performance deficiencies. Steady-state water system modeling can describe system performance under a variety of different demand conditions such as normal conditions of average day or maximum day demands or nonnormal system conditions such as maximum day demand plus fire load, peak hour demand, or low-flow conditions with minimum system demands (e.g., to assess water transfer to storage). The steady-state evaluation is essentially a snapshot of the system under a fixed and defined demand situation. However, steady-state modeling cannot assess what happens before or after these conditions occur and cannot assess how the system responds to changes in conditions. Extended-period simulation is an extension of the steady-state analysis in that it conducts a series of evaluations that verify system performance. Performance indicators can include storage utilization, pump capacity, flow reversal, energy usage, and water quality. Starting with the steady-state model, extended-period simulations can be set up by establishing initial conditions, describing system demands, and describing the rules governing the operation of the system during the simulation period. This section discusses these additional requirements required for extended-period simulation and calibration.

6.3.2. Initial Conditions

The initial hydraulic conditions of an extended-period simulation are the same as the boundary conditions used in a steady-state analysis in terms of tank levels and pump run status. Care must be taken regarding the time of day at which the model simulation begins. Normally, extended-period simulations begin at midnight when the system conditions can be markedly different than during the middle of the day when steady-state simulations are run. The initial conditions used will be based on the operational constraints and philosophy that is in place or is to be tested by the modeling efforts.

Additional efforts are required to determine the initial conditions for water quality modeling that are discussed in chapter 7.

6.3.3. System Operating Conditions

System operational constraints, either imposed by physical conditions or by policy designs, must be identified so that their impact on the system operation and performance can be evaluated. There are a number of different types of operating philosophies that can be evaluated using extended-period analyses. This system operation, in the form of information requirements, is described in the following sections.

6.3.4. Overall Operational Philosophy

Utilities establish an acceptable level of service to satisfy customer demands. Most utilities have a goal of providing continuous delivery of high-quality drinking water to all customers within a reasonable range of pressures. This goal assumes that adequate pumping, storage, and redundant critical system components are available so this level of service can be achieved during normal or maximum demand conditions and even during emergency operations (e.g., firefighting, equipment maintenance, or natural disasters). Regular water outages or other lengthy interruptions in water service are not tolerated in most US communities. However, the acceptable level of service may differ between normal and emergency conditions. Extended-period analyses may help identify several combinations of operating conditions that could lead to subpar levels of service.

6.3.4.1. Peak Day. Often, a water system is designed to sustain continuous maximum day demands, requiring that the supply over a 24-hour period equals the total system demand over that period. Water storage facilities are capable of supplying instantaneous peak demands in excess of maximum day supply and of storing water during low demand periods. This philosophy assures that the system meets the system demand under any conditions, minimizes fluctuations in supply, and optimizes the storage requirements. Conditions are deemed acceptable if the storage tank volumes at the end of the 24-hour period return to their initial levels.

6.3.4.2. Peak Week. If the supply is constrained or if the possibility of a peak day event experienced over a series of consecutive days is not great, the system should be designed for a longer period of recovery. Again, the supply over a 7-day or greater period equals the total system demand over that period while the storage is capable of supplying instantaneous demands over the available supply and storing supply in excess of low-usage periods. This philosophy allows for a smaller percentage of supply relative to peak day demand but does require additional storage. As will be discussed in chapter 7, this philosophy may have adverse impacts on water quality.

6.3.4.3. Time-of-Day. Other operating philosophies allow for variations in conditions according to the time of day. Some utilities only operate their treatment facilities for one or two shifts during a day. Some power suppliers make more attractive rates available to the water utility if the power demand occurs during the power utility's off-peak hours. The water utility can then respond by operating more power-using facilities, such as treatment plants or pumps, during off-peak hours. These time-of-day philosophies can realize operating cost savings for labor and power, but they may possibly require additional, possibly significant, capital cost outlays to provide water during these specified times of days through additional storage volumes or additional pumping capacity.

6.3.4.4. Minimum and Average Day. Extended-period simulations of minimum and average demand conditions are often used to plan daily operations or to assess water quality. As will be discussed in chapter 7, water quality analyses generally require simulations of very long duration that exhibit repeating cycles of tank filling and draining as supply sources are turned on and off on a regular schedule.

6.3.4.5. Seasonal. Many water systems have sources of water that are used differently in different times of the year. These operating conditions could require as many separate extended-period simulation configurations as there are different operations. With multiple operating configurations, the most extreme configurations in terms of the performance criteria being evaluated should be selected. Many times the variations are dictated by differences in water quality or availability of the source flows

6.3.5. Reservoir or Service Storage Data

Steady-state distribution system models are concerned with water surface elevation at a particular point in time. One of the most important pieces of information required for extended-period simulations is the amount of service storage available and the characteristics of that storage. While steady-state analyses are based on a fixed water level or pressure at a tank, the extended-period model must account for much more information. Vital information on storage facilities needed for extended-period simulation models includes:

- Overflow elevation—the maximum water surface elevation
- Minimum water level—required to maintain adequate system pressures
- Reservoir physical attributes—uniformly circular, rectangular, irregular at bottom or top
- Storage/unit of height—the amount of water contained per unit of height (if the tank is not circular or rectangular)
- Effective height—the difference in elevation between the maximum and minimum water surface elevation
- Piping configuration of inlets and outlets—some tanks have a single pipe that serves both as an inlet and an outlet, Other tanks have separate and sometimes multiple inlet and outlet pipes. If a tank has an inlet pipe that discharges into the tank above the maximum water level, a node can be added with an elevation above the overflow elevation of the tank with the node connected to the tank by means of a check valve to allow flow into the tank only. Some modeling programs can now accommodate differences in tank inlet and outlets and even simulate flow into the tank above the high water level.

6.3.6. Pump Station Information

Pump station input requirements are also significantly more detailed in the extended-period model than in the steady-state model. Accurate individual pump curves are essential for an accurate extended-period model, because minor fluctuations in suction pressure or service reservoir levels greatly affect model predictions. The actual station operating parameters, or control set points and location of the controlling point in the system (reservoir or system junction point), must be input into the model, depending again on the accuracy required. The pertinent information for the station includes the following:

- Certified pump curves for each pumping unit within the station pump; controlling point location
- Controlling elements, generally tank levels or node pressures but sometimes flow through a pipe or valve
- Control set points to turn a pump on or off, adjust a pump's speed or valve's setting, or to open or close a pipe or valve
- Initial status (on/off) for each pump
- Detailed piping configuration of suction and discharge piping, including estimates of minor loss coefficients for fittings

6.3.7. Well Data

The well pump level and pump curve are generally input into the extended-period simulation model. The extended-period simulation model also requires the location of the controlling point in the system (reservoir or system junction point), the actual on/off set points, and initial condition be input to allow the well to function in the model as it would in the real world. Often, there is a significant difference between the static water level in the well and the drawdown level in the well when the well pump is operating. The drawdown water level in the well can be used as a fixed reservoir elevation or a head loss can be imparted through a control valve to simulate the variation of well drawdown with flow rate.

6.3.8. Control Valve Data

Control valves are defined by the type of function they perform. Some types of control valves operate automatically based on pilot settings at the valve reflected by conditions directly up or downstream of the valve. These include pressure-reducing valves, pressure-sustaining valves and pressure-breaking valves. In most modeling programs these settings are input directly with the valve's characteristics. Other types of valves can have more complex functions. These include flow control valves and throttle control valves, which can have a fixed setting but can also be controlled by other system components using logical rules in the model.

6.3.9. SCADA Information

Extended-period simulations are designed to mimic water system operation in response to system demands over time. The more precise and detailed the information included in the model, the greater the model accuracy and greater the potential rewards of the evaluation. Essential to accurate extended-period system analysis is obtaining two basic components of information. The first is an accurate description of system demand fluctuations over a defined time period. Optimally, these periods are recorded at regular intervals of an hour or less. The second is a detailed description of the operating parameter changes (pump on/off) and reaction of the water system to those system demand fluctuations (storage volume changes) over the same finite time. System status information obtained every 24 hours is essential in describing system peak day events for a water system and useful in describing static model evaluations; however, 24-hour system status information is insufficient in developing an extended-period simulation model. Central in obtaining system operation parameter changes and reaction to the water system is a SCADA system that monitors operation parameters as they change and stores information in an easily retrievable form for use in the model. This information, broken into finite time intervals, such as hourly changes (at a minimum), is the standard to measure system performance versus model predictions. The following is the basic SCADA information required for extended-period simulations:

- Reservoir fluctuations over time
- Pump run times and flow variations over time
- System pressure fluctuations over time
- System control valve status, pressures and flow, if measured

6.3.9.1. Using SCADA Data in EPS Models. In most cases, SCADA data will be used in several ways for extended-period modeling analysis (Figure 6-1). These include establishing the initial status of pumps, valves, and tanks; defining the time varying patterns at system boundaries such as tank levels or flows through pumps or inflow points; defining or confirming controls that govern pump or valve operation; and, most important, comparing modeling results with actual field measurements.

Prior to obtaining SCADA data for use in EPS models, it is important to know the time intervals and run durations to be used in the model. The next step is to assess the types of data and the format of the data that are available in SCADA because most SCADA systems were not set up with modeling in mind. Many SCADA systems have historians or other types of report generators that can provide information for the models. Coordination with the SCADA system operators or integrators is important as is knowing the format used by modeling software. Often, the SCADA data can be exported in a file format that can be opened in a spreadsheet or database program. The data can then be reformatted to match the needs of the modeling software.

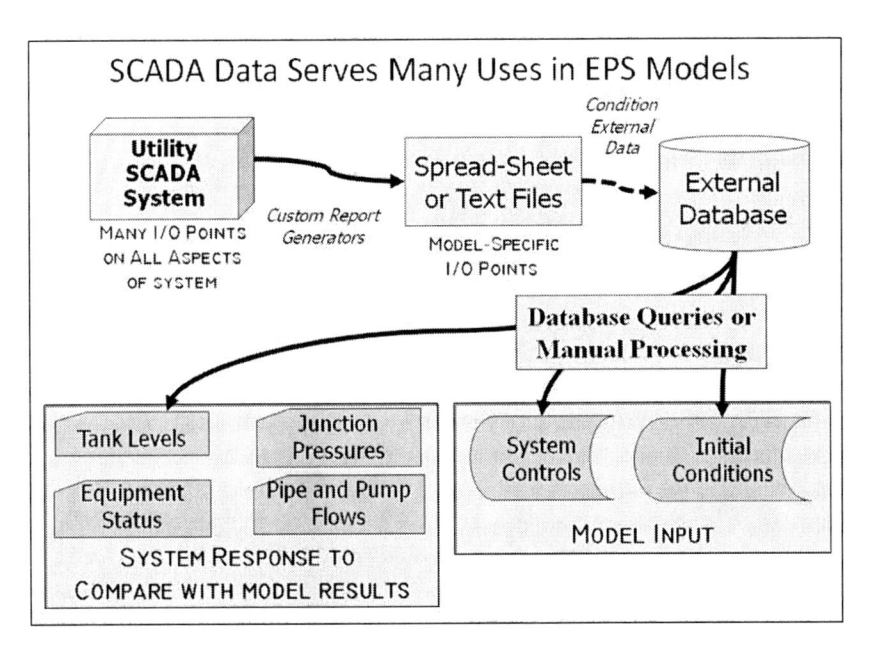

Figure 6-1 Using SCADA data in EPS models

Many of the important considerations for using SCADA data are discussed in chapter 2. When using SCADA data for EPS models, the time interval of the data used is one of the most important issues to consider. This is because the time intervals at which SCADA data are polled can be very frequent, usually in periods of minutes or seconds. Except in specialized cases, EPS models only require time intervals of several minutes or hours. Furthermore, there may be considerable noise in the data recorded by field instruments. Rather than using SCADA data at intervals that match the model's intervals, it may be helpful to use more frequent SCADA intervals and derive a running average of values at the model intervals to filter out the noise. In other cases, the SCADA system may only record data on change in status of the variable being recorded. In these cases, you may have to extrapolate the data for time intervals without recordings.

6.3.9.2. EPS Modeling Without SCADA Data. While using SCADA data certainly makes EPS modeling more convenient, lack of SCADA data does not mean EPS modeling is impossible. There is likely some means of measuring flows and tank levels in place, usually in the form of circular or strip charts. Hourly data can be extracted from these charts and tabulated for use in EPS models. Generally, daily operations logs indicate when pumps were turned off or on, tank levels at the end of the day, and total water production. The degree of accuracy of the model results may be reduced, but the results can be of some value. In absence of permanent SCADA instruments, flowmeters and pressure sensors can be installed on a temporary basis with the system information from these instruments saved on local data loggers. Stored data can be downloaded into portable computers and copied for use in the model.

6.3.10. Diurnal Curve Characterization

Derivation of diurnal curves (see chapter 3) for major customer types is important for all types of distribution system analyses, especially extended-period simulations. The accuracy of an extended-period simulation model is highly dependent on the quality of the demands and of the diurnal patterns assigned to the model. To create an accurate extended-period simulation model, it is not sufficient to broadly characterize zones or areas based on assumed demands. Specific water demands and diurnal patterns are needed to ensure the accuracy of the model. Ideally, the specific water-use characteristics of the utility's customers will be accurately described in order to predict water use over time with validity. For residential customers, homes with large landscaped lots demonstrate outside watering characteristics far different than homes with water-efficient landscapes or multifamily residential complexes. Different types of businesses and industries exhibit widely different usage patterns. These differences in water use, however subtle, can translate into inaccuracies in the model if not included in the diurnal curves for each customer classification. Figure 6-2 presents a few different types of diurnal demand patterns that may be seen in a distribution system with multiple user types.

Few water-system service areas consist of a homogeneous customer type. However, it is probable that most contain a few areas that are characterized as homogeneous and about which the actual system demand is described. For example, if a residential area can be found whose water use is segregated from the remaining system and whose water-use characteristics are described using SCADA records or field measurements using temporary flowmeters, a diurnal curve for this residential type can be described. A similar residential area, but combined with some multifamily or commercial use, is evaluated using the previously determined residential diurnal to filter out the residential demand, thereby describing the multifamily or commercial diurnal.

Classification of customer types can be performed in various ways. Some billing departments maintain information on the category of customers while some assessors' departments contain information on the land use in each property parcel. In some cases, distinctions can be made based on the size of the customers' meters or based on land use zoning. Once diurnal curves are derived for the various customer types, it is necessary to apply base demands, either as individual users or as a block or blocks of similar individual users, to the developed diurnal curves and populating the node influence areas in the model.

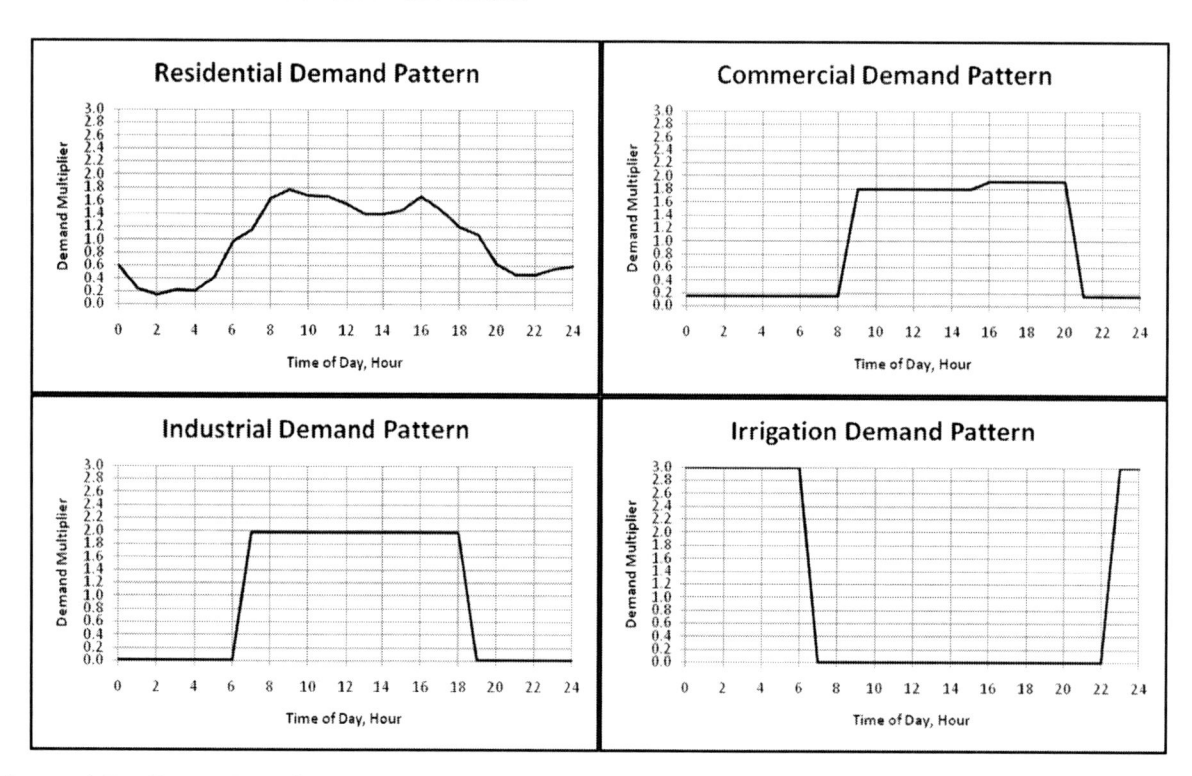

Figure 6-2 Examples of typical diurnal demand patterns for different use categories

Developing a good set of demand data for all customer types significantly improves the accuracy, and therefore the value, of a hydraulic model. Yet few utilities have the personnel or the budget to install a significant number of water meters sophisticated enough to monitor water usage for the broad type and number of customer types required to characterize water demand over small increments of time concurrent with SCADA information. However, equipment can be purchased to attach to existing meters that collects diurnal demand information. If necessary, meters can be manually inspected over a 24-hour period to gather information. Placement of these meters should include a good representation of each customer type and cover the entire geographic extent of the utility's service area.

Aside from taking an exhaustive series of actual field measurements representative of the various customer types throughout the entire service area, the remaining option involves using data available from billing records and SCADA information; the meter usage data are classified by customer type or geographical area and assigned a diurnal curve based on flow information from SCADA.

In addition to collecting and deriving diurnal patterns for specific classes of customers based on metering records, there are a number of citations in the literature that describe typical usage patterns for different categories that can be used in the absence of meter data. At a minimum, patterns for residential, commercial, industrial, and institutional users should be developed as well as estimates of the quantities and patterns of unaccounted-for uses whether based on actual data or derived from other sources. Further complicating matters are pressure-dependent demands, such as irrigation, which will vary as system pressures fluctuate. Systems with significant qualities of these types of demands should consider more detailed analysis of these demands.

Ultimately, the model's various diurnal demand patterns must be composited and compared with the systemwide average production patterns. When using several diurnal patterns to represent different categories of use, it is helpful to have the average of each hourly demand multiplying factor equal one to better manage the overall system demand. A typical composite demand pattern is presented in Figure 6-3 illustrating each component's demands added to the previous.

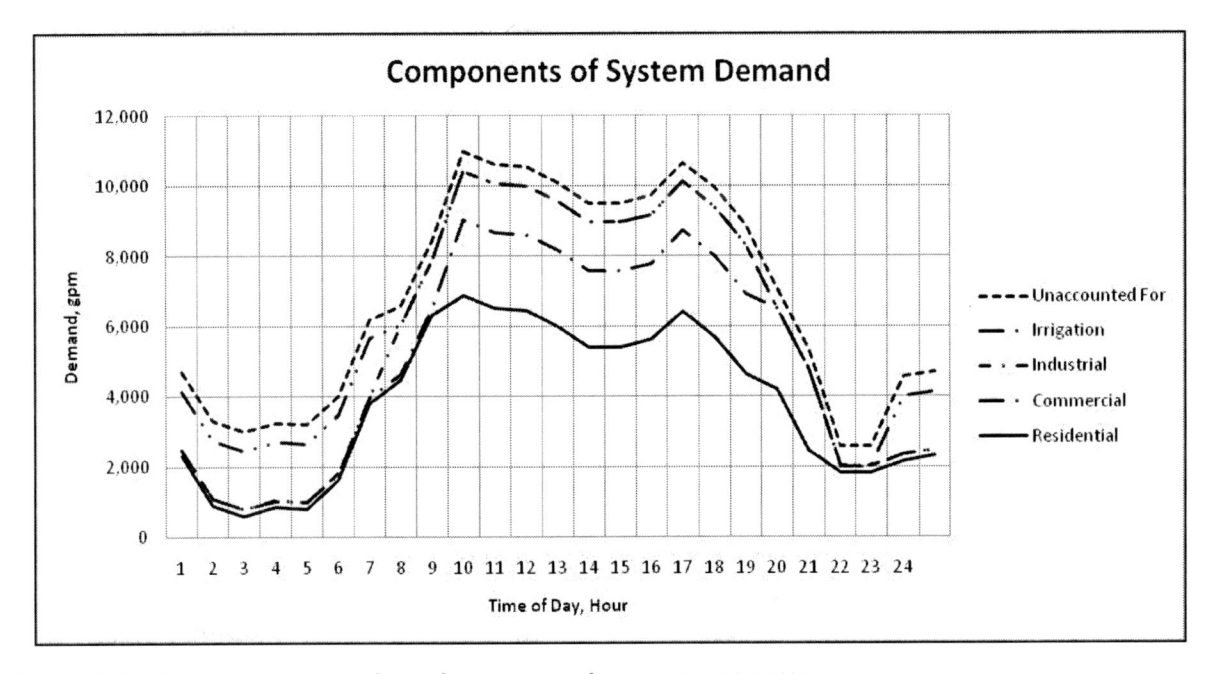

Figure 6-3 Example system diurnal pattern and component patterns

6.4. EXTENDED-PERIOD MODEL CALIBRATION

System components requiring calibration include piping configuration, size, demands, and coefficients of friction to accurately describe friction losses associated with transporting vast quantities of water through the system. Extended-period simulation calibration involves simulating a certain series of conditions to represent a dynamic exhibition of the system through the demand fluctuations that occur over a 24-hour, or even greater, time period. System calibration requirements for this evaluation include the steady-state calibration requirements but extend to additional information mandatory for accurate system evaluations. Further discussion of techniques for extended-period simulation can be found in chapter 4.

The extended-period model calibration process compares the results of extended-period model runs with recorded data. Four basic components of models directly impact the results of extended-period simulations: system configuration, system condition, actual demands, and the system's reactions to demand. Two of these components, system configuration and condition, should have been calibrated during the steady-state water model calibration process and are not normally changed during extended-period calibration. The water system demand component varies over time and is represented by the diurnal demand patterns. The final component is the reaction of the water system to those demands. Some of this reaction is prescribed by the operating rules governing the operation of pumps and valves. Other elements, such as valve position and pump status, help define the initial status of the system on which the remaining results are based. The rest of the system reaction, namely flows and pressures, is dependent on the other components and can be gauged from data contained in the SCADA information.

6.5. TYPES OF EXTENDED-PERIOD SIMULATION ANALYSES

Extended-period simulation analysis techniques, performed after the model is calibrated, are tailored based on the objectives of the analysis. Modeling criteria must first be described and then the actual analysis technique is determined based on those criteria. In general, the same criteria used in steady-state analyses are valid for extended-period simulations for hydraulic conditions alone. Water quality modeling, which is used for different types of extended-period analyses, uses similar hydraulic criteria in addition to the water quality criteria discussed in chapter 7. The system must maintain adequate pressures over the entire duration of the simulation. If the modeling results indicate low pressures, these deficiencies are probably caused by inadequate capacity in a pipe or pump to deliver water where it is needed, by high elevations relative to pump output or storage tanks, or some combination of the two. Evaluating the resulting extended-period simulation pressures over a 24-hour period can help identify the reasons for the deficiencies. Low pressures due to poor capacity generally occur during a period of high demand while the more persistent low pressures are likely found with customers at higher elevations. Several of the more typical analysis objectives for extended-period simulations are discussed in some detail in the following sections.

6.5.1. Storage Versus Production

The most basic extended-period simulation evaluation involves determination of the adequacy of system storage relative to production. The main premise is that the water system's storage facilities should return to their initial conditions (levels) within a prescribed time frame. The following example illustrates some of the aspects involved in evaluating storage versus production. In these types of analyses, the primary objective is that initial tank levels must be returned to in a 24-hour period. That means the total system production over a 24-hour period equals the total system demand over that same time frame, with storage either making up for hourly differences in demand over production or storing excess production over demand. Storage and production must combine to provide service while returning the system to its initial status at the end of the 24-hour period. Because the demands are not symmetric with respect to storage of production, the combination of storage volumes plus water system piping must be sized appropriately.

The results of a real water system analysis are provided to illustrate the impact these analyses have in a real-world situation. The example system involves interaction of two reservoirs, each with a well-defined service area and containing two very distinct diurnal demand curve types.

The system demand in this example contains two separate components, a residential demand and a golf course demand. Figure 6-4 represents a peak-day diurnal curve for this system.

In this figure, the diurnal curve for the entire residential population (shown as a single dot dashed line) is based on the residential diurnal curve established for the residential population, adjusted to the actual residential population served. The diurnal curve for the golf course load over the same time period (shown as a double-dot dashed line) is based on the course's diurnal curve and recordings of the actual instantaneous demand. In an extended-period simulation evaluation, these two loads are added to form the total system diurnal curve for the period, shown in the figure as a solid line.

The two reservoirs each have a well supply on-site, plus a single well within the service area. The system's physical parameters are described in Table 6-1.

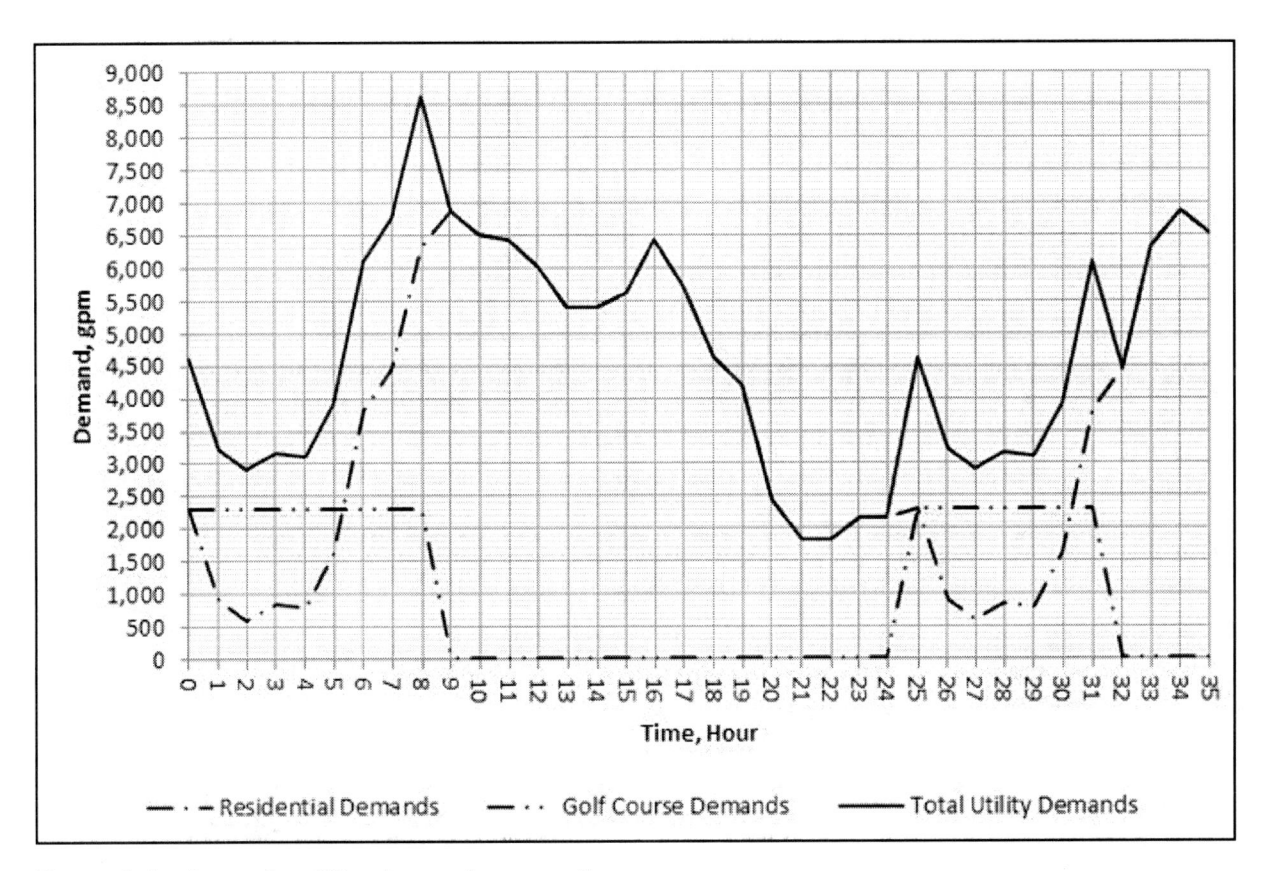

Figure 6-4 Example utility demands versus time

138 COMPUTER MODELING OF WATER DISTRIBUTION SYSTEMS

Table 6-1　System physical parameters for extended-period simulation analysis

Reservoir No. 1		Reservoir No. 2	
Overflow elevation	5,485.00 ft	Overflow elevation	5,485.00 ft
Bottom elevation	5,453.00 ft	Bottom elevation	5,450.63 ft
Effective storage	2.0 MG	Effective storage	3.08 MG
Diameter, ft	103 ft	Diameter, ft	128 ft
Storage/ft	62,500 gal/ft	Storage/ft	96,250 gal/ft
Well No. 1	1,800 gpm @ 563 ft of lift		
Well No. 2	2,075 gpm @ 563 ft of lift		
Well No. 3	2,075 gpm @ 563ft of lift		
Fire Protection requirements:	5,000 gpm for 5 hr or 1.5 MG		

The example utility in its existing configuration includes only reservoirs 1 and 2 and wells 1 and 2. An analysis was performed using the 24-hour peak day diurnal curve followed by an additional 12 hours of 90 percent of peak day demand to determine the adequacy of storage and production. The results of the analysis with this existing configuration (Case 1) are shown in Figure 6-5. The figure indicates that the water system could not completely recover for the second day, resulting in a situation in which the reservoir would come close to storage set aside for fire protection. In some cases, the lower reservoir levels may lead to unacceptably low pressures at some locations in the system.

This situation would be alleviated by the introduction of additional production, as shown in Figure 6-6. An identical evaluation was prepared using Well 3 as a supplemental well to the system. Case 2 (Figure 6-6) illustrates the effect on the system of Well 3, set to operate when the reservoirs approach a water level 10 ft (3 m) below overflow. The reservoir water elevations returned to a 24-hour full situation by providing the additional supply from Well 3.

6.5.2. Vulnerability Analysis

Another key evaluation that can be performed using extended-period simulation is reliability or vulnerability analysis. Types of this analysis would include loss of pumping, supply, or storage facilities; loss of critical transmission mains; planned maintenance outages; and pump cycling studies. This analysis is used to identify critical weak points within the system, assess the extent of those weak points and effect on water service, and determine system modifications that provide some relief from the problem. This relief might lead to identifying the need for some redundant production, storage, or a combination. For example, the loss of a well at the Reservoir-1 site results in reservoir elevations as shown in Figure 6-7 (case 3), assuming all other assumptions stay constant. The system that was balanced previously is now unbalanced to the point that Reservoir-1 level compromises fire storage and, if a second peak day event occurred, would drain.

6.5.3. Emergency System Operations

An extremely valuable benefit of extended-period simulations is the evaluation of system response under emergency conditions, such as simulated major fires during a maximum-day demand condition or simulated power outages. These "what if" analyses provide operations personnel important information on how water system facilities respond when stressed. This enables advanced planning for better response in emergency situations, if and when, they occur.

EXTENDED-PERIOD SIMULATION 139

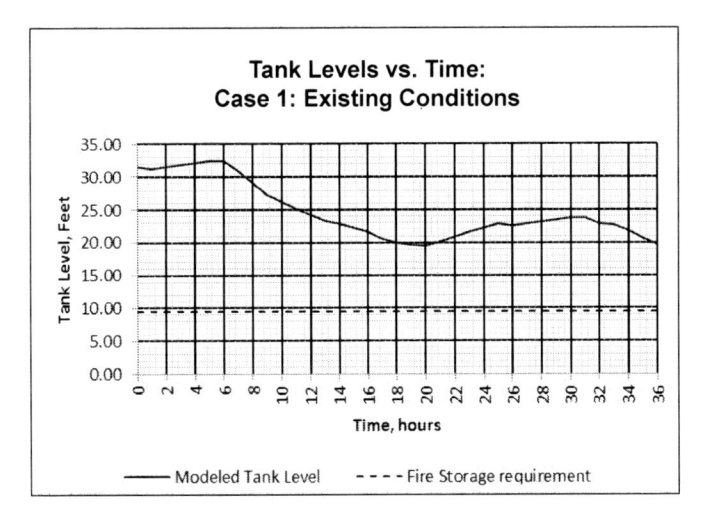

Figure 6-5 Example of storage versus production for existing conditions, Case 1

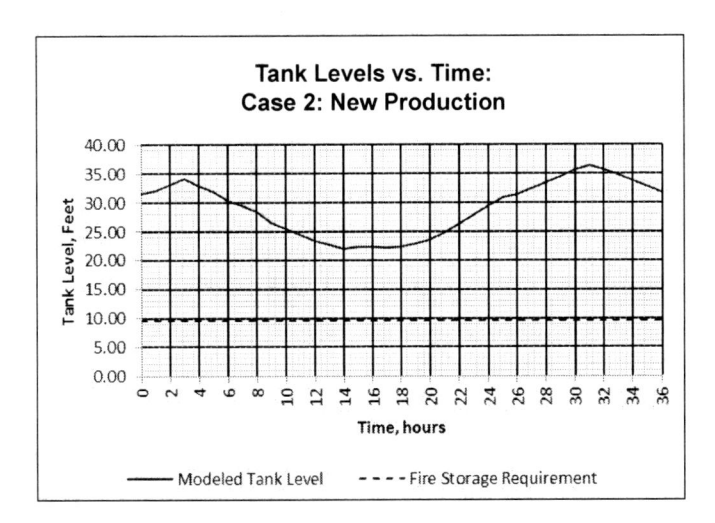

Figure 6-6 Example of storage versus production with new production, Case 2

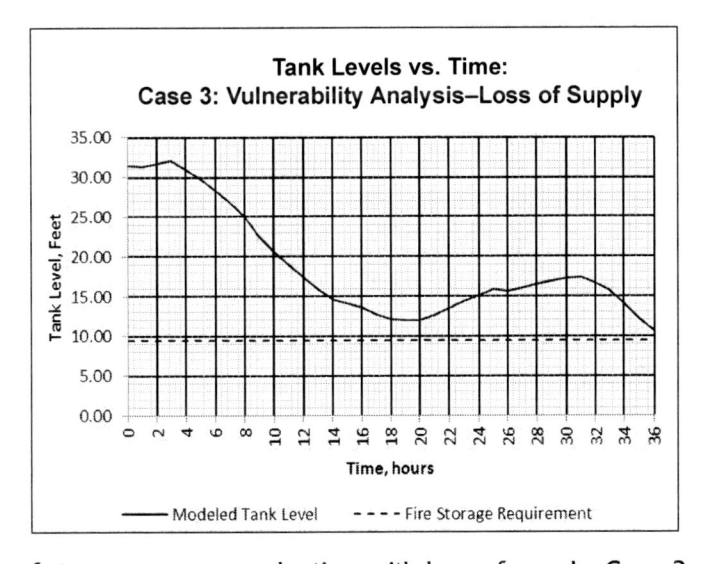

Figure 6-7 Example of storage versus production with loss of supply, Case 3

Major industrial areas typically require very large fire flows. To simulate system response under worst-case conditions, analyses assume that these large fire flows occur during early morning hours, coinciding with high morning demands. Similar to a simulated fire, system power outages are also assumed during critical demand periods to determine system response. Under such circumstances, wells and pumping stations without standby power sources would be out of service. Therefore, the only water available comes from supplies having emergency power generators, system storage, or emergency interconnections with other systems. This type of analysis can be used to determine how long the system can reliably operate under these conditions. For example, a simulated fire in the system results in a drop reservoir elevations as shown in Figure 6-8 (case 4), assuming all other assumptions stay constant. It is important that the tank levels stay above those required for minimum levels of service at all times during a fire.

During emergency simulations, system response should be analyzed for areas having below standard pressures (less than 20 psi [138 kPa]), negative pressures, and low or empty reservoir levels. For multizone systems, interzone water transfers should be analyzed to ensure that the whole water system responds to emergency conditions. These simulations are useful tools in changing operating procedures to improve system performance.

6.5.4. Energy Optimization

Energy optimization is very important for efficient water system operation because of the rising cost of energy. Energy optimization involves a delicate balance between supply, pumping, and storage based on cost differentials between various energy sources, either by type or by time of service.

The potential energy savings from energy optimization can be as much as 10 or 20 percent of the total energy cost. For this reason, evaluating energy requirements and developing a power-management program are key steps in the development of distribution system improvements.

To assess potential energy savings, the electrical rate structure should be analyzed to determine the least expensive times to pump. Electrical rates typically consist of two components, the demand charge and the energy charge. The demand charge is the maximum instantaneous power usage at a facility. In some rate structures, the demand charge is based on the highest demand in the last 12 months, so if the peak was an isolated event, the utility pays a substantial amount of money for the one occurrence. The energy charge represents the total amount of energy used over the billing period. The cost for energy varies at different times of the day, so if the utility schedules reservoir-filling activities during the lowest cost period, additional savings are found.

Some of the items to consider in developing an energy optimization plan are as follows:

- Power consumption by system components
- Alternate system operations to reduce energy consumption
- Peak shaving through standby power
- Water demand management
- Construction of additional pipes, pumps, or storage to reduce overall pumping requirements

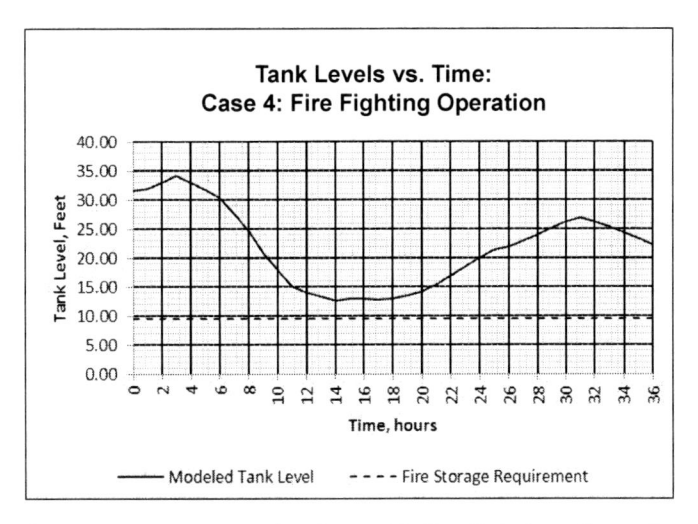

[FIGURE: Tank Levels vs. Time: Case 4: Fire Fighting Operation — plot of Tank Level (Feet) vs Time (hours), showing Modeled Tank Level and Fire Storage Requirement]

Figure 6-8   Example of storage versus production with fire fighting, Case 4

Many of the software packages include some sort of energy optimization capabilities that can help a utility determine the pumping routine to achieve the best energy management.

Several types of energy optimization techniques are described in the following sections to illustrate the effects of time-of-day rates and blending energy types on water service or system capital requirements.

6.5.4.1. Time-of-Day Electric Rates. Electricity generation, unlike water supply, must meet or exceed instantaneous electric demand. This fact results in significant capital expenditures for peak electric generation equipment that is idle the vast majority of time. Some electric generation companies attempt to smooth their system's electricity load over nonpeak times, targeting large users by rewarding them for off-peak electric demands and penalizing them for energy demand that coincides with the system peak electric demand. With few exceptions, water utilities comprise a community's larger—if not largest—electric user, and water system demand peaks usually mirror electric demand peaks; in fact, a significant percentage of water system demand occurs during a 12-hour peak electric demand time frame. Figure 6-9 illustrates the same utility example using a 12-hour pumping curtailment from 8 a.m. to 8 p.m. Reservoir 1 would actually drain into fire storage during the 36-hour period while Reservoir 2 would be well into its fire storage. The result would be a lack of service to some and insufficient fire protection to all.

Water utilities using only electric power have options available to them to curtail on-peak electric consumption. The first involves a balance of additional storage and supply to make up the deficit. The second involves an analysis to determine the actual number and extent of system peaks expected to occur and balance the cost of the additional infrastructure versus the additional energy costs incurred in violating the on-peak pumping for those periods of the year.

6.5.4.2. Driver Types/Blend. One way to avoid peak electric demand rates is to supplement electric power needs by using other pump drivers. A natural gas or diesel engine connected to a well pump can provide water during on-peak electric periods and supplement electric drivers off-peak. In addition, peak water demand occurs during the growing season at a time when natural gas energy requirements for heat are traditionally at low levels. Some electric utilities may credit water utilities for running emergency generators during peak electricity demand periods and delivering the excess electricity back into the power grid. Figure 6-9 (case 5) illustrates the impact of changing Well 3 to a natural gas driver. Clearly, the additional production of Well 3 during the on-peak electric period is not sufficient to completely offset the storage losses caused by the other wells being off, but there is a significant improvement over the reservoir levels illustrated in Figure 6-10 (case 6).

Variable speed pumps are another way of improving energy efficiency. Constant speed pumps that use a throttling control valve to adjust flows use the same amount of energy regardless of the flow rate. Adjusting the speed of the pump allows operation closer to the pump's best efficiency point and will use less energy when lower flow rates are pumped. Many modeling programs now have the capability to simulate system operation with variable speed pumps based on maintaining pressures or flows.

The electrical industry throughout North America has seen substantial changes over the years. The modeler should meet with power utility representatives to review the impact of different operating strategies on the power cost of the water utility. Generally, these different operating strategies require that the water utility have additional pumping and storage capabilities to allow the operation to take advantage of power cost savings.

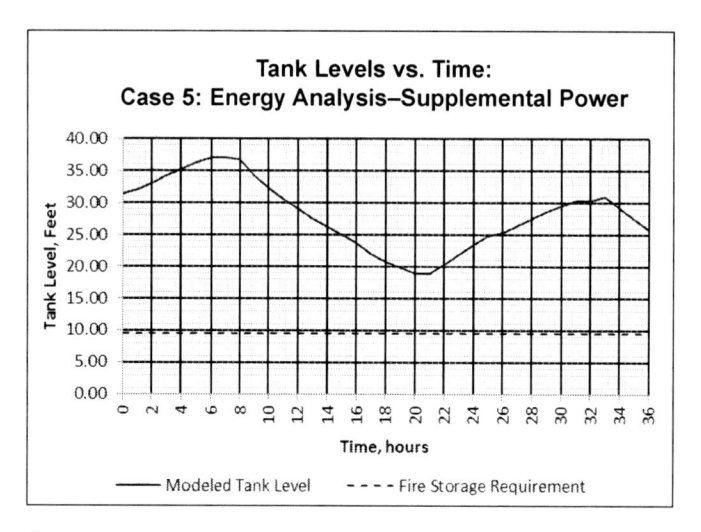

Figure 6-9 Example of storage versus production with supplemental power, Case 5

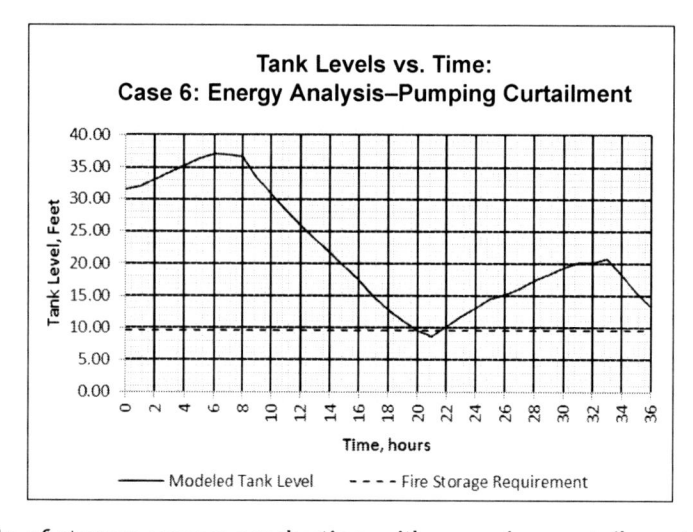

Figure 6-10 Example of storage versus production with pumping curtailment, Case 6

6.5.5. Sizing Storage Tanks

Extended-period simulations are best suited for assessing and designing storage facilities. Based on steady-state modeling, the desired tank overflow elevation can be selected based on providing sufficient pressures in the area served by the tank. Once the overflow level is set, extended-period simulations are used to determine the range of water levels in the tank for various demand conditions including minimum day, average day, and maximum day (including maximum day with fire flows). Different locations of the tanks can be evaluated to assess the levels of service throughout the day and to assess how well the tank will operate with the pumps used for filling. Storage tanks' volumes, as discussed in chapter 5, involve a number of components. Extended-period modeling is used to determine the volume requirements for demand equalization.

6.5.6. Operator Training

Operator training is an excellent application of distribution system modeling within a utility. The models are used to train new staff on the unique operational characteristics of the water distribution system. The models are also useful in simulating different system failures and identifying the most effective response to reduce the impact on customers.

The operators are also the best resource for determining if the model accurately represents flow and pressure measurements based on the utility's SCADA system. This allows the modeler and the operator to work together to identify system deficiencies or problems in the model assumptions.

6.6. CASE STUDY: CITY OF FULLERTON, CALIFORNIA

In a study conducted for the city of Fullerton, Calif. (shown in Figure 6-11), extended-period simulations were conducted with the city's computer model to identify necessary new facilities and operational improvements to the water supply system. These extended-period simulations enabled the city to save considerable operating costs by identifying ways to maximize low-cost well water, and to reduce the amount of pumping (and energy required) throughout the water system.

The city of Fullerton receives its water supply from wells in the local Orange County Groundwater Basin and from water imported by the regional water supplier, Metropolitan Water District of Southern California (MWD). City wells pump to forebays or reservoirs, and booster pumping stations pump the water into the system. A few wells pump directly into the system. MWD delivers imported, treated water from the Colorado River and California State Water Project, which draws from several sources across the state. MWD's transmission system provides metered connections to the city's distribution system. These connections are controlled by pressure-regulating and flow-control facilities.

The city has a complex distribution system composed of 7 MWD connections, 10 storage facilities, 12 wells, 11 booster pumping stations, and 28 pressure-regulating stations. The service area is divided into 13 pressure zones. The system is divided into four main pressure zones and nine smaller, subpressure zones. Storage reservoirs and booster pumping stations equalize flows between and maintain system pressures within each zone. All zones are interconnected through pressure-regulating stations that maintain minimum zone pressures systemwide.

The basic operational scheme involves well water pumped upward and imported MWD water generally flowing downward through the system. Well water is boosted into lower-elevation pressure zones and is pumped again into successively higher-pressure zones. Most MWD water enters the distribution system through higher-pressure zones. Imported water is allowed to flow into lower pressure zones through pressure-regulating stations based on pressure drops in response to system demands.

The time-dependent model included all major water system facilities: transmission and distribution pipelines (typically 8-in. [20.3 cm] diameter and larger), storage reservoirs, booster pumping stations, wells, MWD connections, and the more actively used pressure-regulating stations. Physical features of these water system facilities were defined in mathematical form and entered into the computer model. Numerous time-dependent (24-hour time period) computer simulations were performed under various demand conditions (including existing and future demand conditions) to evaluate the behavior of the network.

6.6.1. Existing Inefficiencies

Computer simulations pinpointed a number of potential operating improvements within the existing system. For example, a reservoir in Zone 1, the lowest pressure zone, was maintained at an unusually high water level throughout maximum demand conditions and therefore, never fully contributed to the system. Investigation revealed that the pressure gradient setting at a nearby pressure-regulating station was set too close to the reservoir's high water level. This setting caused the pressure zone to be served primarily by the pressure-regulating station. The reservoir was continually kept full from this higher pressure zone. The reservoir cannot provide flow to the pressure zone and thus experiences inadequate water circulation. Lowering the pressure setting reduced the flow through pressure-regulating station and allowed much more flow from the reservoir during high demands.

In another example, the computer analysis showed that water pumped from another Zone-1 reservoir into Zone 3, a higher zone, was recirculating through a nearby pressure-regulating station to supply an intermediate pressure zone, Zone 2. More water should have been pumped from the reservoir directly to Zone 2 where it was needed, rather than to Zone 3. By adjusting the settings of booster pumps that pump from Zone 1 to Zone 2 to run more often, the recirculation between pressure zones and associated wasted pump energy was eliminated.

6.6.2. Operational Improvements

Having identified deficiencies in the existing system, future demand conditions were analyzed. Once future water system improvements were selected, time-dependent analysis was used to determine the most efficient water supply system operation under future conditions. Basic study objectives were to (1) minimize interzone water transfer and resultant energy lost through pressure-regulating stations; (2) maximize lower-cost groundwater production while minimizing the purchase of higher cost MWD water; and (3) determine reservoir storage capacity needed to meet regulatory, emergency, and fire flow requirements of the future system.

6.6.3. Pressure-Regulating Stations

The computer model showed that interzone water transfers could be greatly reduced without adversely affecting system pressure or supply by adding to or modifying supply facilities as well as lowering downstream pressure gradient settings at pressure-regulating stations. With these changes, an analysis of future demand conditions showed much less water would be transferred during days of maximum demand. Overall, interzone water transfers would be reduced approximately 80 percent compared with existing system operation.

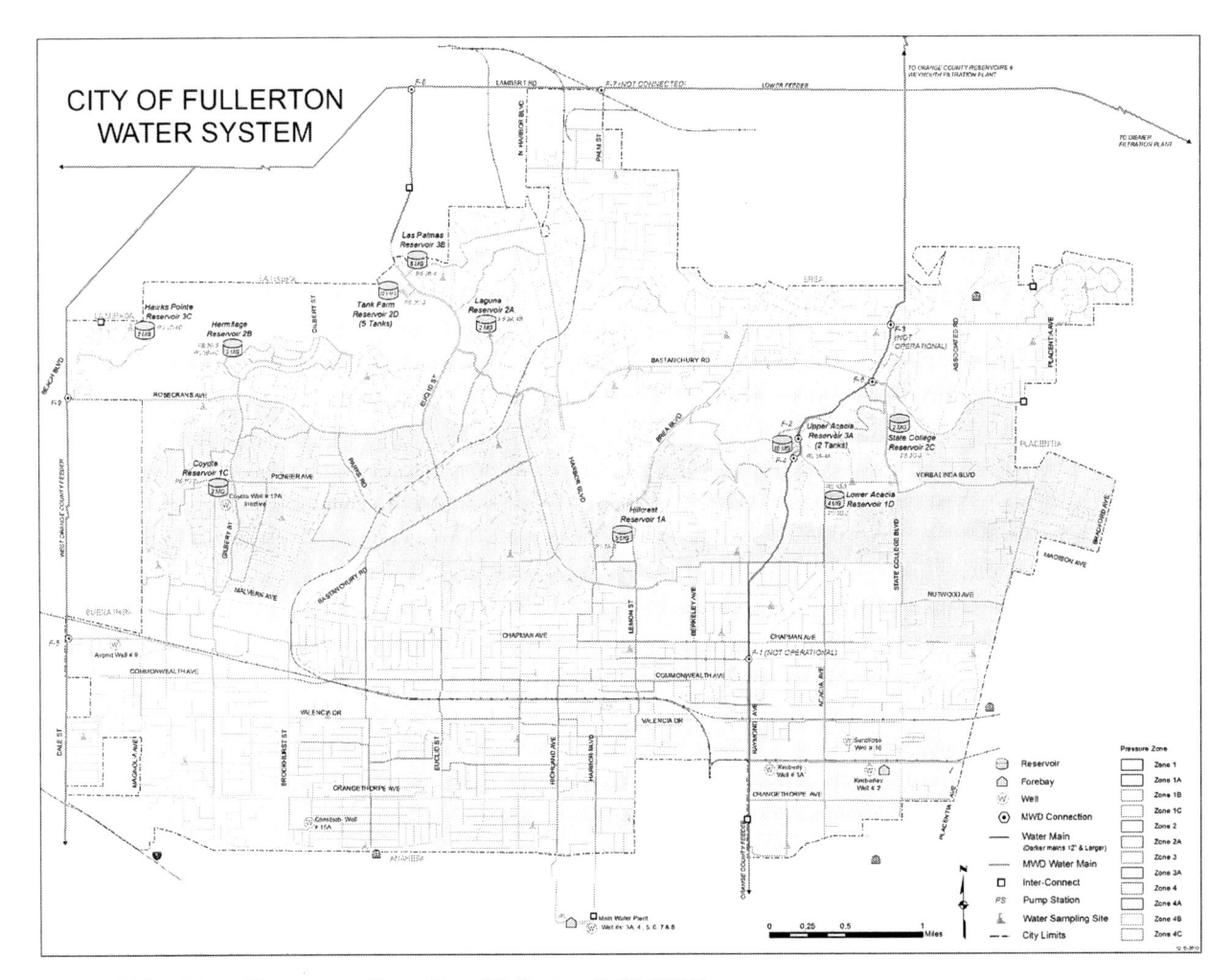

Source: Urban Water Management Plan, City of Fullerton, Calif. (2005)

Figure 6-11 Location map for Fullerton case study

6.6.4. Booster Stations and Wells

A major objective of the extended-period simulation was to identify ways to meet most of the city's water demand from groundwater sources instead of imported MWD water. To explore conditions for optimum groundwater production, new wells and additional booster pumping capacity at existing wells were added to the model and the results were studied. The computer model showed that two new wells and an increase in booster pumping capacity to match well production capacity could produce enough groundwater to meet most of the system's winter and nighttime demands. Also, an increased amount of groundwater could be mixed with MWD water to meet summer demand. A new booster station to pump water from Zone 1 to Zone 3 was also recommended to increase the amount of well water pumped to higher pressure zones and help solve a fire flow deficiency.

6.6.5. Reservoirs

To determine reservoir storage capacity requirements, future needs of the city were studied, as well as individual needs of each pressure zone. Generally, citywide storage capacity was adequate to meet future needs. However, some individual pressure zones required additional storage capacity because of limitations in interzone water transfers. Here, time-dependent computer simulations showed that pressure Zone 2 required additional emergency storage to meet ultimate demands. Also, although Zone 3 had adequate storage capacity, there were physical constraints in the distribution system piping that, in effect, isolated a large part of the zone without adequate emergency storage capacity. Therefore, more emergency storage was proposed for that area.

6.7. REFERENCES

Bhave, P.R. 1991. *Analysis of Flow in Water Distribution Systems.* Lancaster, Pa.: Technomic.

Cross. H. 1936. *Analysis of Flow in Networks of Conduits or Conductors.* Urbana, Ill.: Univ. of Illinois.

Eggener, C.L., and Polkowski, L. 1976. Network Modeling and the Impact of Modeling Assumptions. *Jour. AWWA,* 68:4:189–196.

Gupta, R., and Bhave, P. 1996. Comparison of Methods for Predicting Deficient Network Performance. *Journal of Water Resources Planning and Management,* 122:3:214.

Jeppson, T.W. 1976. *Analysis of Flow in Pipe Networks.* Ann Arbor, Mich.: Ann Arbor Science Publishers.

Jordan., R.A, Priest, M., Jain, D.K., and Jacobesen, L.B. 1999. Master Planning Utilizing H$_2$0NET Extended-Period Simulation and Microsoft Access Generated Summary Reports. In *Proceedings of the Information Management and Technology Conference.* Denver, Colo.: AWWA.

Shamir, U., and Howard, C.D.D. 1968. Water Distribution Systems Analysis. *Journal of the Hydraulics Division,* ASCE, 94:1:219.

Stone, K., Barbato, L.M., and Koval, E.J. 2002. Colorado Springs Utilities: A Case Study for Extended-Period Simulation. In *Proceedings of the Information Management and Technology Conference.* Denver, Colo.: AWWA.

Todini, E., and Pilati, S. 1987. A Gradient Method for the Analysis of Pipe Networks. In *International Conference on Computer Applications for Water Supply and Distribution.* Leicester, UK.

Walski, T.M., Gessler, J., and Sjostrom, J.W. 1990. *Water Distribution—Simulation and Sizing.* Ann Arbor, Mich.: Lewis Publishers.

Wood, D.J. 1980. *Computer Analysis of Flow in Pipe Networks.* Urbana, Ill.: University of Illinois.

Wood, D.J., and Charles, C.O.A. 1972. Hydraulic Analysis Using Linear Theory. *Journal of the Hydraulics Division American Society of Civil Engineers,* ASCE, 98:7:1157–1170.

Wood, D.J., and Rayes, A.G. 1981. Reliability of Algorithms for Pipe Network Analysis. *Journal of the Hydraulics Division,* ASCE, 107:10:1247–1248.

WRc Plc. 1989. *Network Analysis—A Code of Practice.* Swindon, UK: Water Research Centre.

AWWA MANUAL M32

Chapter 7

Water Quality Modeling

7.1. INTRODUCTION

A primary goal of water distribution systems is to deliver potable water when and where it is needed. Ideally, there would be no change in the quality of water from the time it leaves the source until the time it is consumed. In reality, significant changes occur as water travels through a distribution system and on to the customer's tap.

Under normal operating conditions these changes occur for several reasons. Chemical reactions that began at the treatment plant continue in the distribution system. Some of the most important of these are the reactions involving disinfectant residual, usually a form of chlorine. These reactions reduce the concentration of the disinfectant over time limiting its ability to control pathogens and may also result in production of undesirable by-products, such as chloroform or taste-and-odor-producing compounds. Another example is the formation of calcium carbonate or other coatings along pipe walls. Blending of waters from different sources in the distribution system produces changes in water chemistry such as pH and chloramine speciation.

Water also reacts with the walls of the pipe through which it flows. Corrosion and subsequent buildup of tubercles and oxide coatings are typical examples. Pipe walls also support the growth of attached colonies of microorganisms or thin biofilms. When this growth is excessive, it often increases disinfectant demand, produces taste-and-odor-causing compounds, and offers protection for opportunistic pathogenic organisms. In addition to the normal chemical reactions occurring in the system, water quality degradation through accidental or intentional contamination is an important concern.

7.2. NEED FOR WATER QUALITY MODELING

The spatial and temporal variation of water quality in a distribution system involves very complex phenomena because of both the hydraulic mixing patterns and the water quality transformations that occur within the network. The topological complexity of pipe networks in distribution systems and the ever-changing patterns of water usage rates over a day and over different seasons create myriad pathways that water travels on its journey from treatment plant to consumer. Individual parcels of water lose their identity as they mix with other parcels of different residence time and quality throughout the system. Thus, the water quality reaching a tap will vary significantly over the course of a day or between seasons, and the water quality at two different taps, even if they are in close proximity, can display significantly different water quality because of the travel pathways and transformations between the treatment plant and the taps.

The complexity of flow paths and transformations makes it difficult, if not impossible, for a person experienced in a particular distribution system to determine travel times, blending, and subsequent water quality throughout the system. Mathematical models of the hydraulics and water quality of a system, after careful application, can provide a method for estimating the movement, transformations, and water quality in a network.

7.3. USES OF WATER QUALITY MODELING

Water quality models are used to predict the spatial and temporal distribution of a variety of constituents within a distribution system. These constituents include the following:

- The portion of water originating from a particular source (tracing)
- The age of water (i.e., how long since it left the treatment plant)
- The concentration of a nonreactive tracer compound either added to or removed from the system (e.g., fluoride or sodium)
- The concentration and loss rate of residual disinfectant (e.g., chlorine or chloramines)
- The concentration and growth rate of disinfection by-products (DBPs) (e.g., trihalomethanes or haloacetic acids)

The ability to model the transport and outcome of these parameters helps system managers perform a variety of water quality studies. Examples include the following:

- Calibrating and verifying hydraulic models of the system through the use of chemical tracers
- Locating and sizing storage tanks and modifying system operation to reduce water age
- Identifying ways to reduce water age and stagnation
- Determining the degree of intermixing of multiple water sources
- Modifying system design and operation to provide a desired blend of waters from different sources
- Designing a cost-efficient routine monitoring program to identify water quality variations and potential problems
- Identifying the most effective sampling locations in a distribution system
- Determining the best location and dosage for rechlorination stations
- Assessing and minimizing the risk of consumer exposure to DBPs
- Assessing system vulnerability to incidents of external contamination
- Identifying response times for various contamination scenarios
- Finding the best combination of (1) pipe replacement, relining, and cleaning; (2) reduction in storage holding time; and (3) location and injection rate of booster stations to maintain desired disinfectant levels throughout the system

Recent developments in water quality modeling simulation include estimation of the growth or decay of multiple constituents simultaneously. This application is useful for evaluation disinfection by-product formation, planning corrosion control programs, controlling secondary nitrification in chloraminated systems, and simulation of sedimentation in pipes to help plan flushing programs. Efforts to extend water quality models to include such features as predicting biofilm buildup and growth of microorganisms are still in the research stage (Powell et al. 2004). Curently, water quality modeling to directly assess such distribution system problems as minimizing biofilm growth, controlling coliform outbreak occurrences, and avoiding red water problems will be handled in future advances in software.

7.4. WATER QUALITY MODELING TECHNIQUES

At the current time, the three main uses of water quality modeling include water age, source/contaminant tracing, and constituent growth/decay.

7.4.1. Water Age

Changes in the age of water throughout a distribution system can be simulated. Water age is the time spent by a parcel of water in the network. New water entering the network from reservoirs or source nodes enters with age of zero. Water age provides a simple, nonspecific measure of the overall quality of delivered drinking water. In simple terms, age is treated as a reactive constituent whose growth follows zero-order kinetics with a rate constant equal to 1 (i.e., each second the water becomes a second older).

7.4.2. Source/Contaminant Tracing

Tracing the sources of water can also be performed. Source tracing tracks over time the percent of water from a particular source node that reaches any junction node in the network. The source node can be any tank or reservoir in the network. In simple terms, the source node is treated as a constant source of a nonreacting constituent that enters the network with a concentration of 100. Source tracing is a useful tool for analyzing distribution systems drawing water from two or more different raw water supplies. It can show to what degree water from a given source blends with that from other sources, and how the spatial pattern of this blending changes over time.

7.4.3. Constituent Growth/Decay

Of greatest importance in water quality modeling is the simulation of the growth or decay of chemical or biological constituents in the water system. The following section describes the governing principles of these phenomena and the data requirements for modeling water quality.

7.5. GOVERNING PRINCIPLES OF WATER QUALITY MODELING

Just as hydraulic models conform to the conservation laws of fluid mass and energy, water quality models are based on conservation of constituent mass. These models represent the following phenomena occurring in a distribution system (Clark and Grayman 1998).

7.5.1. Advective Transport of Mass Within Pipes

A dissolved substance travels down the length of a pipe with the same average velocity as the carrier fluid while at the same time reacting (either growing or decaying) at a given rate. Longitudinal dispersion is usually not an important transport mechanism under most operating conditions. This means that the modeler can assume that there is no intermixing of mass between adjacent parcels of water traveling down a pipe. The governing equation for advective transport of mass within pipes is

$$\frac{\partial C_i}{\partial t} = -u_i \frac{\partial C_i}{\partial x} + r(C_i) \tag{7-1}$$

Where:

- C_i = concentration (mass/volume) in pipe i as a function of distance x and time t
- u_i = flow velocity (length/time) in pipe i
- r = rate of reaction (mass/volume/time) as a function of concentration

Recent studies have found that pipes in low flow zones, such as dead-end pipes, which are common in municipal water-distribution systems, experience a degree of longitudinal dispersion (Tzatchkov et al. 2009). Further research is needed to assess the full extent of this issue and to develop appropriate modifications to network models.

7.5.2. Mixing of Mass at Pipe Junctions

Most water quality models assume that at junctions receiving inflow from two or more pipes, the mixing of fluid is complete and instantaneous. Thus, the concentration of a substance in water leaving the junction is simply the flow-weighted sum of the concentrations in the inflowing pipes. The governing equation for mixing of mass at pipe junctions is

$$C_{i|x=0} = \frac{\sum_{j \in I_k} Q_j C_{j|x=L_j} + Q_{k,ext} C_{k,ext}}{\sum_{j \in I_k} Q_j + Q_{k,ext}} \tag{7-2}$$

Where:

- j = pipe with flow leaving node k
- I_k = set of pipes with flow into k
- L_j = length of pipe j
- q_j = flow (volume/time) in pipe j
- $q_{k,ext}$ = external source flow entering the network at junction node k
- $C_{k,ext}$ = concentration of the external flow entering at junction node k

The notation $C_{i|x=0}$ is the concentration at the start of pipe i, while $C_{i|x=L}$ is the concentration at the end of the pipe.

Recent research has indicated that under some circumstances, mixing may be incomplete at junctions (Romero-Gomez et al. 2008; Austin et al. 2008). Further study and field verification are needed to determine the extent of incomplete mixing and modifications to water quality models to account for this phenomenon.

7.5.3. Mixing of Mass Within Storage Tanks

Most water quality models can model four different types of mixing characteristics within storage tanks

- Complete mixing
- Plug flow (first in/first out [FIFO])
- Short-circuiting (last in/first out [LIFO])
- Compartment

The characteristics of each of these types of models are described in chapter 9.

7.6. REACTIONS WITHIN PIPES AND STORAGE TANKS

While a substance moves down a pipe or resides in storage, it undergoes reaction. The rate of reaction, measured in mass reacted per volume of water per unit of time, depends on the type of water quality constituent being modeled. Some constituents, such as fluoride, do not react and are termed *conservative*. Others, such as chlorine residual, decay with time, while DBPs, such as trihalomethanes (THMs), grow with time. Some, such as chlorine again, react with materials in both the bulk liquid phase and at the liquid–pipe wall boundary.

Distribution-system water quality models represent these phenomena (transport within pipes, mixing at junctions and storage tanks, and appropriate reaction kinetics) with a set of mathematical equations. These equations are then solved according to an appropriate set of boundary and initial conditions to predict the variation of water quality throughout the distribution system. Even though water quality calculations are made throughout the length of each pipe, output reporting of quality is usually made only for nodes. The set of equations that comprises a typical water quality model is described in section 7.5.

7.7. COMPUTATIONAL METHODS

In most situations, water quality models are run using extended-period modeling conditions. Extended-period models take explicit account of how changes in flows through pipes and storage tanks occurring over an extended period of system operation affect water quality. These models predict both spatial and temporal variations in water quality throughout a network and provide a realistic picture of system behavior.

The solution of steady-state water quality models involves solving a set of simultaneous linear equations that express the conservation of mass expressions at network nodes in terms of the unknown nodal concentrations (Wood and Ormsbee 1989). The coefficients in these equations contain information from the hydraulic behavior of the network, which is determined separately.

Several solution methods are available for extended-period water quality models (Rossman and Boulos 1996). They require that a hydraulic analysis be run first to determine how flow quantities and directions change from one time period to another throughout the pipe network. Usually, the time steps used for water quality analyses are shorter than those used in hydraulic analyses—usually less than 5 minutes. Such small time steps can result in very long run times but are necessary to achieve accurate results. At the start of each new hydraulic time period, each pipe is divided into a number of subsegments, and water is either routed between subsegments (the Eulerian approach) or the subsegments are moved downstream (the Lagrangian approach) over a succession of smaller time steps. At the same time, account is taken of any reactions occurring in each subsegment. At the end of each water quality time step, the water entering each node from inflowing pipes is blended to determine a new concentration value for that node. This concentration is released from the node into its outflowing pipes at the start of the next time period. When a new hydraulic time step begins, the water quality routing begins again under a new set of hydraulic conditions.

7.8. DATA REQUIREMENTS

7.8.1. Hydraulic Data

Water quality models use the same physical and operational characteristics as hydraulic models, incorporating the length and diameter of each pipe and how a distribution system's nodes and links are connected to one another. In addition, the water quality model uses the flow solution of a hydraulic model as part of its input data. An extended-period water quality model uses a time history of flows in each pipe and flows to and from each storage tank as determined from an extended-period hydraulic analysis. Water quality models do not normally use hydraulic data such as pipe roughness or nodal elevation, head, or pressure for water quality calculations, although the water level in tanks figures prominently in these calculations.

7.8.2. Water Quality Data

A water quality model needs to incorporate the quality of all external inflows into the network. An example would be the concentration of chlorine leaving the clearwell of a water treatment plant or at a source of supply from an adjacent utility as well as the quantity of chlorine injected at any booster stations in the system. Extended-period models also need to include how the concentration levels of external inflows change over time. These kinds of data are obtained from existing source-monitoring records when simulating existing operations, from laboratory investigations, or they can be set to desired values to investigate potential operational changes.

Dynamic water quality models are supplied with a set of initial water quality values for all nodes and storage tanks throughout the network. These initial values are established in several ways. One way is to use the results of a field monitoring study if they are available. A second method is to equilibrate the model by running it for a sufficiently long enough period of time from an arbitrary set of initial conditions under a repeating pattern of source and demand loadings until the model reaches a repeating pattern of outputs (dynamic equilibrium). This equilibration approach is limited by the calibration of the hydraulic model and the assumption that the demands repeat every day. The results of this approach can be used as the initial conditions for subsequent water quality analyses. The duration of the simulation for this approach will vary based on the type of water quality analysis being run, the size of the system, and the initial water quality estimates. The modeler should note an initial rise in water quality parameter followed by leveling off at the equilibrium value as shown in Figure 7-1.

7.8.3. Reaction Rate Data

7.8.3.1. Chlorine Decay Coefficients. Modeling the fate of residual disinfectant is one of the most common applications of distribution system water quality models. The two most frequently used disinfectants in distribution systems are chlorine and chloramines (a combination of chlorine and ammonia). These two disinfectants have a residual that can be simulated using most modeling programs. The term *free chlorine* denotes the amount of uncombined chlorine that consists of both hypochlorous acid (HOCl) and hypochlorite ion (OCl$^-$). *Combined chlorine* refers to the amount of chloramines (primarily monochloramine) present. The *total chlorine* is the sum of

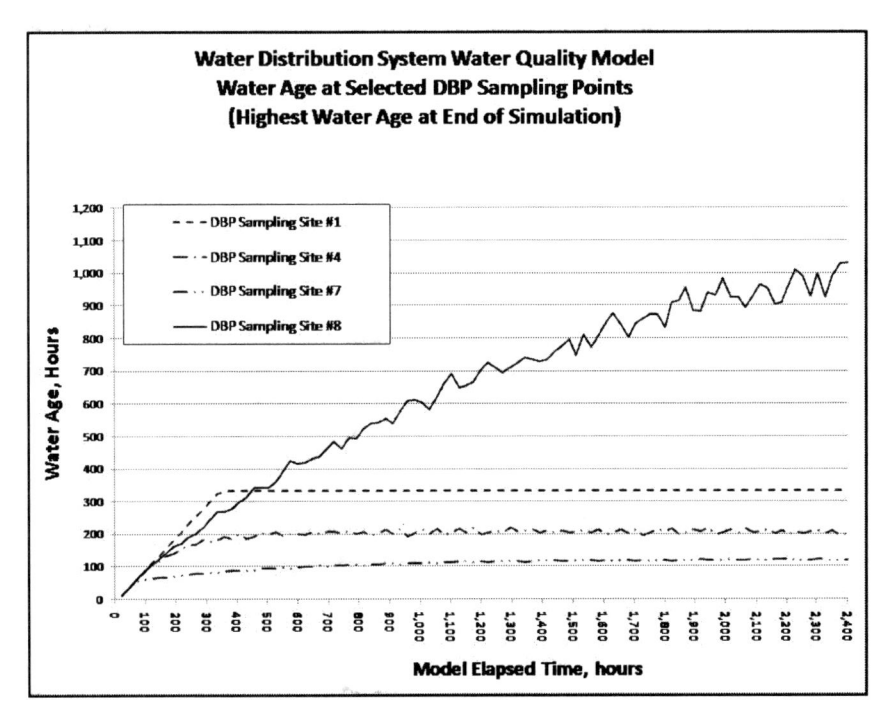

Figure 7-1 Illustration of water quality model equilibration

these two. Free chlorine is more reactive than chloramines, and its reaction kinetics have been studied more extensively. Most research in this area has shown that there is more loss of chlorine observed within the distribution system than for the same amount of water stored in an inert container over the same period of time. This implies that there are several separate reaction mechanisms for chlorine decay—one involving reactions within the bulk fluid and other involving reactions with material on or released from the pipe wall (Vasconcelos et al. 1997). A combination of these mechanisms may be occurring as the water passes through storage tanks or reservoirs, and separate reaction rates are often assigned to tanks in most modeling programs.

7.8.3.2. Bulk Decay Coefficients. The rate of free chlorine decay in the bulk fluid depends on the chemical nature of the water and its temperature. Chlorine reacts rather quickly with such inorganic compounds as iron and manganese. These types of reactions are probably completed during the time spent in a treatment plant's clearwell. Reactions with organic matter proceed at a slower pace and may not have been completed by the time the water enters the distribution system. The actual rate depends on both the amount and type of both inorganic and organic matter present at the entry point.

Free chlorine in the bulk water phase is typically represented by a first-order decay rate, which can be expressed as an exponential decay over with time as indicated by:

$$C = C_o \, e^{(-k_b t)} \qquad (7\text{-}3)$$

Where:

C_o = the initial concentration at the head of the pipe

t = the time of travel spent in the pipe

k_b = a bulk reaction coefficient with units of 1/time

Laboratory methods for measuring k_b are described later in this chapter. The k_b value can vary widely for different waters or for the same source water subjected to different treatment processes or at different times of year. Highly reactive waters have coefficients above 1.0/day, while slightly reactive waters show coefficients below 0.1/day (Hart et al. 1986; Vasconcelos et al. 1996). For the same water, k_b can roughly double for each 10-degree rise in water temperature. It is often necessary to assign different k_b values to different pipes throughout a distribution system to account for the blending of source waters with different reaction rates.

Source tracing analyses can be used to identify which parts of the system are influenced by which source. Reaction rates may be adjusted to account for blending of the source waters. Multi Species eXtension (MSX) modeling (see section 7.9) may be used in multisource systems to dynamically calculate time varying bulk decay coefficients for individual pipes based on the fraction of water emanating from different sources (Grayman et al. 2011). Bulk decay coefficients in tanks deserve careful consideration because daily temperature changes in a tank's contents affect the bulk decay coefficient and because residence times in tanks can be substantial. The way the contents within a tank are mixed affects tank residence time, which in turn affects the chlorine concentration of water leaving the tank.

7.8.3.3. Wall Demand Coefficients. Pipe wall demand for free chlorine typically occurs in unlined metallic pipes or in pipes where a significant growth of biofilm has occurred. In the former case, the demand is caused mainly by corrosion resulting in release of iron from the pipe wall, while in the latter case, the organic material is associated with the biofilm slime layer that exerts a chlorine demand. There are two approaches used for modeling wall demand. With the first approach, the value of k_b is increased to account for the increased chlorine demand. The second approach introduces a second rate coefficient, k_w, so that the overall coefficient K becomes (Rossman et al. 1994)

$$K = k_b + 2k_w/r \tag{7-4}$$

The term $2/r$, where r is the pipe radius, is the pipe surface area per unit of volume available for reaction.

The wall coefficient, k_w, with units of length divided by time, depends on the surface characteristics of the pipe, such as the amounts of exposed iron surface and biofilm buildup. Because older pipes tend to exhibit more chlorine demand than newer ones and these pipes also tend to have lower hydraulic roughness coefficients, k_w is hypothesized to be inversely proportional to a pipe's Hazen-Williams C-factor. In this case, the overall chlorine decay coefficient becomes

$$K = k_b + \alpha/(r\, C_{HW}) \tag{7-5}$$

where α is a factor that is system-dependent.

Both k_w and α are very site-specific. A study of portions of four distribution systems where wall demand was exerted found that k_w values ranged between 0.1 to 5.0 ft/day (0.03 to 1.5 m/day) and that α ranged from 10 to 650 between the different systems (Vasconcelos et al. 1996).

Some research studies have found anomalous behavior in wall demand that is not described by the methodologies previously outlined (Powell et al. 2004). Additional research is needed in this arena to refine the wall demand processes.

7.8.3.4. THM Formation Coefficients. As free chlorine reacts with organic material in water, classes of compounds known as *disinfection by-products* that include trihalomethanes (THMs) and sum of haloacetic acids (HAA5s) are formed. Members of these classes are suspected carcinogens, and the USEPA has set a maximum limit of 80 mg/L of total THM in water and a maximum limit of 60 mg/L of HAA5s in water. USEPA requires compliance sampling for these constituents in the distribution system. The amount of sampling required can be reduced through use of water quality modeling of the distribution system.

The kinetics of THM formation are rather slow, with reactions proceeding for many hours after water leaves the treatment plant. It is not uncommon for more than half the total THM production to occur in the distribution system. There is a maximum THM level that forms for each water, depending on the nature and amount of organic matter and on the chlorine level present. This amount is called the THM formation potential (THMFP). Under constant conditions of temperature and pH, THM formation with time is approximated by the following equation:

$$\text{THM} = \text{THM}_0 + (\text{THMFP} - \text{THM}_0)(1 - exp^{-k_{\text{THM}}t}) \qquad (7\text{-}6)$$

Where:

THM_0 = an initial THM concentration

k_{THM} = a formation rate constant

t = elapsed time

This constant, k_{THM}, is determined from a laboratory test similar to that used to determine the decay of residual chlorine that also determines the THMFP. This method is described later in this chapter.

Researchers have suggested alternative models for representing the formation of THMs in distribution system. Speight et al. (2009) summarize some of the alternative methodologies that have been studied.

The growth of HAA5 in a distribution system initially follows similar kinetic rates as THMs, but they are likely to follow a more complex reaction with growth in the distribution system as HAA5s have been observed to subsequently decay in some parts of distribution systems. Simulation of HAA5s was not feasible until recently with the advent of simulation of multiple species.

7.9. MODELING OF MULTIPLE SPECIES

Recent developments in modeling software allow users to define and model complex reaction schemes between multiple chemical and biological species in both the bulk flow and at the pipe wall (Shang et al. 2008a). This is referred to as an MSX. Examples of processes that could be modeled using MSX include the auto-decomposition of chloramines to ammonia, the formation of disinfection by-products like HAA5, biological regrowth including nitrification dynamics, combined reaction rate constants in multi-source systems, and mass transfer of limited oxidation pipe-wall adsorption reactions.

These developments recognize that many constituents can be found in two different physical phases within a distribution system. These are the bulk water phase relative to the motion of water in the pipes and a surface phase relative to the fixed pipe surface. Bulk phase species can be chemical or biological components that are carried in the bulk volume of water moving at the average velocity. Surface phase species are integral with or attached to the pipe wall and react with the bulk water as it passes by a fixed point on the pipe. An example of this phenomenon is the decay of residual chlorine in the bulk fluid as it reacts with natural organic matter while it also decays as it deactivates biofilms attached to the pipe wall. Other examples of bulk phase species include dissolved ions or compounds, suspended matter, and chemicals adsorbed onto suspended or colloidal particles. Other examples of surface species include ferrous iron ions on iron or steel pipes, oxidized forms of iron, calcium carbonate scale, phosphate species used for corrosion control, particulate matter that has settled out of solution or become attached to the pipe wall, or organic compounds diffusing from or to the walls of plastic pipe materials. Some of these components can exist in both phases and can be transferred to or from either phase. These are cases for the use of multiple species modeling.

Software algorithms developed to model multiple species employ these integration techniques using a choice of different methods that are applicable to different types of constituents or combinations of constituents. The user can usually specify the mathematical expressions that govern the reaction dynamics of the system being studied, which affords users the flexibility to model a wide range of chemical reactions of interest to water utilities, consultants, and researchers. For further details, please refer to the specific software user manuals, e.g., Shang et al. (2008b). Examples of use of MSX include Klosterman et al. (2009) and Helbing and VanBriesen (2009).

7.10. OBJECTIVES OF WATER QUALITY TESTING AND MONITORING

Field measurements serve multiple purposes when used to support water quality modeling:

- Model development: as a means of developing water quality reaction and transformation relationships
- Model calibration: to develop the best values for reaction and transformation parameters used by existing water quality models
- Model validation: to ascertain that a previously calibrated water quality model performs appropriately for the situation being studied
- Calibration and testing of the underlying hydraulic model

Five general types of water quality tests and measurements are discussed in this section: distribution system water quality surveys, use of previously collected (historical) field data, tracer studies, tank and reservoir field studies, and laboratory kinetic studies. Frequently, multiple types of studies are conducted in tandem to fully parameterize, calibrate, and validate a hydraulic and water quality model.

7.11. MONITORING AND SAMPLING PRINCIPLES

Water quality field testing procedures are centered on the collection and analysis of water samples. When used for regulatory purposes, sampling and analysis must follow standards prescribed by federal and state agencies. Water quality sampling done exclusively in support of water quality modeling typically is not required to follow these standards. However, to insure useful results, a quality assurance/quality control (QA/QC) program specifying methods and practices should be established and followed.

There are a variety of mechanisms for collecting representative water samples and analyzing the samples. These mechanisms are summarized in the following section.

7.11.1. Sample Collection

The objectives of sample collection are to (1) collect samples that are representative of the location or facility being sampled; and (2) avoid contamination of the sample. There are four general methods for collecting water samples.

- In-situ measurements: Instruments are inserted in the water to make direct measurements. This procedure is used for taking measurements in pipes, tanks, and reservoirs. Measurements may be taken automatically and logged or sent remotely to a receiver, or taken manually such as reading a thermometer inserted in water.

- Manual grab samples: A sample of water is manually taken from the distribution system for analysis. Dedicated sampling taps, hose bibs, sinks, and fire hydrants are generally used for manual collection into sampling containers.

- Automated grab samples: Automated sampling devices can collect multiple samples at specified time intervals and/or in response to a signal. Samples can be collected into separate containers or into a single container to collect a composite sample.

- Continuous sample collection: A continuous side stream can be extracted from a distribution system tap for subsequent automated or manual analysis.

7.11.2. Sample Analysis

There are several options for analyzing field samples. Selection of an appropriate method depends on the specific parameters being monitored, availability of equipment, required level of accuracy, cost of analysis, required expertise, and other factors.

- Manual analysis in the field: Field kits are available for manually analyzing many parameters in the field. For example, chlorine residual, pH, and temperature are typically measured in the field using such kits.

- Automated analysis in the field: For longer term analysis or as part of an intensive field study, automated instrumentation may be connected to a continuous source of flow in order to measure water quality at a prescribed frequency rate. Parameters that are frequently measured in the field using on-line monitors include chlorine, pH, temperature, conductivity, oxidation–reduction potential (ORP), turbidity, and total organic chlorine.

- Laboratory analysis: For many analyses, it is either necessary or preferred to analyze the field collected samples in the laboratory. Proper methods must be used in collecting, labeling, preserving, and transporting the samples to the laboratory.

7.11.3. Monitoring Parameters

The parameters to be monitored depend on the specific uses of the monitoring data. When used as part of a model calibration/validation process, parameters should correspond to the specific parameters being modeled and related parameters required within the calibration/validation process. The most frequently modeled water quality parameter in distribution systems is chlorine. In this case, free and/or total chlorine residual measurements should be taken in addition to temperature measurements. For modeling studies of disinfection by-products, the specific disinfection by-products should be monitored. In the case of tracer studies, the specific tracer being used (i.e., fluoride) or a surrogate parameter such as conductivity for measuring sodium or calcium chloride or UV-254 for measuring organic content should be measured. Other parameters that are sometimes taken in conjunction with a modeling study include bacteriological samples, organic compounds, and specific chemicals.

7.11.4. Monitor Locations

Selection of monitoring locations depends on a number of factors including ultimate use of the data, availability and cost of equipment and resources, and accessibility of alternative sites. For on-line equipment, additional factors include the availability of electricity, protection of the environment and a drain for side-stream water. Various mathematical methods have been employed in selecting monitor locations that best fit a set of stated criteria. These methods have been used in siting monitors or sampling locations for both water security purposes and for routine water quality sampling (Grayman 2008).

7.12. WATER QUALITY SURVEYS

Water quality surveys encompass a wide range of water-quality field data collection and analysis activities. Surveys can range in time and complexity from the collection of data at a few stations within the distribution system to intensive multi-day or longer term data collection activities at many stations and involving significant labor and equipment resources. Though the scale of these activities varies significantly, many of the planning and implementation activities for the survey are similar.

There are five major steps that apply to all water quality surveys: (1) development of a detailed sampling plan and other preparations for sampling such as procurement of materials (e.g., sampling bottles, reagents, colorimeters), development of forms, obtaining preapproval for staff overtime, and so on; (2) performance of the field sampling study; (3) post-sampling data analysis; (4) preparation of a report documenting the field study; and (5) use of the field data in the development (parameterization) of the distribution system model.

As part of the preparation of the sampling plan, the distribution system/water quality model is applied to the sampling area using the operating conditions likely to be in effect during the field study. Based on predicted results, sampling locations and sampling frequency are established. In some cases, alternative operating conditions are simulated to determine their effect on the planned sampling strategy. Even though the model is not fully calibrated and, in fact, is only an approximation of the actual system, the information gained by predicting the behavior of the system under the expected sampling conditions is frequently helpful. The following issues should be addressed in developing a sampling plan, preparing for sampling, conducting the actual sampling campaign, and the post-sampling analysis process (Clark and Grayman 1998).

7.12.1. Sampling Locations

Sampling sites should be carefully and accurately delineated in the plan. Examination of the results of the preliminary modeling runs is an important factor in selecting general locations for sampling. Samples are taken from permanent sampling taps, public buildings, businesses, residences, or water utility facilities. The locations are selected to reflect places where water quality data are desired including water sources. Other selection criteria include accessibility throughout the entire study, required flushing times, safety issues, liability, and the degree to which the sample site is representative of some portion of the distribution system. Customer plumbing greatly affects water quality, so selecting a site close to the water main is preferable. There are differing opinions on usage of hydrants for samples. While hydrants, with proper flushing, provide a convenient method for obtaining samples from a main, stagnant water in the hydrant or excessive biofilm/corrosion products typical of hydrant risers can confound the water quality data.

7.12.2. Sampling Frequency

Samples are taken manually or by using automated collection or analysis instruments. The frequency of sampling is generally governed by the availability of personnel, the number of samples to be analyzed in the laboratory, and most importantly, the expected temporal variation in water quality in the study area. For manual sampling, a sampling *circuit* is generally established in which samples are taken sequentially by a sampling crew. The circuit should be clearly marked on a map. The time it takes to sample at a single station and over the entire circuit is estimated by a preliminary test of the circuit or by estimating the times associated with each aspect of sampling. (e.g., time to flush the sampling tap, time to take and analyze a sample, time to travel between sites, etc.). The duration of a sampling program varies depending on the goals of the study, travel time across the network, and whether or not a tracer is used in the study.

7.12.3. System Operation

The operation of a water system (e.g., plant or well flow rates, tank levels, pumps, etc.) can impact the resulting water quality in the system. During a sampling study, system operation conditions are viewed as a controlled set of variables rather than a random set of occurrences. This requires close cooperation between the sampling team and the system's operations group. The system operation during the sampling event either reflects normal operating conditions or reflects a desired set of operational conditions. Operational changes affect the time required to establish a steady-state water quality condition and, therefore, the length of the sampling study.

7.12.4. Preparation of Sampling Sites

Prior to actual sampling, the sampling sites should be evaluated and prepared. Preparation could include installation of sampling taps, calculation of required flushing time, notification of owners, and marking the sites for easy identification. Valves and appurtenances are checked to ensure that they are in good working order.

7.12.5. Sample Collection Procedures

Procedures for collecting samples should be specified in the plan. These procedures include required flushing times (generally 2 to 6 minutes), methods for filling and marking sample containers, listing reagents or preservatives to be added to selected samples, methods for storing samples, and data logging procedures. The sample is only as good as the collection methods, therefore USEPA, AWWA, or other references should be consulted.

7.12.6. Collection of Ancillary Data

Use of the sampling data in conjunction with a model requires extensive knowledge of the system. Preparations should be made to collect information during the sampling study on flows, pump operations, valve settings, and any other information required by the model.

7.12.7. Analysis Procedures

Samples taken in the field are analyzed at the sampling site, at a field laboratory located in the sampling area, or in a centralized laboratory. Procedures to be followed for each type of analysis should follow commonly accepted practice and/or *Standard Methods for the Examination of Water and Wastewater* (APHA 2012) and be specified in the plan.

7.12.8. Logistical Arrangements

Logistics refers to the practical arrangements that are necessary to organize something successfully, especially something involving many people or many pieces of equipment. Logistical arrangements include the coordination of lodging for nonresident participants, provision for meals, transportation, and all the other minor details associated with any field study.

An important element of the logistics is personnel scheduling. A detailed personnel schedule should be developed for the sampling study. Ideally, the schedule includes crew assignments and a work schedule for each study participant. The personnel scheduling should include coordination with utility staff and outside personnel such as consultants and equipment providers, as appropriate.

Other logistical arrangements include coordination related to safety issues and ensuring that all necessary equipment is available. These items are discussed separately and in more detail in the following section.

7.12.9. Safety Issues

The following safety-related concerns should be addressed in preparation for the sampling campaign: notification of police and other governmental agencies; public notification; notification of customers who are directly affected; issuance of safety equipment, such as flashlights, vests, and so on; use of marked vehicles and uniforms identifying the participants as official water utility employees or contractors; and issuance of official identification cards or letters explaining their participation in the study. Whenever possible, samplers should not work alone, especially during nighttime or in dangerous areas. If sample collectors will be using hose bibs from homes that are not part of the routine monitoring programs, homeowners should be notified.

7.12.10. Data Recording

An organized method for recording all data should be devised including the design of a data recording form. Data are recorded in ink using military (24-hour) time, and the individual data sheets are numbered sequentially and transferred to a central location at frequent intervals. A date-time-location identifier should be included on both the recording sheet and on the sample containers' labels.

7.12.11. Equipment and Supply Needs

Equipment includes field sampling equipment, safety equipment, laboratory equipment, and so on. Expendable supplies include sampling containers, reagents, marking pens, batteries, and so on. As part of the sampling plan, the needs and availability of equipment and supplies are identified and alternative sources for equipment investigated. Some redundancy in equipment should be planned because equipment malfunction or loss is possible.

7.12.12. Training Requirements

Training of sampling crews is essential and should be specified as part of the sampling plan. Even though crews are composed of experienced workers, it is important that all crews be trained in a consistent set of procedures. Topics include the following: the sampling locations, sample collection and analysis procedures, data recording procedures, and contingencies. All crew members should read the sampling plan prior to the training sessions.

7.12.13. Contingency Plans

The old adage that "if something can go wrong during a sampling study, it will" serves as a good basis for contingency planning. Contingency planning includes equipment malfunction, illness of crew members, communication problems, severe weather, unexpected system operation, and customer complaints. The sampling crew should brainstorm what could go wrong during the training period and assist in developing contingencies. Backup sampling sites are a part of a contingency plan.

7.12.14. Communications and Coordination

During the sampling study, key personnel are often distributed throughout the area as follows: samplers at single stations or riding a circuit; the laboratory analysts in a permanent or field lab; operations personnel at central control stations; and the study supervisor at any of many locations. In order to coordinate actions during the study or to respond to unexpected events, some means of communication are needed. Alternatives include radios (normally available in utility vehicles), cellular phones, walkie-talkies, or the low-tech solution of a person circulating in a vehicle. Ideally, a study is coordinated through a central location where a study supervisor is located. The supervisor functions as the central coordinator, receiving information from the field, laboratory, and operations center, and making decisions and disseminating information back to the field. Decisions on changes in sampling schedules or operating procedures are made out of this center based on full knowledge of all aspects of the ongoing study.

7.12.15. Calibration and Review of Analytical Instruments

A specific plan for reviewing and calibrating all instruments is necessary, and this task must be carried out prior to the study. When multiple instruments are used to measure the same parameter, each instrument is calibrated against a standard and compared. A plan to check the calibration of instruments during the study is also essential.

7.12.16. Preparation of Data Report

Immediately following the sampling campaign, the data are examined and transferred to a computer format using spreadsheet or database management software and a comprehensive data report is prepared. The data form should have been constructed with data input in mind (i.e., rows and columns to match spreadsheet format). The purpose of the data report is to contain all pertinent information on the field study and subsequent lab analysis. This includes information on the sampling study, the study area, and the results of the lab analysis organized in a usable manner.

7.13. USE OF HISTORICAL DATA

Historical distribution system water quality data may be available, usually resulting from regulatory sampling required by the Total Coliform Rule (TCR), the Disinfectants and Disinfection By-products Rule, and other federal or state regulations. The efficacy of using such historical data in support of water quality modeling should depend on many factors, including the following:

- Assurance that the data have been correctly collected and reported

- Knowledge of the exact location and time that the data were collected

- Knowledge of the state of the distribution system operation (i.e., boundary conditions) at the time of data collection

Unlike field studies specifically designed in support of water quality modeling, the modeler has no control over practices associated with past field studies. In many cases, complete information is not known concerning the historical data. As a result, subsequent use of the data in the calibration or validation of a water quality model may lead to erroneous results.

7.14. TRACER STUDIES

Tracer studies in distribution systems involve the observation of the movement of a substance within the distribution system. The information gained from a tracer study is used to characterize the travel times (velocity) within the system and to aid in the testing and calibration of a hydraulic model.

Tracers that are used in such studies include (1) constituents that are normally added to a distribution system, such as fluoride, which are turned off for some period and traced; (2) nontoxic, conservative substances, such as sodium chloride, calcium chloride, or lithium chloride, which are injected at a known concentration; or (3) naturally occurring substances, such as hardness or UV-254, which vary between different sources.

The underlying concept behind tracer studies for use in the characterization/calibration/testing process is as follows: When modeling a conservative substance, there are essentially no water quality parameters that may be adjusted. In other words, if the hydraulic parameters are correct and the initial conditions and loading conditions for the substance are accurately known, the water quality model should provide a good estimate of the concentration of the substance throughout the network. The use of the water quality model and conservative tracer as a means of calibrating the hydraulic model is based on this relationship.

The calibration process using water quality modeling is summarized as follows:

1. An appropriate conservative tracer is identified for a distribution system. Factors that affect this decision include cost, analysis requirements, local or state regulations, and ease in handling the tracer.

2. A controlled field experiment is performed in which either (a) the conservative tracer is injected into the system for a prescribed period of time; (b) a conservative substance that is normally added, such as fluoride, is shut off for a prescribed period; or (c) a naturally occurring substance that differs between sources is traced.

3. During the field experiment, the concentration of the tracer is measured at selected locations in the distribution system along with other parameters that are required by a hydraulic model, such as tank water levels, pump operations,

flows, and so on. The tracer study is conducted in conjunction with a water quality survey. Depending on the selected tracer, measurement of the tracer is done manually using in-field analysis, is done in the laboratory on samples collected in the field, or involves the use of automated, continuous chemical analyzers and data loggers.

4. Standard means are used to adjust the parameters in the hydraulic model to represent the operations of the distribution system. The water quality model is used to model the conservative tracer.

5. If the model adequately represents the observed concentrations, this indicates the likelihood of a good calibration of the hydraulic model for the conditions being modeled. Significant deviation between the observed and modeled concentrations indicates that further calibration of the hydraulic model is required. Various statistical and directed search techniques are used in conjunction with the conservative tracer data to aid the user in adjusting the hydraulic model parameters so as to better match the observed concentrations.

Tracer studies have been successfully conducted at many test sites around the United States and in other countries (USEPA 2006a). Recently, tracer studies have been employed more frequently in support of the Initial Distribution System Evaluation (IDSE) requirement of the Stage 2 Disinfectants and Disinfection Byproducts Rule (USEPA 2006b). Because tracer studies generally require a significant amount of resources to perform, prior to embarking on tracer studies a detailed study plan should be developed and the costs and resource requirements evaluated.

7.15. TANK AND RESERVOIR FIELD STUDIES

Field studies of distribution system tanks and reservoirs may be used to both characterize the flow and mixing behavior within the facility and to assist in the calibration or validation of mathematical models of the storage facility. Field testing may be categorized as: water quality studies, tracer studies, and temperature studies. Water quality studies provide data on the temporal and spatial variation of water quality parameters within the storage facility and in the inflow and outflow. Tracer studies provide information on the mixing behavior in the facility. Temperature studies are used to gather information on how the temperature may vary at different locations and depths within the facility and over both short-term and long-term periods of time. The three types of studies may be performed by themselves or in tandem to develop a better understanding of how the reservoir behaves.

To understand the mixing phenomena in a tank or reservoir, a conservative tracer may be added to the inflow and the resulting concentration measured over time at various locations within the facility and in the outflow. When a tracer study is performed for distribution system reservoirs and tanks and to understand the internal mixing phenomena, it is best to measure the tracer at available locations within the facility in addition to the outflow. Substances that are used as tracers should be conservative (not decay over time) and must conform to local, state, and federal regulations. Fluoride, calcium chloride, sodium chloride, and lithium chloride are the usual candidates for such studies.

Temperature variations in a tank or reservoir can significantly affect the mixing characteristics in the facility. Water temperatures may be measured manually from a sample tap or water may be drawn from different depths using a pump or sampling apparatus. Alternatively, an apparatus composed of a series of thermistors and a data logger may be placed in a reservoir or tank to measure temperature at preset intervals at different depths. Figure 7-2 illustrates the results produced by three thermistors

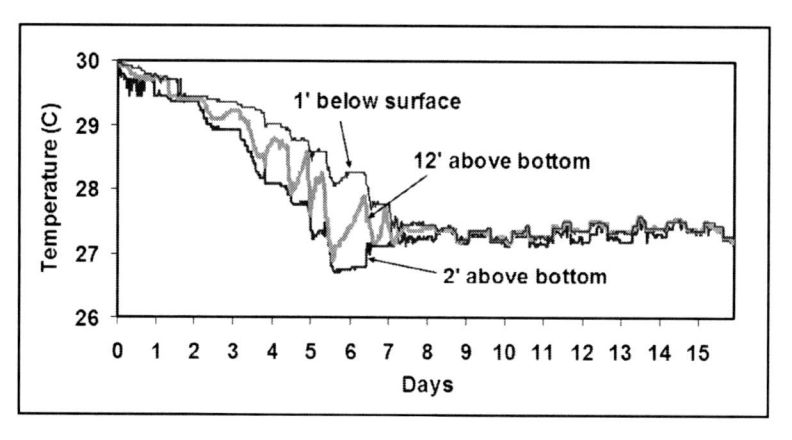

Figure 7-2 Example results from thermistor study showing temperature variation in tank

placed in a large reservoir at different elevations for a 16-day period. As illustrated, the tank displays a slight degree of stratification with a maximum temperature difference of 1.5°C that forms and then dissipates over a 5-day period.

7.16. LABORATORY KINETIC STUDIES

Laboratory tests are available for estimating the kinetics of two types of reactions represented in distribution system water quality models—bulk chlorine decay and THM formation.

7.16.1. Chlorine Decay Bottle Test

The chlorine decay bottle test measures the rate of chlorine decay that occurs in the bulk flow without any influence from the pipe wall. The test is generally conducted on finished water leaving a treatment plant or on the water that enters the zone of the distribution system being modeled. Waters from different sources undergo separate tests. Figure 7-3 lists the steps used to conduct the test. Results of the test provide an estimate of the bulk decay rate coefficient, k_b, that is used to model chlorine disappearance in a distribution system.

Similar tests can be conducted for chloramine decay. In this case, the duration of the test is likely much longer than for chlorine because of the lower reactivity of monochloramine.

7.16.2. THM Formation Test

A similar bottle test can be used to estimate rate parameters for the growth of THMs. This is a modified form of the standard SDS (simulated distribution system) test (APHA 2005) where instead of a single sample being incubated for 3 days, multiple samples are incubated together, each for a different period of time. The test is run long enough so that THM levels plateau to a constant level. This level becomes the formation potential THMFP, and the formation rate k_{THM}, is estimated by plotting the natural logarithm of THM at time t minus THM at time zero against time and computing the slope of the straight line fit through these points. Alternatively, nonlinear least squares are used to estimate k_{THM} directly from the equation

$$THM = THM_0 + (THMFP - THM_0)(1 - exp(-k_{THM}t))$$

where THM is the measured THM value at time t and THM_0 is the measured value at time zero.

```
This procedure should be followed to generate data that can be used to estimate the
rate of chlorine decay in finished drinking water.

   1. Split a single sample of finished water into 24 to 36 amber bottles of 250 ml
      size or larger. Fill the bottles so they are headspace free and cap them.

   2. Place the bottles in a small bucket or wire cage and immerse them in a water
      bath in a sink with running water so that the initial water temperature of
      the sample can be maintained.

   3. Starting from time zero, periodically remove three bottles at a time and
      analyze the contents for free chlorine. Discard bottle contents after taking
      measurements.

   4. For each group of bottles analyzed, record the time (in hours from the start
      of the test), the free chlorine concentration of their contents, and the
      temperature of the water bath.

An ideal schedule for analyzing the bottles would provide at least 10 sets of
chlorine values that cover a range from 100% down to about 25% of the initial
chlorine concentration. A typical schedule might analyze bottles at 0, 0.5, 1, 2,
3, 6, 12, 18, 24, 36, 48, 60, and 72 hours. However some adjustments might have to
be made if the intermediate results indicate that chlorine is decaying either much
more rapidly or much more slowly than anticipated.

After the test data are generated, a bulk decay coefficient can be estimated by
plotting the natural logarithm of chlorine concentration versus time, and fitting a
straight line through the points. The use of three separate chlorine readings at
each time can be used to eliminate obvious outliers. The slope of the line is the
bulk decay coefficient in units of 1/time (e.g., if time is in hours then the
coefficient has units of 1/hours). When fitting the line it is best to force it to
pass through the initial measurement at time zero. Alternatively, one could use a
nonlinear curve fitting routine, available in many commercial spreadsheet and curve
plotting software packages, to estimate k_b directly from the equation

        C = C_o exp(-k_b t)

where C is the chlorine concentration measured at time t and C_o is the measured
concentration at time zero.
```

Figure 7-3 Protocol for chlorine decay bottle test

7.17. WATER QUALITY MODELING AND TESTING CASE STUDY

The following case example describes a landmark water quality modeling study conducted in 1993. Though the study was conducted in the relatively early days of distribution system water quality modeling, the methods are still highly relevant. At the end of the case example description, there is a brief discussion on how more recent developments in testing and modeling could be used to improve this historical study.

The study illustrates how water quality data collected from an intensive short-term sampling program are used to calibrate both an extended-period hydraulic model and a dynamic water quality model of a distribution system. It also demonstrates some of the complexities involved in modeling chlorine decay in multisource systems where both bulk and pipe wall demands are present.

7.17.1. Background

At the time of the study, the North Marin Water District (NMWD) served a suburban population of approximately 53,000 people in the northern portion of Marin County, Calif. In 1993, as part of an Awwa Research Foundation project, a water quality sampling and modeling study was conducted to characterize and model chlorine decay in the distribution system (Vasconcelos 1996).

NMWD was served by two sources of water: the Russian River, via an aqueduct used throughout the year, and Stafford Lake, which provided water from approximately May through October. Russian River water entered the system with a chlorine residual in the range of 0.3 to 0.4 mg/L and a THM concentration in the range of 10 to 20 µg/L. Stafford Lake had a high humic content, and following conventional treatment and prechlorination, the water left the clearwell with a chlorine residual of approximately 0.5 mg/L and THM levels that sometimes exceeded 100 µg/L. Sodium hydroxide was added in the treatment process, resulting in average sodium levels of 23 mg/L, as compared to sodium levels in the Russian River of about 9 mg/L. The significant difference in sodium concentrations in the two sources provided an easy method for identifying the source of the water as it blended within the distribution system.

The NMWD system was divided into a series of zones and satellite systems. For the sampling and modeling study, only the primary zone (Zone I) was represented with transfers to other zones represented as external demands. Zone I contains three tanks and several pump stations used to lift water into higher pressure zones. The export to the higher zones is represented in the model by demands assigned to nodes located at the point of export.

7.17.2. Sampling Study

A 42-hour sampling study was performed in July 1993 to gather data for use in the assessment and modeling aspects of the project. Prior to conducting the study, a network model was applied to determine the projected response of the system and to aid in selecting sampling stations and frequency. A skeletonized representation of the distribution system network is presented in Figure 7-4, showing the location of the sampling stations. Flows from the two sources and tank water levels were recorded every 15 minutes by a SCADA system.

During the sampling study, samples were taken manually at approximately 2-hour intervals at 16 stations. Each sample was analyzed for temperature, sodium, and free and total chlorine. Selected samples were also analyzed for a wide range of parameters in order to characterize the water and to study THM formation. Bottle decay tests were run to determine the chlorine bulk decay coefficient for the two sources of water and for an even mixture of the two sources. All sampling results were stored in a database management system for later graphical and statistical analysis.

7.17.3. Hydraulic Validation

Following the completion of the sampling study, the initial modeling task was to test an existing hydraulically calibrated model of Zone I of the NMWD system for the conditions encountered during the study. The model was first updated to reflect recent hydraulic modifications in the network. The time history of SCADA recorded tank levels and pump station flows were used to compute total demand in the network for each hour of the sampling period. Total demands were disaggregated to individual network nodes by prorating them against demands used in the original model calibration.

Testing of the hydraulic model was done by comparing predicted and observed sodium concentrations over time throughout the system because of the blending of water from the two sources. Initial sodium levels at all nodes in the model were assigned by interpolating from the initial samples taken at the monitoring stations. During the period of sampling, the Stafford Treatment plant operated only during the day for 8- to 9-hour periods, resulting in widely fluctuating sodium concentrations throughout the system. Figure 7-5 illustrates the observed and modeled sodium concentrations at three stations: M003, which received Russian River water at all times;

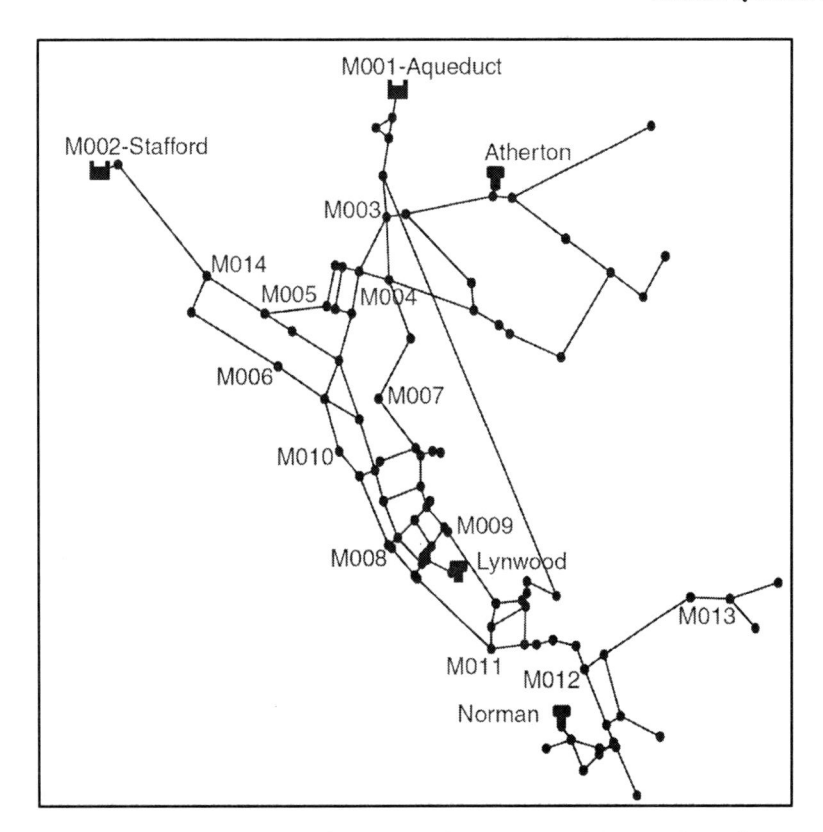

Figure 7-4 Skeletonized representation of Zone I of the North Marin Water District

M005, which received Stafford Lake water when that treatment plant was operating and Russian River water at other times; and M012, which received blended water. The study relied on visual comparison (as opposed to statistical analysis) of the modeled and observed sodium concentrations, along with a comparison of the storage tank water elevation trajectories, to determine that an acceptable level of hydraulic calibration had been obtained. For systems that do not have multiple sources of water that contain significantly different water quality "signatures," similar calibration is performed by the addition of a conservative tracer, such as fluoride.

7.17.4. Water Quality Calibration

The sampling results were used to provide an understanding of chlorine decay kinetics in the NMWD system and to develop a calibrated chlorine model of the system. The two sources of water were very different in terms of their chlorine decay kinetics, as illustrated by their bulk chlorine reaction coefficients:

Russian River:	1.32/day
Stafford Lake:	17.7/days
50/50 blend:	10.8/day

This complicated the modeling of chlorine decay within the network because most water quality models use a single bulk decay coefficient for each pipe that is applied to all water flowing through the pipe regardless of its source of origin.

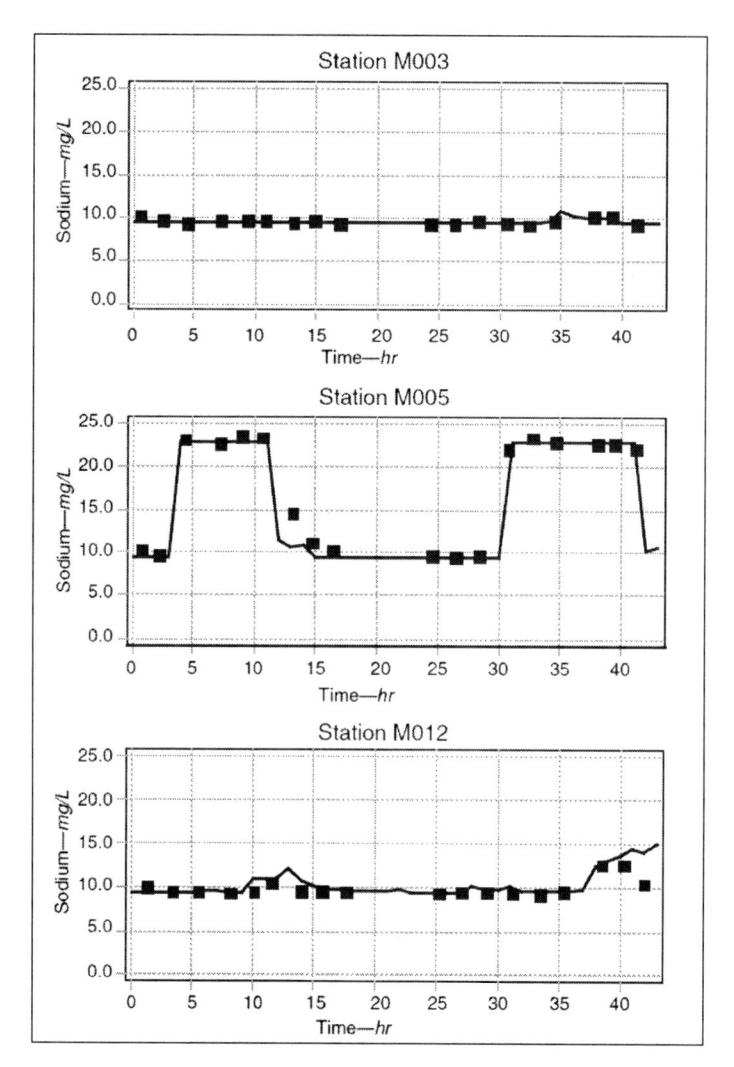

168 COMPUTER MODELING OF WATER DISTRIBUTION SYSTEMS

Figure 7-5 Comparison of observed and modeled sodium concentrations in the North Marin Water District

    To resolve this situation, a model run was made to determine what the average contribution of flow to each of the monitoring stations from Stafford Lake was over the 42-hour sampling period. The results, portrayed in Figure 7-6, show that the Stafford source serves mainly the western edge of the system. This suggested dividing the system into two zones, as shown in the figure. The pipes in the eastern zone, which receive mostly Aqueduct (Russian River) water, were assigned the bulk decay coefficient associated with the Aqueduct (Russian River) water. Because, on average, the water in the Stafford-zone pipes was close to a 50/50 blend of Aqueduct (Russian River) and Stafford water, these pipes were assigned the decay coefficient determined for the 50/50 blended water. The main line connecting Stafford to station M014 was assigned the bulk coefficient for 100 percent Stafford water because no Aqueduct (Russian River) water ever travels through this pipe. As with sodium, initial concentrations of chlorine throughout the network were assigned by interpolating from the initial sampled values at each of the monitoring stations. Inputs of chlorine from the two sources were kept at their recorded values during the duration of the sampling study.

    Initial runs of the model were made with no pipe wall demand for chlorine. When simulated chlorine concentrations were compared with measured values, it became

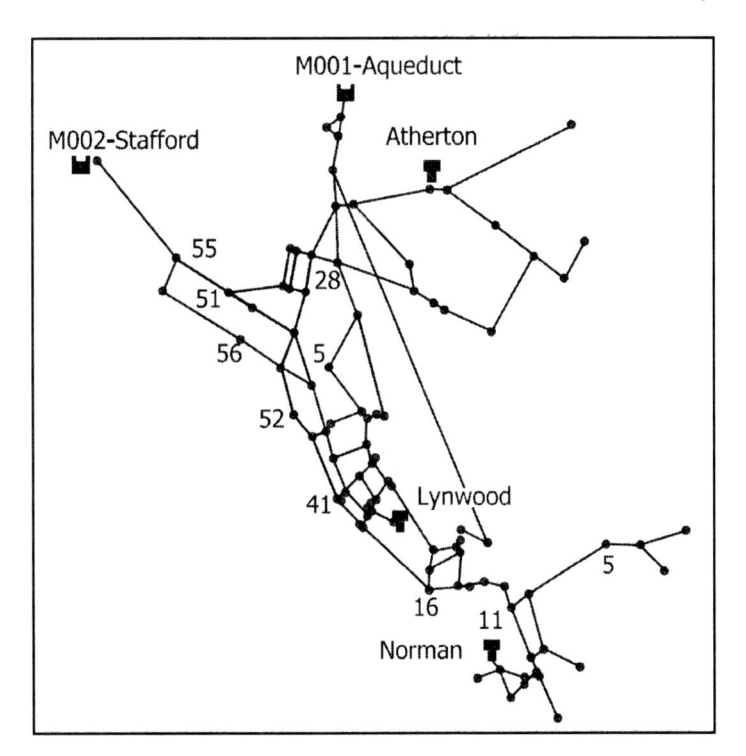

Figure 7-6 Average percent of Stafford Lake water in the North Marin Water District

apparent that chlorine levels in the Stafford zone were overpredicted. This result implied that the pipes in this portion of the system were exerting an additional chlorine demand from the pipe wall. This was logically consistent with the age and material of these pipes (40–50 year-old mortar-lined steel or cast iron) when compared to the newer asbestos–cement pipes found in the southerly and easterly portions of the system.

A wall demand was then introduced into the model for the pipes in the Stafford zone of the system. The wall coefficient was systematically varied over a range of values until the overall mean absolute error between observed and predicted chlorine values at all of the sampling locations was minimized. This occurred at a wall coefficient of 5 ft/day (1.5 m/day) producing a mean absolute error of 0.05 mg/L. Time series plots comparing the model with observed chlorine levels for several of the monitoring stations are presented in Figure 7-7.

7.17.5. Conclusions

This exercise demonstrated that a distribution system water quality model could be calibrated to reproduce conditions observed within the North Marin system during the July 1993 sampling period. Additional testing of the model, using data collected from other operating periods, was needed before it could be applied within the district.

In the period since the case study was performed, there have been some significant improvements in the testing and modeling methods available. Boccelli et al. (2004) and Sautner et al. (2005) describe the use of continuous monitors for measuring both conductivity and chlorine. Conductivity can be used to measure the presence of chemicals such as sodium chloride and calcium chloride that are injected as tracers in a distribution system or natural tracers such as sodium that differ significantly in the two sources in North Marin.

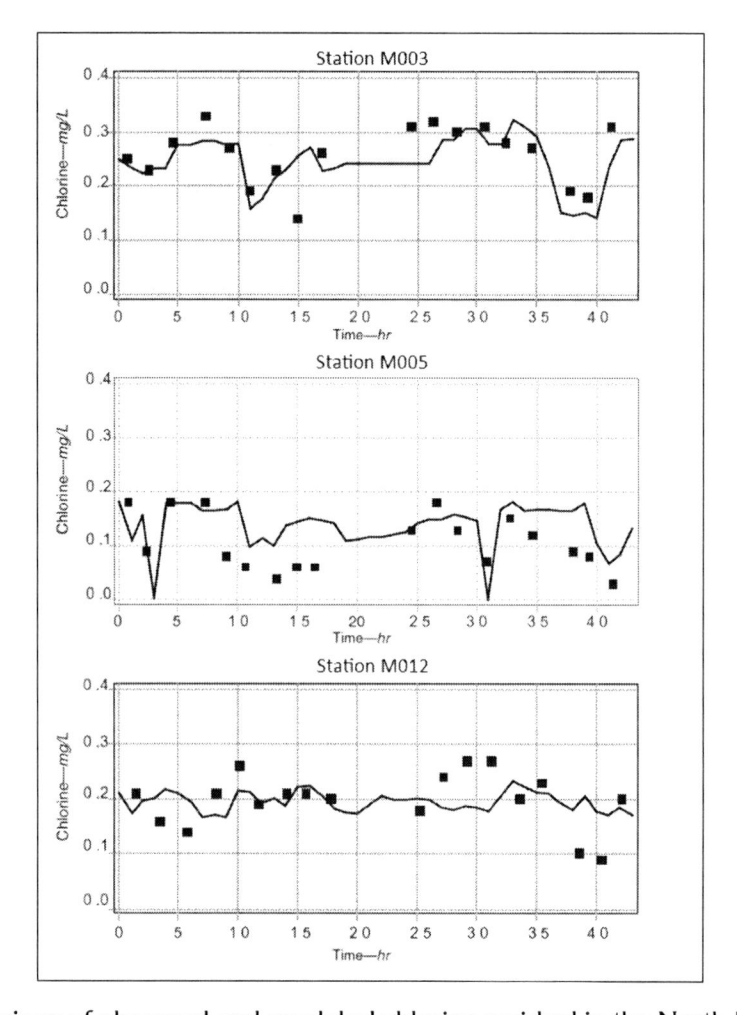

Figure 7-7 Comparison of observed and modeled chlorine residual in the North Marin Water District

There have been recent advances in water quality modeling using the Multi Species eXtension (MSX) that can lead to more accurate chlorine modeling when there are two sources with differing bulk decay coefficients (Grayman et al. 2011). In place of the "zoned" approach used in the original North Marin study to assign bulk decay coefficients to parts of the system, MSX can be used to dynamically calculate the fraction of water coming from each source to each pipe at each time step and to calculate the resulting bulk decay coefficient for the blended water.

7.18. REFERENCES

APHA. 2005. *Standard Methods for the Examination of Water and Wastewater.* 21st Edition. American Public Health Association, American Water Works Association, Water Environment Federation.

Austin, R.G., van Bloemen Waanders, B., McKenna, S., and Choi, C.Y. 2008. Mixing at Cross Junctions in Water Distribution Systems – Part II. An Experimental Study. *ASCE Journal of Water Resources Planning and Management,* 134:3:295–302.

Boccelli, D.L., Shang, F., Uber, J.G., Orcevic, A., Moll, D., Hooper, S., Maslia, M., Sautner, J., Blount, B., and Cardinali, F. 2004. Tracer Tests for Network Model Calibration. In *Proceedings of the 2004 World Water and Environmental Resources Congress, American Society of Civil Engineers* [CD ROM document]. Salt Lake City, Utah.

Clark, R.M., and Grayman, W.M. 1998. *Modeling Water Quality in Drinking Water Distribution Systems.* Denver, Colo.: American Water Works Association.

Grayman, W., Kshirsagar, S., Rivera-Sustache, M., and Ginsberg, M. 2011. An Improved Water Distribution System Chlorine Decay Model Using EPANET MSX. In *Modeling of Urban Water Systems*, Monograph 20. Ed., W. James. Guelph, Ont.: CHI.

Grayman, W.M. 2008. Designing an Optimum Water Monitoring System. In *Wiley Handbook of Science and Technology for Homeland Security*. Ed., J.G. Voeller. New York: John Wiley & Sons.

Hart, F.L., Meander, J.L., and Chiang, S.M. 1986. CLNET—A Simulation Model for Tracing Chlorine Residuals in a Potable Water Distribution Network. In *Proc. Distribution System Symposium.* Denver, Colo.: AWWA.

Helbing, D.E., and VanBriesen, J.M. 2009. Modeling Residual Chlorine Response to a Microbial Contamination Event in Drinking Water Distribution Systems. *ASCE Journal of Environmental Engineering*, 135(10):918–927.

Klosterman, S., Hatchett, S., Murray, R., Uber, J., and Boccelli, D. 2009. Comparing Single- and Multi-Species Water Quality Modeling Approaches for Assessing Contamination Exposure in Drinking Water Distribution Systems. In *Proceedings, World Environmental and Water Resources Congress, Kansas City, MO, May 17–21, 2009.* Reston, Va: American Society of Civil Engineers (ASCE).

Powell, J., Clement, J., Brandt, M., Casey, R., Holt, D., Grayman, W., and LeChevallier, M. 2004. *Predictive Models for Water Quality in Distribution Systems.* Denver, Colo.: Awwa Research Foundation.

Romero-Gomez, P., Ho, C.K., and Choi, C.Y. 2008. Mixing at Cross Junctions in Water Distribution Systems—Part I. A Numerical Study. *ASCE Journal of Water Resources Planning and Management* 134:3:284–294.

Rossman, L.A., Clark, R.M., and Grayman, W.M. 1994. Modeling Chlorine Residuals in Drinking-Water Distribution Systems. *Journal of Environmental Engineering, ASCE Journal of Water Resources Planning and Management*, 120:4:803–820.

Rossman, L.A., and Boulos, B.F. 1996. Numerical Methods for Modeling Water Quality in Distribution Systems: A Comparison. *ASCE Journal of Water Resources Planning and Management*, 122:2:137–146.

Sautner, J.B., Maslia, M.L., Valenzuela, C., Grayman, W.M., Aral, M.M., and Green, Jr., J.W. 2005. Field Testing of Water-Distribution Systems at U.S. Marine Corps Base, Camp Lejeune, North Carolina, in Support of an Epidemiologic Study. In *Proceedings, ASCE/EWRI Congress 2005, May 15–19, 2005, Anchorage, Alaska.*

Shang, F., Uber, J.G., and Rossman, L.A. 2008a. Modeling Reaction and Transport of Multiple Species in Water Distribution Systems. *Environ. Sci. Technol.*, 42(3):808–814.

Shang, F., Uber, J.G, and Rossman, L.A. 2008b. *EPANET Multi-Species Extension User's Manual.* USEPA Office of Research and Development—National Homeland Security Research Center. EPA/600/S-07/021. Revised October 2008. Washington, D.C.: USEPA.

Speight, V., Uber, J., Grayman, W., Martel, K., Friedman, M., Singer, P., and DiGiano, F. 2009. *Probabilistic Modeling Framework for Assessing Water Quality Sampling Programs.* Denver, Colo.: Water Research Foundation.

Summers, R.S., Hooper, S.M., Shukairy, H.M., Solarik, G., and Owen, D. 1996. Assessing DBP Yield: Uniform Formation Conditions. *Jour. AWWA*, 88:6:80–93.

Tzatchkov, V.G., Buchberger, S.G., Li, Z., Romero-Gomez, P., and Choi, C. 2009. Axial Dispersion in Pressurized Water Distribution Networks—A Review. In *Proc. International Symposium on Water Management and Hydraulic Engineering. Ohrid/Macedonia, 1–5 September 2009.*

USEPA. 2006a. *Water Distribution System Analysis: Field Studies, Modeling and Management—A Reference Guide for Utilities.* Washington, D.C.: USEPA. http://www.epa.gov/nrmrl/pubs/600r06028/600r06028.pdf

USEPA. 2006b. *Evaluation Guidance Manual for the Final Stage 2 Disinfectants and Disinfection Byproducts Rule.* Washington, D.C.: USEPA. http://www.epa.gov/safewater/disinfection/stage2/pdfs/guide_idse_full.pdf

Vasconcelos, J.J. 1996. *Characterization and Modeling of Chlorine Decay in Distribution Systems.* Denver, Colo.: Awwa Research Foundation.

Vasconcelos, J.J., Rossman, L.A., Grayman, W.M., Boulos, P.F., and Clark, R.M. 1997. Kinetics of Chlorine Decay. *Jour. AWWA,* 89:7:54–65.

Wood, D.J. and Ormsbee, L.E. 1989. Supply Identification for Water Distribution Systems. *Jour. AWWA,* 81:7:74–80.

AWWA MANUAL M32

Chapter **8**

Transient Analysis

8.1. SYNOPSIS

Transients can introduce large pressure forces and rapid fluid accelerations into a piping system. These disturbances may result in pump and device failures, system fatigue or pipe ruptures, and backflow/intrusion of untreated and possibly hazardous water. Many transient events can lead to water-column separation, which can result in catastrophic pipeline failures. Thus, transient events can cause health risks and can lead to increased leakage, interrupted service, decreased reliability, and breaches in the piping system integrity. Transient flow simulation has become an essential requirement for assuring safety and the safe operation of drinking water supply and distribution systems.

This chapter introduces the concept and fundamentals of hydraulic transients, including the causes of transients, general rules to help determine whether or not the system may be exposed to unacceptable conditions under a transient event, governing transient equations, numerical solution methods, guidelines for control and suppression of transients, transient modeling considerations, and transient data requirements. Illustrative examples are also discussed and conclusions are stated.

The chapter is therefore geared toward engineers involved in the planning, design, and operation of water supply and distribution systems, and engineers who need an insight into the most common causes of hydraulic transients and suitable methods that can be applied to alleviate their consequences. Such capabilities will greatly enhance the ability of water utilities to evaluate cost-effective and reliable water supply protection and management strategies for preserving system hydraulic and water quality integrity, preventing potential problems, and safeguarding public health.

8.2. INTRODUCTION

Most people have been in an older house with pipes that rattle when someone turns off a faucet. When the faucet handle turns, closing the valve almost instantaneously, the pipes rattle against the walls. This is called *water hammer*, which is also referred to as a *surge* or as a *hydraulic transient*. Water hammer refers to rapid and often large pressure and flow fluctuations resulting from transient flow conditions in pipes transporting fluids. Transient flow analysis of the piping system is often more important than the analysis of the steady-state operating conditions that engineers normally use as the basis for system design. Transient pressures are most significant when the rate of flow is changed rapidly, such as rapid valve closures or pump stoppages. Such flow disturbances, whether caused by design or accident, may create traveling pressure waves of excessive magnitude. These transient pressures are superimposed on the steady-state conditions present in the line at the time the transient occurs. The total force acting within a pipe is obtained by summing the steady-state and transient pressures in the line. The severity of transient pressures must thus be accurately determined so that the pipes can be properly designed to withstand these additional shock loads. In fact, pipes are often characterized by pressure ratings (or pressure classes) that define their mechanical strength and have a significant influence on their cost.

Transient events may be associated with equipment failure, pipe rupture, separation at bends, and the introduction of contaminated water into the distribution system via unprotected cross-connections or intrusion. High-flow velocities can remove protective scale and tubercles and increase the contact of the pipe with oxygen, all of which will increase the rate of corrosion. Uncontrolled pump shutdown can lead to the undesirable occurrence of cavitation and water-column separation, which can result in catastrophic pipeline failures due to severe pressure rises following the collapse of the vapor cavities. Vacuum conditions can create high stresses and strains that are much greater than those occurring during normal operating regimes. They can cause the collapse of thin-walled pipes or reinforced concrete sections, particularly if these sections were not designed to withstand such strains (e.g., pipes with a low pressure rating).

Cavitation occurs when the local pressure is lowered to the value of vapor pressure at the ambient temperature. At this pressure, gas within the liquid is released and the liquid starts to vaporize. When the pressure recovers, liquid enters the cavity caused by the gases and collides with whatever confines the cavity (i.e., another mass of liquid or a fixed boundary) resulting in a pressure surge. In this case, both vacuum and strong pressure surges are present, a combination that may result in substantial damage. The main difficulty here is that accurate estimates are difficult to achieve, particularly because the parameters describing the process are not yet determined during design. Moreover, the vapor cavity collapse cannot be effectively controlled. In less drastic cases, strong pressure surges may cause cracks in internal lining or damage connections between pipe sections and, in more serious cases, can destroy or cause deformation to equipment such as pipeline valves, air valves, or other surge protection devices. Sometimes the damage is not realized at the time but results in intensified corrosion that, combined with repeated transients, may cause the pipeline to collapse in the future. Transient events in pipelines also damage seals that often lead to increase leakage and significant water loss.

Transient events can have significant water quality and health implications. These events can generate high intensities of fluid shear and may cause resuspension of settled particles as well as biofilm detachment. Moreover, low pressure caused by transients may promote the collapse of water mains; leakage into the pipes at loose joints, cracks, and seals under subatmospheric conditions; backsiphonage at cross-connections; and potential intrusion of untreated, possibly contaminated groundwater in the distribution system. Pathogens or chemicals in close proximity to the pipe can become potential contamination sources, where continuing consumption or leakage can pull contaminated water into the depressurized main.

Recent studies have confirmed that soil and water samples collected immediately adjacent to water mains can contain various levels of microorganisms, an indicator of fecal pollution (fecal coliforms, *E. coli*, *Clostridium perfringens*, coliphages) and in some cases enteric viruses (Besner et al. 2008; Karim et al. 2003; Kirmeyer et al.

2001). This is especially significant in systems with leaking pipes below the water table. Problems with low or negative pressure transients have been reported in the literature (Walski and Lutes 1994; LeChevallier et al. 2003). Gullick et al. (2004) studied transient pressure occurrences in actual distribution systems and observed 15 surge events that resulted in a negative pressure. Hooper et al. (2006) and Besner et al. (2007) also reported such events in full-scale systems.

The most often identified cause for the events reported in the literature was the sudden shutdown of pumps, either unintentional (power failure) or intentional (pump tests). Using a pilot-scale test rig, Friedman et al. (2004) confirmed that negative pressure transients can occur in the distribution system and that the intruded water can travel downstream from the site of entry. Locations with the highest potential for intrusion were identified as sites experiencing leaks and breaks, areas of high water table, and flooded air-vacuum valve vaults. Preliminary results by Besner et al. (2007) showed that significant concentrations of indicator microorganisms could be detected in the water found in flooded air-vacuum valve vaults. In the event of a large intrusion of pathogens, the chlorine residual normally sustained in drinking water distribution systems may be insufficient to disinfect contaminated water, which could lead to damaging health effects. A recent case study in Kenya (Ndambuki 2006) showed that in the event of a 0.1 percent raw sewage contamination, the available residual chlorine within the distribution network would not render the water safe.

Transient events that can allow intrusion to occur are caused by sudden changes in the water velocity due to loss of power, sudden valve or hydrant closure or opening, a main break, fire flow, or an uncontrolled change in on/off pump status (Boyd et al. 2004). Transient-induced intrusions can be minimized by knowing the causes of pressure surges, defining the system's response to surges, and estimating the system's susceptibility to contamination when surges occur (Friedman et al. 2004). Therefore, water utilities should never overlook the effect of pressure surges in their distribution systems. Even some common transient protection strategies, such as relief valves or air chambers, if not properly designed and maintained, may permit pathogens or other contaminants to find a "backdoor" route into the potable water distribution system. Any optimized design that fails to properly account for pressure surge effects is likely to be, at best, suboptimal, and at worst completely inadequate.

Pressure transients in water distribution systems are inevitable and will normally be most severe at pump stations and control valves, in high-elevation areas, in locations with low static pressures, and in remote locations that are distanced from overhead storage (Fleming et al. 2006; Friedman et al. 2004). All systems will, at some time, start up, switch off, undergo unexpected flow changes, and will likely experience the effects of human errors, equipment breakdowns, earthquakes, or other risky disturbances. Although transient conditions can result in many abnormal situations and breaches in system integrity, the engineer is most concerned with those that might endanger the safety of a plant and its personnel, that have the potential to cause equipment or device damage, or that result in operational difficulties or pose a risk to the public health.

Transient pressures are difficult to predict and are system dependent, including specific system layout, configuration, design, and operation. Engineers must carefully consider all potential dangers for their pipe designs and estimate and eliminate the weak spots. They should then perform a detailed transient analysis to make informed decisions on how best to strengthen their systems and ensure safe, reliable operations (McInnis and Karney 1995; Karney and McInnis 1990).

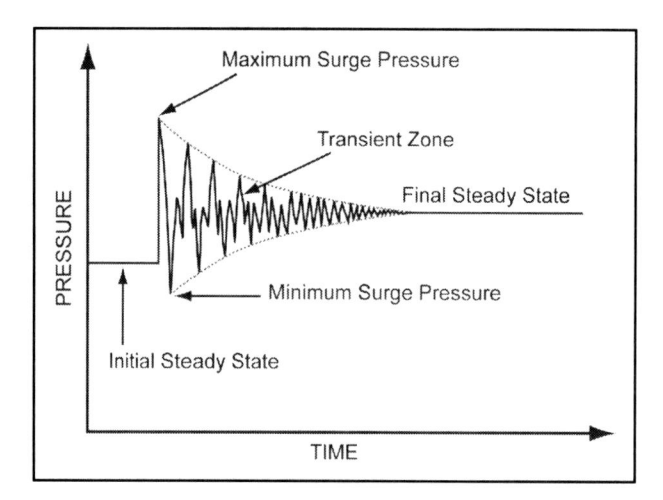

Figure 8-1 Example steady-state transition after a period of rapid transients

8.3. CAUSES OF TRANSIENTS

Transient events are disturbances in the water flow caused during a change in operation, typically from one steady-state or equilibrium condition to another (Figure 8-1). The principal components of the disturbances are pressure and flow changes at a point that cause propagation of pressure waves throughout the distribution system. The pressure waves travel with the velocity of sound (acoustic or sonic speed), which depends on the elasticity of the water and that of the pipe walls. As these waves propagate, they create transient pressure and flow conditions. Over time, damping actions and friction reduce the waves until the system stabilizes at a new steady state. Normally, only extremely slow flow regulation can result in smooth transitions from one steady state to another without large fluctuations in pressure or flow.

8.3.1. Basics—Rapid Changes in Velocity

In general, any disturbance of the flow of water generated during a change in mean flow conditions will initiate a sequence of transient pressures (waves) in the pipe system. Disturbances will normally originate from changes or actions that affect fluid devices or boundary conditions. Typical events that require transient considerations include:

- Pump shutdown or pump trip (loss of power)
- Pump startup
- Valve opening or closing (variation in cross-sectional flow area)
- Changes in boundary pressures (e.g., losing overhead storage tank, adjustments in the water level at reservoirs, pressure changes in tanks, etc.)
- Rapid changes in demand conditions (e.g., hydrant flushing)
- Changes in transmission conditions (e.g., main break)
- Pipe filling or draining—air release from pipes
- Check valve or regulator valve action

In municipal water systems, most surge problems occur as a result of closing (or opening) valves too rapidly or when pumps trip due to an unplanned power failure. Both high and low surge pressures may cause problems. If special precautions are not taken, the magnitude of the resulting transient pressures can be sufficient to cause severe damage. Figures 8-2 to 8-5 describe four typical hydraulic transient problems. The problem of shutting down a pump is illustrated in Figure 8-2. When the pump is suddenly shut down, the pressure at the discharge side of the pump rapidly decreases and a negative pressure wave (which reduces pressure) begins to propagate down the pipeline toward the downstream reservoir causing low pressures at the pump and elsewhere in the system. When the negative pressure wave reaches the high point (which already has a relatively low pressure due to the higher elevation) in the pipe, the pressure can drop below atmospheric to reach vapor pressure. At this pressure, gas within the water is gradually released and the water starts to vaporize (water-column separation). On subsequent cycles of the transient when the pressure recovers, the cavity can collapse generating a large pressure surge spike. On the suction side of the pump, the solid sloping line represents the initial hydraulic grade and the dashed straight line depicts the final hydraulic grade, while startup transients are not shown.

It should be noted that when the pipeline velocity reverses and the water column returns toward the pump, it is suddenly stopped by the check valve on the discharge side of the pump causing very high pressures. If no check valve exists, the pump can spin backward, perhaps reaching speeds that can be damaging to the equipment. During normal pump operation, transients can typically be controlled by using slow closing and slow opening pump control valves, or with variable frequency drives (VFDs) and soft start/stop controllers on the pump motors.

The problem of pump startup transient is illustrated in Figure 8-3. When a pump is started, the pressure at the discharge side of the pump rises, sending a positive pressure wave (which increases pressure) down the pipeline toward the downstream reservoir. The resulting peak pressure can cause the pipe to collapse if the pressure rating of the pipe is less than the maximum surge pressure. When the initial positive pressure wave reaches the downstream reservoir, it is converted into a negative pressure wave that propagates back to the pump and may induce cavitation. On the suction side of the pump, the solid straight line represents the initial hydraulic grade and the dashed sloping line depicts the final hydraulic grade, while shutdown transients are not shown.

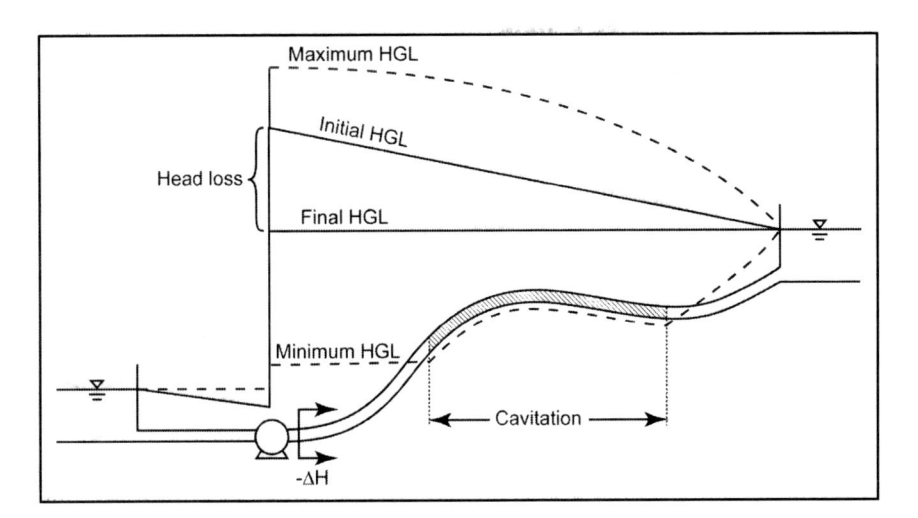

Figure 8-2 Transient caused by pump shutdown

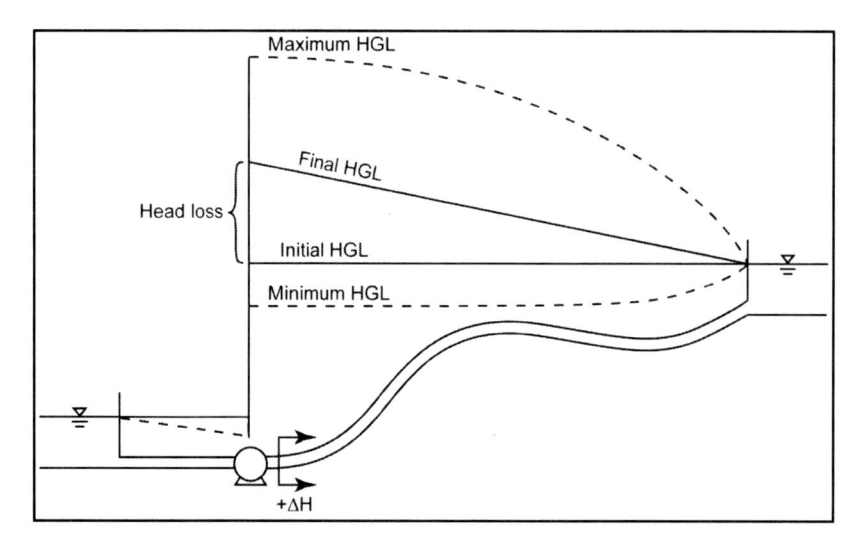

Figure 8-3 Transient caused by pump startup

Pipe filling is especially vulnerable to pressure transients when the pipeline is empty and large amounts of air are expelled during startup. Pump startup transients in pressurized pipelines are not commonly a problem, because the pumps can be started against a closed pump control valve, and then the valve can be slowly opened, or variable frequency drives can be used on the pump motors.

Opening a valve and closing a valve too fast can also result in severe hydraulic transients and are illustrated in Figures 8-4 and 8-5, respectively. When the valve in Figure 8-4 is rapidly opened, a negative pressure wave is initiated at the valve and propagates upstream toward the reservoir decreasing the pressure in the pipe. Similar to the pump shutdown scenario, the initial negative surge can drop to vapor pressure causing cavitation in the pipe. In the second example (Figure 8-5), rapidly closing the downstream valve generates a positive pressure wave at the valve that propagates toward the upstream reservoir increasing the pressure in the pipe.

In municipal systems, opening and closing of hydrants too quickly will sometimes cause unacceptable transients. In Puerto Rico, the testing of a 60-in. (152 cm) butterfly valve caused the 72-in. (183 cm) pipeline to rupture due to a rapid valve closure (Figure 8-6a). Walski (2009) showed that increasing hydrant-closing time can make a dramatic difference in pressure surges in the distribution system, greatly decreasing the impacts of hydrant closure.

Sudden and complete failure of a single pipe may also cause other unacceptable pressures elsewhere in the water system. A pipeline rupture sends low pressure waves propagating in both directions from the break. This can produce negative pressures and cavitation at higher elevations. Sometimes a pipe rupture on thin-walled pipeline will cause a pipe collapse elsewhere in the pipeline. The low pressure waves will get reflected from nearby tanks and reservoirs as positive pressure waves that in turn can also damage pipes in other locations. Other real life examples of catastrophic failures due to surge pressures are shown in Figures 8-6b and 8-6c.

Pipe systems must be designed to handle both normal and abnormal operating conditions. If an analysis indicates that severe transients may exist, the main solution techniques generally used to mitigate transient conditions are (Boulos et al. 2006; Wood et al. 2005a; Walski et al. 2003) as follows:

- Installation of stronger (higher pressure class) pipes
- Rerouting of pipes
- Improvement in valve and pump control/operation procedures

TRANSIENT ANALYSIS 179

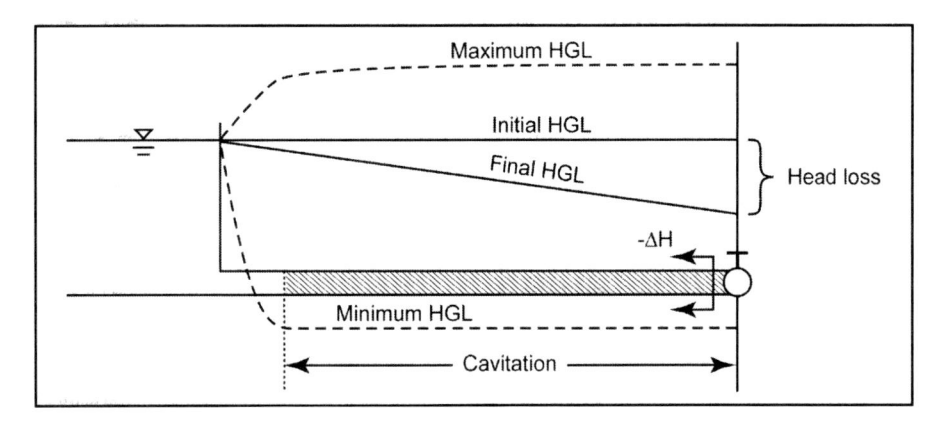

Figure 8-4 Transient caused by rapid valve opening

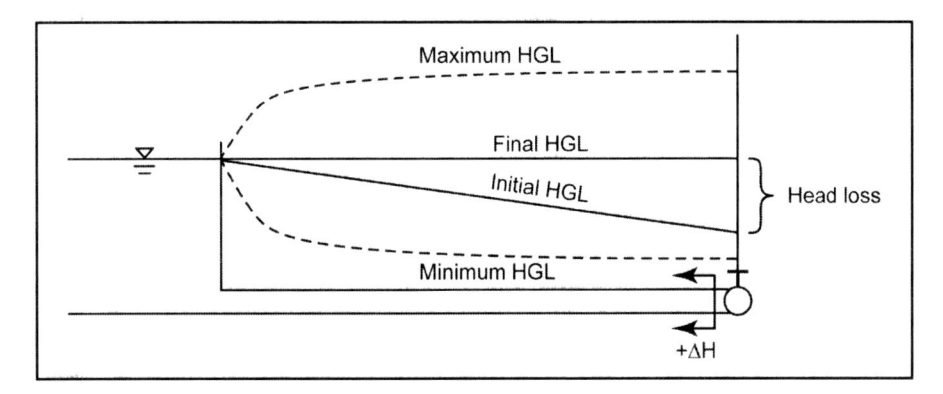

Figure 8-5 Transient caused by rapid valve closure

Figure 8-6a Rupture caused by valve closure (Superaqueduct of Puerto Rico)

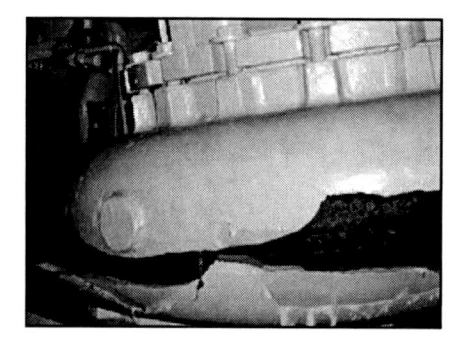

Figure 8-6b Damaged pump bowl

Figure 8-6c Broken air admission valve

- Limiting the pipeline velocity
- Reducing the wave speed (e.g., different pipe material)
- Increasing pump inertia (e.g., fitting a flywheel between the pump and motor)
- Design and installation of surge protection devices

8.3.2. Contributing Factors

The severity of and potential for surge problems depend on contributing factors such as the length of pipeline, the shape of the pipeline profile, the location of pipeline "knees," the static head, and the initial or steady-state velocities. Figure 8-7 compares varying pipeline profiles, showing the best profile with the least potential for surge problems compared to pipeline profile shapes that have increasingly higher potential for surge problems.

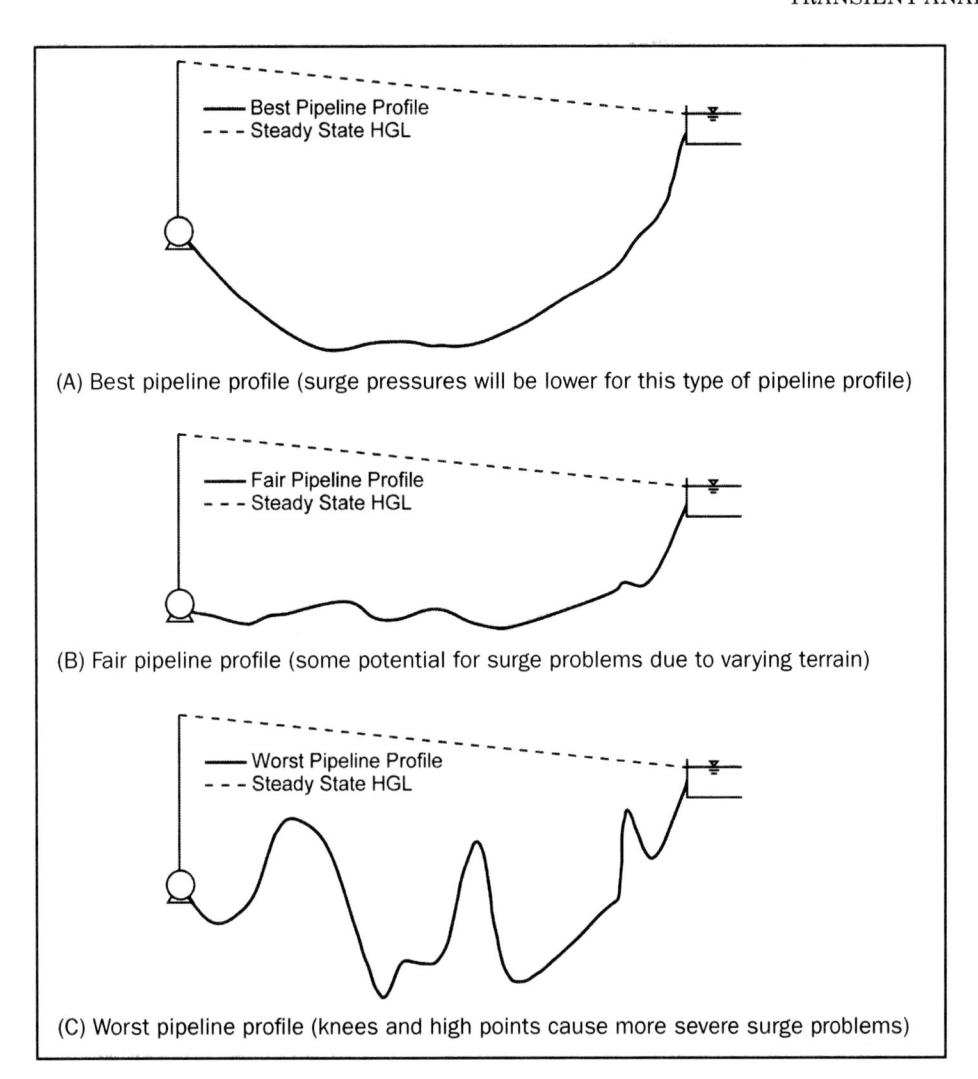

Figure 8-7 Varying pipeline profiles

8.3.3. Rules of Thumb for Identifying Vulnerable Systems

The following rules of thumb are intended as an aid in identifying one or more possible conditions that tend to cause surge problems in a water distribution system:

- Pipelines without demands or water takeoffs with lengths greater than 1,000 ft (305 m)
- Static head greater than 40 ft (12 m)
- Steady-state velocities greater than 2 ft/sec (0.6 m/sec)
- Pipeline profiles with "knees" and high points
- In-line booster pump stations with long suction lines
- Some air valves, float operated, allowing rapid discharge of air during power failure

Systems with high velocities (> 5 ft/sec or 1.5 m/sec) can also be vulnerable even with much shorter pipelines. It should be noted that exceeding any one of the general rules of thumb indicates a potential for surge problems, and that prudent design should include a detailed surge analysis. Jung et al. (2007a) studied the need for comprehensive transient analysis of water distribution systems and concluded that only systematic and informed surge analysis can be expected to resolve the complex transient characterizations and adequately protect water distribution systems from the vagaries and challenges of rapid transient events.

8.3.4. Dead-End Pipelines

Dead-end pipelines are generally needed to feed into new developments. Although initial velocities are relatively low, these dead ends can sometimes be a source of severe transients as hydrants are opened and closed for flushing. Dead ends, which may also be caused by closure of check valves, lock pressure waves into the system in a cumulative fashion, and wave reflections will double both positive and negative pressures. For example, when a surge wave of approximately 300 ft (91 m) reaches a dead end, that dead end will cause a positive pressure wave reflection of 300 ft (91 m). Because the incoming and outgoing waves are additive, the dead end experiences the doubling of the surge pressure or 600 ft (183 m). Therefore, dead ends constitute some of the most vulnerable locations for objectionable pressures and should be carefully considered in a surge analysis. Additionally, if the ground elevation of the dead end is sufficiently high, cavitation can occur under a transient event. Figures 8-8 and 8-9 illustrate this situation. The system shown in Figure 8-8 has a varying terrain with very low flow rates and long lengths of dead-end pipes. There are approximately 15,000 ft (4,572 m) between the pumping station and the hydrant, with 364 ft (111 m) of static head. All pumps in the pump station have quick-closing check valves on their discharge sides. Without any surge control, a normal pump shutdown can cause column separation, vapor pressures, and unacceptable pressure fluctuations as shown in Figure 8-9. The effects of dead ends on surge analysis were studied in detail by Jung et al. (2007b).

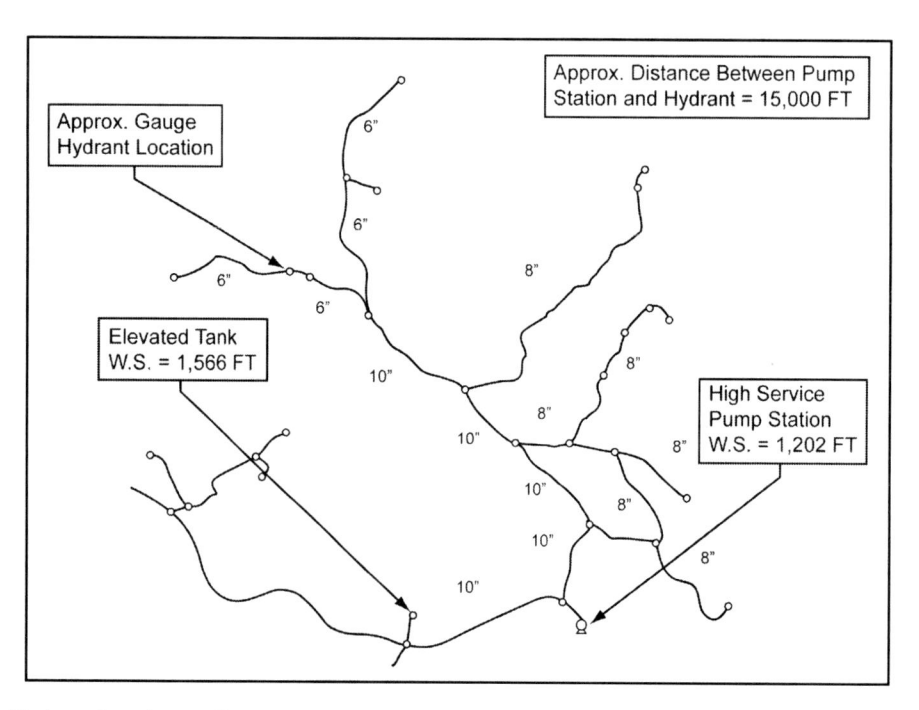

Figure 8-8 Network schematic

8.4. BASIC PRESSURE WAVE RELATIONS

8.4.1. Wave Action in Pipes

The relationship between pressure change (ΔP) and flow change (ΔQ), which is associated with the passage of a pressure wave, defines the transient response of the pipe system and forms the basis for the development of the required mathematical expressions (Boulos et al. 2006; Wood et al. 2005a; Walski et al. 2003; Wylie and Streeter 1993; Chaudhry 1979). Figure 8-10 shows flow and pressure conditions, which exist a short time Δt apart, as a pressure wave of magnitude ΔP propagates a distance Δx in a liquid filled line.

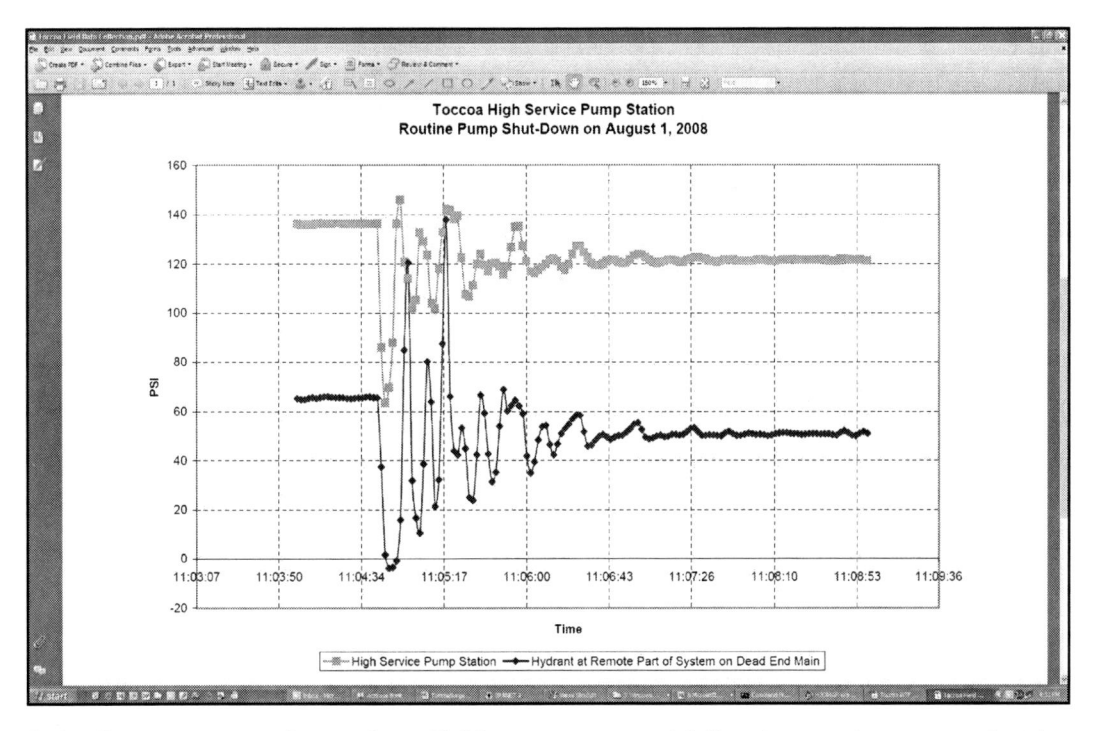

Figure 8-9 Pressure surge fluctuations (field measurements) following routine pump shutdown

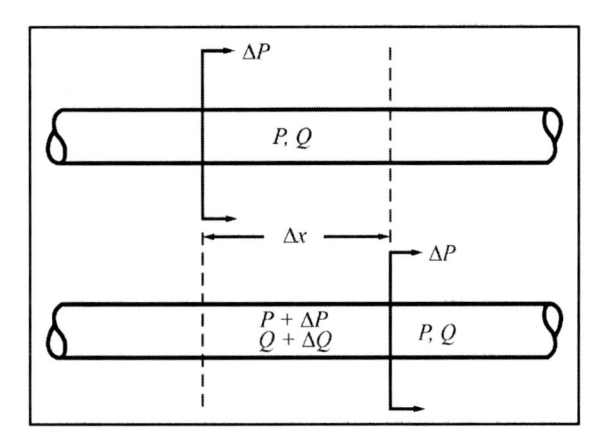

Figure 8-10 Pressure wave propagation in a pipe

During the short time Δt, the pressure on the left side of the wave front is $P + \Delta P$ while the right side of the wave front is P. This unbalanced pressure causes the fluid to accelerate. The momentum principle is:

$$(P+\Delta P-P)A = \rho\Delta x \frac{\Delta Q}{\Delta t} \tag{8-1}$$

where A is the pipe cross-sectional area; and ρ is the liquid density. Canceling and rearranging give:

$$\Delta P = \rho \Delta Q \frac{\Delta x}{A \Delta t} \tag{8-2}$$

The term $\Delta x/\Delta t$ is the propagation speed of the pressure wave. The wave speed is equal to the sonic velocity (c) in the system if the mean velocity of the liquid in the line is neglected. Because the mean velocity of the liquid is usually several orders of magnitude smaller than the sonic velocity, this is acceptable. Thus:

$$\Delta P = \frac{\rho c \Delta Q}{A} \tag{8-3}$$

or in terms of pressure head:

$$\Delta H = \frac{c \Delta Q}{gA} \tag{8-4}$$

or in a more general form

$$\Delta H = \pm \frac{c}{g} \Delta V \tag{8-5}$$

where g is the acceleration of gravity. The resulting head rise equation is called the *Joukowsky relation*, sometimes called the *fundamental equation of water hammer*. The equation is derived with the assumption that head losses due to friction are negligible and no interaction takes place between pressure waves and boundary conditions at the end points of the pipe. The negative sign in this equation is applicable for a disturbance propagating upstream and the positive sign for one moving downstream. Because values of wave speed in many pipelines are in the range of 3,000–4,000 ft/sec (915–1,220 m/sec), typical values of c/g in Eq. 8-5 are large, often 100 or more. Thus, this relationship predicts large values of head rise that highlights the importance of transient analysis. For example, if an initial velocity of 3 ft/sec (0.9 m/sec) is suddenly arrested at the downstream end of pipeline and c/g equals 100 m/sec, a head rise of 300 ft (91 m) will result.

8.4.2. Wave Speed

The wave speed c for a liquid flowing within a line is influenced by the elasticity of the line wall. For a pipe system with some degree of axial restraint a good approximation for the wave propagation speed is obtained using (Thorley 1991):

$$c = \sqrt{E_f/\rho(1+K_r E_f D/E_c t_l)} \tag{8-6}$$

where E_f and E_c are the elastic modulus of the fluid and conduit, respectively; D is the pipe diameter; t_l is the pipe thickness; and K_r is the coefficient of restraint for longitudinal pipe movement. Typically, three types of pipeline support are considered for restraint. These are

Case a: The pipeline is restrained at the upstream end only.

$$K_r = 1 - \mu_p/2 \tag{8-7}$$

Case b: The pipeline is restrained throughout.

$$K_r = 1 - \mu_P^2 \tag{8-8}$$

Case c: The pipeline is unrestrained (has expansion joints throughout).

$$K_r = 1 \tag{8-9}$$

where μ_p is the Poisson's ratio for the pipe material. Table 8-1 lists physical properties of common pipe materials.

8.4.3. Wave Action at Pipe Junctions

In a piping system, junction nodes have a significant impact on the direction and movement of pressure waves in the system. The effects of a pipe junction on pressure waves can be evaluated using conservation of mass and energy at the junction. Energy losses at the junction usually cause only minor effects and are neglected.

A wave of magnitude ΔH impinging on one of the junction legs, jin, is transmitted equally to each adjoining leg (Figure 8-11). The magnitude of the waves is $T_{jin} \Delta H$ where the transmission coefficient, T_{jin}, is given by:

$$T_{jin} = \frac{2\left(\dfrac{g_{jin} A_{jin}}{c_{jin}}\right)}{\sum \dfrac{g_j A_j}{c_j}} \tag{8-10}$$

where the summation j refers to all pipes connecting at the junctions (incoming and outgoing). A reflection back in pipe jin occurs and is of magnitude $R_{jin} \Delta H$ where:

$$R_{jin} = T_{jin} - 1 \tag{8-11}$$

Table 8-1 Physical properties of common pipe materials

Material	Young's Modulus E_c (GPa)	Young's Modulus— E_c (psi) × 10^6 (Typical Values)	Poisson's Ratio μ_p
Asbestos Cement	23–24	3	0.2
Cast Iron	80–170	15	0.25–0.27
Concrete	14–30	3	0.1–0.15
Reinforced Concrete	30–60	6	—
Ductile Iron	172	24	0.3
PVC	2.4–3.5	0.4	0.46
Steel	200–207	30	0.30
High-Density Polyethylene	0.9–1.1	0.13	0.4

For the simultaneous impingement of waves arriving in more than one leg the effects are superimposed.

Eq. 8-10 provides the basis for evaluating the effect of wave action at two special junction cases: dead end junctions and open ends or connections to reservoirs. A dead end is represented as a two pipe junction with A_2 equal to zero. With A_2 equal to zero, T_{jin} equals 2 and R_{jin} is 1, which indicates that the wave is reflected positively from the dead end. This condition implies that the effects of pressure waves on dead ends can be of significant importance in transient consideration. If the pressure wave reaching the dead end is positive, the wave is reflected with twice the pressure head of the incident wave. If the pressure wave reaching the dead end is negative, the wave reflection will cause a further decrease in pressure that can lead to the formation and collapse of vapor cavity. For a reservoir connection, A_2 is infinite so T_{jin} is zero and R_{jin} equals –1, which indicates that a negative reflection occurs at a reservoir.

8.4.4. Wontrol Elements

A general analysis of pressure wave action at a control eleave Action at Cment (e.g., pump, valve, orifice) in a pipe system is described below (Boulos et al. 2006; Wood et al. 2005a). This analysis provides relations to account for a variety of situations.

Figure 8-12 shows a general situation at a control element where pressure waves ΔH_1 and ΔH_2 are impinging. At the same time the characteristics of the control element may be changing. It is assumed that the relationship between flow through the control element, Q, and the pressure head change across the control element, ΔH, always satisfies a head-flow equation for the control element having the general form:

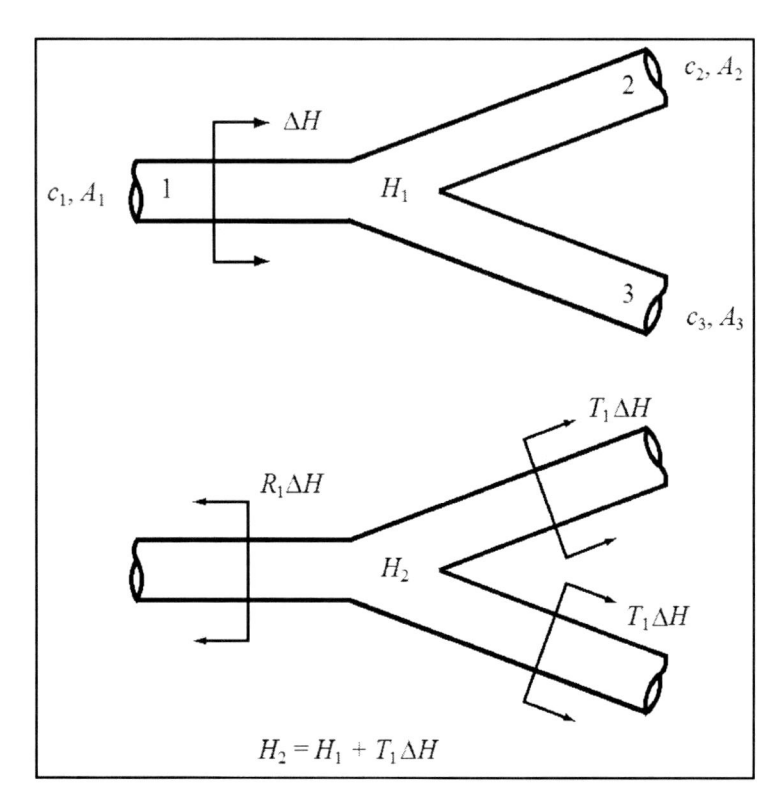

Figure 8-11 Effect of a pipe junction on a pressure wave

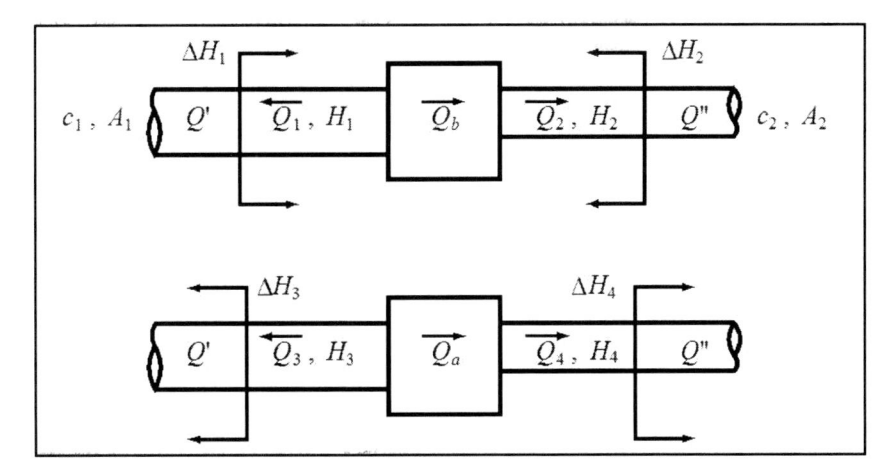

Figure 8-12 Condition at a control element before and after action

$$\Delta H = A(t) + B(t)Q + C(t)Q|Q| \tag{8-12}$$

The terms A, B, and C represent the coefficients for a general representation of the control element head-flow equation. These coefficients may be time dependent but will be known (or can be determined) at all times. The absolute value of Q is employed to make the resistance term dependent on the flow direction. This representation applies to both passive resistance elements such as valves, orifices, fittings, and friction elements and active elements such as pumps.

For passive resistance elements, however, only the coefficient C representing the effect of irreversible loss is not zero. This coefficient represents the ratio of the head loss to the square of the flow through the control element. For hydraulic considerations, this type of square law relationship is appropriate. The sign of the pressure head change is dependent on the direction of flow through the control element that necessitates the use of the absolute value of the flow rate as presented in Eq. 8-12.

In Figure 8-12, subscripts 1 and 2 denote conditions on the left and right side of the control element before the impinging waves arrive, while the subscripts 3 and 4 designate these conditions at the control element after the wave action. Here, Q_b and Q_a are the flows before and after the wave action, respectively.

The basic transient flow relationship for pressure-flow changes is applied to incoming and outgoing waves to yield the following for the outgoing waves:

$$\Delta H_3 = \Delta H_1 + \beta_1 (Q_b - Q_a) \tag{8-13}$$

$$\Delta H_4 = \Delta H_2 + \beta_2 (Q_a - Q_b) \tag{8-14}$$

Where:

$$\beta_1 = \frac{c_1}{gA_1} \quad \text{and} \quad \beta_2 = \frac{c_2}{gA_2} \tag{8-15}$$

Pressure heads after the action are given by:

$$H_3 = H_1 + \Delta H_1 + \Delta H_3 \tag{8-16}$$

and

$$H_4 = H_2 + \Delta H_2 + \Delta H_4 \qquad (8\text{-}17)$$

The characteristic equation relating the pressure head change across and the flow through the control element after the action is:

$$H_4 - H_3 = A(t) + B(t)Q_a + C(t)Q_a|Q_a| \qquad (8\text{-}18)$$

The coefficients of the characteristic equation, $A(t)$, $B(t)$ and $C(t)$, represent the values at the time of the wave action and may vary with time.

Substituting Eqs. 8-16 and 8-17 into Eq. 8-18 and rearranging results in a quadratic relationship for Q_a or:

$$\begin{aligned}C(t)Q_a|Q_a| + (B(t) - \beta_1 - \beta_2)Q_a \\ + A(t) + H_1 + 2\Delta H_1 - H_2 - 2\Delta H_2 + (\beta_1 + \beta_2)Q_b = 0\end{aligned} \qquad (8\text{-}19)$$

Eq. 8-19 can be solved directly for Q_a using the quadratic formula or iteratively using the Newton-Raphson method. Eqs. 8-13 and 8-14 are then solved to give the magnitude of the pressure waves produced by the action, and Eqs. 8-16 and 8-17 yield the pressure head after the action takes place.

This general analysis represents a wide variety of control elements that can be subject to a range of conditions.

8.4.4.1. Control Element Characteristics. The coefficients of the control element characteristic equation (8-12) are determined using head-flow operating data for the control element. Some control elements such as pumps will use all three coefficients to represent the head-flow variation. In some cases, the characteristic equation will be based on data, which represents the head-flow relationship for a relatively small range of operation. For these applications, the coefficients used for the control element analysis will be based on data valid for the operation in the vicinity of the operating point and will be recalculated as the operating point changes. This is true for the analysis of variable speed pumps and for pumps using data representing a wide range of operating conditions, including abnormal situations such as flow reversal.

Many control elements, such as valves, can be modeled using only the C coefficient. These are referred to as *resistive control elements* where the head-flow relation is adequately described by a single resistive term. For this application, the coefficient $C(t)$ is defined as the control element resistance. The term *resistance* is defined as the head drop divided by the square of the flow ($\Delta H/Q^2$). Here, the head drop is in feet (meters), and the flow is in ft^3/sec (m^3/sec).

The control element resistance is directly related to other resistive parameters such as minor loss (K_M), valve flow coefficient (C_v), sprinkler constant (K_s), and others, which characterize the head-flow characteristic of a resistive control element.

8.4.4.2. Wave Propagation With Friction. Because all pipeline systems contain friction, the pressure wave is attenuated as it travels down a line. Line loss can be simulated by concentrating the losses in length L at an orifice as shown in Figure 8-13. This orifice will then partially transmit and reflect pressure waves and account for the effect of wall shear. The friction orifice will therefore attenuate a pressure wave in a manner similar to the total attenuation that will occur as the wave travels the length L in the pipe.

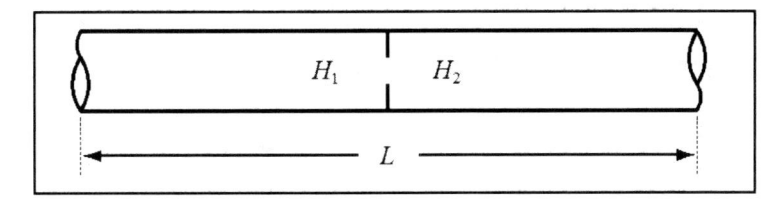

Figure 8-13 Wave propagation in a pipe section considering friction

In this representation, the loss at the orifice is:

$$\Delta H = H_2 - H_1 = \left[-\frac{fL}{2gDA^2}\right]Q^2 = CQ^2 \qquad (8\text{-}20)$$

where f is the friction factor; g is the acceleration of gravity; D is the pipe diameter; and A is the pipe area.

In this case, the coefficients of the characteristic equation (Eq. 8-12) for the line friction orifice are:

$$A(t) = B(t) = 0 \qquad (8\text{-}21)$$

and

$$C(t) = -\frac{fL}{2gDA^2} \qquad (8\text{-}22)$$

The friction factor can be determined using the flow rate through the orifice prior to the wave action. Although it is true that some approximation errors will be introduced if excessively long reaches are used, these errors are generally very small and can be minimized or eliminated using shorter pipe reaches. Ramalingam et al. (2009) developed sound guidelines in the form of error study for selecting the optimal number of friction orifices to ensure accurate results.

8.5. GOVERNING EQUATIONS

The fundamental equations describing hydraulic transients in liquid pipeline systems are developed from the basic conservation relationships of physics or fluid mechanics. They can be fully described by Newton's second law (equation of motion) and conservation of mass (kinematic relation). These equations can incorporate typical hydraulic devices and their interactions with the wave conditions in the pipes.

Applying these basic laws to an elementary control volume, a set of nonlinear hyperbolic partial differential equations can be derived. If x is the distance along the pipe centerline, t is the time, and partial derivatives are represented as subscripts, then the governing equations for transient flow can be written as:

Continuity

$$H_t + \frac{c^2}{gA}Q_x = 0 \qquad (8\text{-}23)$$

Momentum (Dynamic)

$$H_x + \frac{1}{gA}Q_t - f(Q) = 0 \qquad (8\text{-}24)$$

where $f(Q)$ is a pipe resistance (nonlinear) term that is a function of flow rate.

The preceding equations are first-order hyperbolic partial differential equations in two independent variables (space and time) and two dependent variables (head and flow). The solution of these equations with appropriate initial and boundary conditions will give head and flow values in both spatial and temporal coordinates for any transient analysis problem.

Unfortunately, no exact analytical solution exists for these equations except for simple applications that neglect or greatly simplify the boundary conditions and the pipe resistance term. When pipe junctions, pumps, surge tanks, air vessels, and other hydraulic components are included, the basic equations are further complicated. As a result, numerical methods are used to integrate or solve the transient flow equations (Wylie and Streeter 1993; Tullis 1989; Chaudhry 1979).

8.6. NUMERICAL SOLUTIONS OF TRANSIENTS

Several approaches have been taken to numerically model pressure transients in water distribution systems (Boulos et al. 2006; Wood et al. 2005a–b; Walski et al. 2003). The two most widely used and accepted methods are the Lagrangian Wave Characteristic Method (WCM) and the Eulerian Fixed-grid Method of Characteristics (MOC). Each method assumes that a steady-state hydraulic solution is available that gives initial flow and pressure distributions throughout the system. The main difference between the two numerical methods is in the way pressure waves are tracked between the pipe boundaries (e.g., reservoirs, tanks, dead ends, partially opened valves, pumps, junctions, surge control devices, vapor cavities, etc.). The MOC tracks a disturbance in the time-space grid using a numerical method based on characteristics, while the WCM tracks the disturbance based on wave propagation mechanics. Both methods have been well documented in the literature (Ramalingam et al. 2009; Jung et al. 2007; Boulos et al. 2005–2006; Wood et al. 1966, 2005a–b; Walski et al. 2003; Wylie and Streeter 1993) and have been implemented in various computer programs for pipe system transient analysis. The methods will mostly produce the same results at the network nodes when using the same data and model to the same accuracy. A brief description of each method follows.

8.6.1. Method of Characteristics (MOC)

In MOC, the governing partial differential equations are converted to ordinary differential equations and then to a difference form for solution by a numerical method. Solution space comprises two equations called the *characteristic equations* along with two compatibility equations for any point in a space-time grid. The method divides the entire pipeline into a fixed number of segments, writes the characteristic and compatibility equations for every grid location, and then solves these equations for head and flow at all grid locations. The line friction of the entire pipeline is distributed in each of these segments. The various boundary conditions are handled by combining the appropriate characteristic equation with the equations defining the boundary.

8.6.2. Wave Characteristic Method (WCM)

The WCM is based on the concept that transient pipe flow results from generation and propagation of pressure waves that occur as a result of a disturbance in the pipe system. The method essentially tracks the movement of pressure waves as they propagate throughout the system and computes new conditions at either fixed time intervals or only at times when a change actually occurs. It requires the calculation of the effects of these pressure waves impinging on the network junctions and control elements along with the appropriate boundary conditions. The entire line friction is modeled as an equivalent orifice situated at the midpoint of a pipeline or multiple orifices distributed uniformly throughout the pipeline (i.e., distributed friction profile). Pressure and velocity or flow rate time histories are computed for any point in the network by summing with time the contributions of incremental waves.

8.7. METHODS OF CONTROLLING TRANSIENTS

The means of controlling pressure transients in water distribution systems will generally depend on whether the initiating event results in an *upsurge* (e.g., a high pressure event caused by a shutdown of a downstream pump or valve) or a *downsurge* (e.g., a low pressure event caused by the shutdown of an upstream pump or valve). Downsurge events can lead to the undesirable occurrence of water-column separation (cavitation) that can result in severe pressure surges following the collapse of a vapor cavity or intrusion of contaminated water through a leak or other opening.

A number of surge protection devices are commonly used to help control high and low pressure transients in water distribution pipe systems. Small systems are just as vulnerable as large systems. No two systems are completely identical; hence the ultimate choice of surge protection devices and operating strategies will usually differ. Of course, it is always best whenever possible to avoid rapid flow changes. A transient analysis should be carried out to predict the effect of each individually selected device. As a result of the complex nature of transient behavior, a device intended to suppress or fix a transient condition could result in a worsening of the condition if the device is not properly selected or located in the system. Designers must evaluate the relative merits and shortcomings of all the protection devices that they may select. A combination of devices may prove to be the most desirable and economical. A brief overview of various commonly used surge protection devices and their functions is provided in the following discussions. Additional details are available in Boulos et al. (2005, 2006), Wood et al. (2005a), Walski et al. (2003), and Thorley (1991).

8.7.1. Devices and Systems

8.7.1.1. Simple Surge Tank (Open). Open surge tanks or stand-pipes can be an excellent solution to both upsurge and downsurge problems. These tanks can be installed only at locations where normal static pressure heads are small (or tall tanks are acceptable). They serve two main purposes: (1) to prevent high pressures during pump startup conditions or valve shutdown conditions by accepting water; or (2) to prevent cavitation during pump shutdown by providing water to a low-pressure region.

8.7.1.2. Surge Vessel (Air Chamber—Closed Surge Tank—Bladder Tank—Hybrid Tank). Surge vessels (or air chambers), which are pressure vessels partly full of air, have the advantage that they can be installed anywhere along a line regardless of normal pressure head. It should be noted that under very low static conditions or downhill pumping, special care will be required to keep the surge tank from completely draining. These vessels are normally positioned at pump stations (downstream of the pump delivery valve) to provide protection against a loss of power to the pump. They serve the same function as an open surge tank but respond faster and allow a wider range of pressure fluctuation. Their effect depends primarily on their location, vessel size, entrance resistance, and initial gas volume and pressure. Closed surge vessels are normally equipped with an air compressor to control the initial gas volume and to supply make-up gas, which is absorbed by the water. Some closed surge tanks are equipped with a precharged pressurized bladder (bladder surge tanks) that eliminates the need for an air compressor. Hybrid tanks are equipped with an air vent that admits air when the pressure goes below atmospheric pressure.

8.7.1.3. Feed Tank (One-Way Surge Tank). The purpose of feed tanks is to prevent initial low pressures and potential water-column separation by admitting water into the pipe subsequent to a downsurge. They can be either open or closed, will have a check valve to allow flow only into the pipe system, and can be installed anywhere on the line.

8.7.1.4. Pressure Relief Valve. A pressure relief valve ejects water out of a side orifice to prevent excessive high-pressure surges. It is activated when the line pressure at a specified location (not necessarily at the valve) reaches a preset value. The valve opens and closes at prescribed rates over which the designer often has some degree of control. It can eject water into the atmosphere or a pressurized region, or into an open or closed surge tank.

8.7.1.5. Surge Anticipation Valve. A surge anticipation valve is much like a pressure relief valve, but it gets triggered to open on a downsurge in pressure (sensed at a specified location) in anticipation of an upsurge to follow. This valve, when activated, follows and completes a cycle of opening and closing based on valve opening and closing rates. For systems where water-column separation will not occur, the surge anticipation valve can solve the problem of upsurge at the pump due to reverse flow or wave reflection. However, this valve must always be used with caution for it can make low pressure conditions in a line worse than they would be without the valve.

8.7.1.6. Air Release/Vacuum Valve. Air release/vacuum breaking valves are installed at high points in a pipeline to prevent low pressure (cavitation) by admitting air into the pipe when the line pressure drops below atmospheric conditions. The air is then expelled (ideally at a lower rate) when the line pressure exceeds atmospheric pressure. Two-stage air valves release the air through a smaller orifice to prevent the "air slam" that occurs when all the air is released and the water column rejoins. A three-stage air valve can be designed to release the air through a second (smaller) orifice to further reduce the air slam.

8.7.1.7. Check Valve. A check valve allows flow only in one direction and closes when flow reversal is impending. For transient control, check valves are usually installed with other devices such as a pump bypass line as described below. Pumps are often equipped with a check valve to prevent flow reversal. Because check valves do not close instantaneously, it is possible that a substantial backflow may occur before closure that can produce additional and sometimes large surges in the system. Check valve modeling includes a time delay between check valve activation and complete closure of the check valve. The check valve is often treated as a valve closing in a linear fashion that is activated by flow reversal and closes completely over the delay period. Check valves can also be used to isolate high pressure waves from reaching a section of a pipeline. One of the great advantages of a check valve is that it can prevent pipes from draining, and keeping the pipe full of fluid tends to reduce startup transients.

8.7.1.8. Pump Bypass Line. In low-head pumping systems that have a positive suction head, a bypass line around the pumps can be installed to allow water to be drawn into the discharge line following power failure and a downsurge. Bypass lines are generally short line segments equipped with a check valve (nonreturn valve) preventing back flow (from the pump discharge to the suction side) and installed parallel to the pump in the normal flow direction. They are activated when the pump suction head exceeds the discharge head. They help prevent high-pressure buildup on the pump suction side and cavitation on the pump discharge side.

8.7.1.9. Flywheel. Increasing the pump rotational inertia by attaching a flywheel (Figure 8-14), a large-diameter steel plate, to the pump motor is sometimes a useful surge control device, especially for sewage pumping systems, because the flywheel is not in contact with the foul water. When power fails, the rotational energy provided by the flywheel will reduce pump speed gradually, allowing the pumping

Figure 8-14 Flywheels to be installed in a large pump station

system to slowly come to rest, thus avoiding unacceptable transient pressures. Three problems accompany flywheel installations:

- They must be large and heavy enough to provide additional rotating inertia to be effective. Pump and motor bearings and supports must then be designed to accommodate the extra weight, and additional space is required.
- Pump startup requires extra power to overcome the inertia of flywheels.
- Pump startup must be gradual to keep the motors from burning out.

Nonetheless, flywheels are useful type of surge control and can be found in many pumping installations.

8.7.1.10. In-Line Pump Control Valve. For large pumping systems with large-diameter headers and high flows, in-line pump control valves are a viable option. In-line pump control can be operated similarly to surge anticipator valves, which open when line pressure drops below a specified set point, remain open for a preset period of time, and then close in a manner that prevents high pressures resulting from rapid valve closure. Properly installed, these valves prevent high pressures at pump stations, but the valve movements must be set with care; otherwise, more severe problems can occur. Four quadrant pump curves have to be analyzed in conjunction with the other transient parameters to develop a solution. Of course, these valves have no ability to prevent low pressures and column separation from occurring in the downstream pipe lines.

8.7.2. Choice of Surge Protection Strategy

A number of techniques can be used for controlling/suppressing transients in municipal water distribution systems. Some involve system design and operation while others are related to the proper selection of surge protection devices. For example pressure relief valves, surge anticipation valves, surge vessels, surge tanks, pump bypass lines, or any combination of them can be used to control high pressures. Low pressures can be controlled by increasing pump inertia or by adding surge vessels, surge tanks, air release/vacuum valves, pump bypass lines, or any combination of that group. The overriding objective is to reduce the rate at which changes to the flow occur.

Surge protection devices are normally installed at or near the point where the disturbance is initiated such as at the pump discharge or by the closing valve (with the exception of air relief/vacuum breaking valves and feed tanks). Figure 8-15 illustrates typical locations for the various surge protection devices in a water distribution system. A comprehensive transient flowchart for considering the transient protection of the system is shown in Figure 8-16. This flowchart is discussed in detail by Boulos et al. (2005, 2006). When developing a protection strategy, it must be recognized that no two systems are hydraulically the same; hence, no general rules or universally applicable guidelines are available to eliminate unacceptable pressures in a water distribution system. Any surge protection devices and/or operating strategies must be chosen accordingly (Boulos et al. 2005, 2006; Wood et al. 2005a; Walski et al. 2003; Thorley 1991).

The final choice will be based on the initial cause and location of the transient disturbance(s), the system itself, the consequences if remedial action is not taken, and the cost of the protection measures themselves. A combination of devices may prove to be the most effective and most economical. Final checking of the adequacy and efficacy of the proposed solution should be conducted and validated using a detailed transient analysis.

8.7.2.1. Pump Station and Downstream Pipeline Protection. Because pump stations are aboveground, surge failures and remedial solutions can be readily observed. However, failures to buried downstream pipelines are sometimes undetected and become subject to unseen and overlooked failures. It is important to recognize that some surge control solutions protect only the pump station but do not necessarily protect the downstream pipelines from column separation and negative pressures. Both the pump station and the downstream pipelines should be protected.

8.7.2.2. Float-Operated Air Valves Versus Surge Resistant Air Valves. A number of vacuum relief valves are often needed for filling and draining pipelines. Also, additional air valves may be required to protect the pipelines from extreme low pressures due to surge transients. The total number of air valves depends on pipeline profile and initial pipeline velocities. Of course, the number of air valves should be limited and carefully considered, because adding large volumes of air to pipelines can create other types of operational problems, such as air binding. If pipeline profiles are too steep and the initial velocities are too high, numerous air valves would be required. In that case, other types of surge control devices should be considered instead.

Because float-operated air valves require water to close the float, some pipe systems with high static head require surge resistant air valves to prevent damaging the float. Damage to the air valve and high pressures occur when water column slams the float closed, pushing the low density air out around the float.

A safer air valve solution would be the so called *surge-resistant air valves* that trap the air inside the air valve body, using it as a cushion, and then slowly releasing the air to prevent high pressures and damage to the valves.

8.7.2.3. Normal Pump Operation Protection Versus Power Failure Protection. Normal pump operation for a surge-vulnerable system can include slow closing and slow opening pump control valves or variable frequency drives on the pump motors to eliminate undesirable surges during day-to-day operation.

Power failure on surge-vulnerable systems requires the surge-control devices. Backup power and generators do not come on-line quickly enough to prevent transients from occurring. It should be noted that the use of surge chambers requires quick-closing check valves immediately downstream of the pumps. The quick-closing check valves prevent the surge chamber from spilling water back through the pumps during

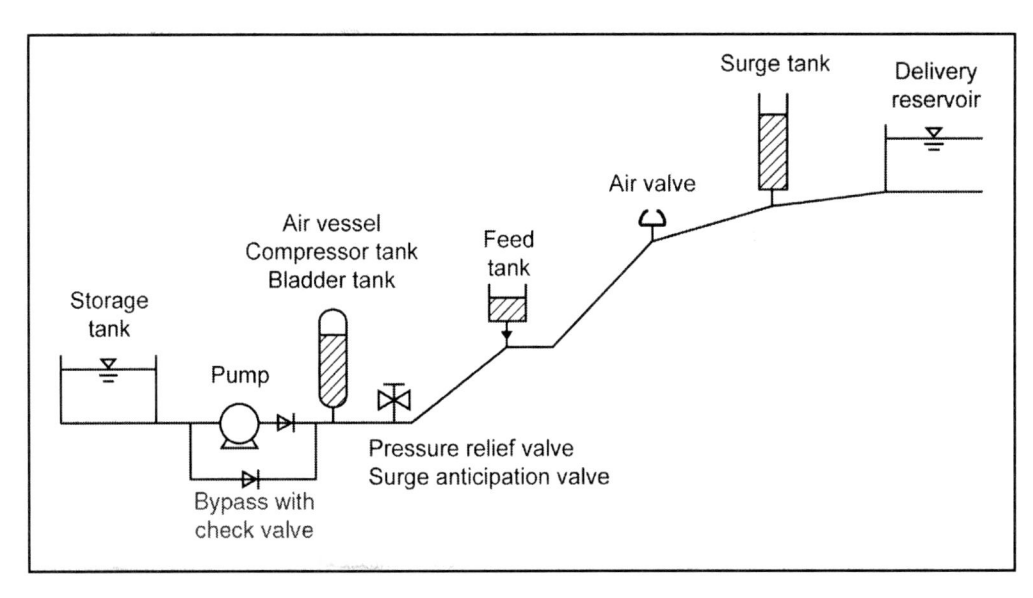

Figure 8-15 Typical locations for various surge protection devices

pump shutdown. Slow-closing pump control valves would not be used in conjunction with a surge chamber. If a surge chamber is required for surge control, it would also be used for normal pump shutdown and startup and pump control valves would not be required.

This section focused on pumps. There are other control strategies such as education of operators to slowly open/close valves, adjusting speed of mechanically operated valves, and not operating tanks such that the altitude valve closes and cannot absorb the transient.

8.8. TRANSIENT MODELING CONSIDERATIONS

Transient analysis is essentially based on equations governing the movement of pressure waves throughout the municipal water system. This type of network analysis requires a significant number of calculations and is an extremely demanding computational exercise. This is usually carried out using transient modeling software. Although the system should be described in full, some simplifications and assumptions will be needed for its model representation. This is necessary to reduce the network model complexity and computational run times, and also because some of the basic transient flow data required will not be available (Thorley 1991).

Having considered some of the detail of transient analysis and protection, it is perhaps helpful to include some tips or guidelines that would assist in preparing computer simulation files. In essence, the good news is that transient modeling uses much of the same data required for steady-state modeling. A steady-state analysis of the initial conditions for the transient analysis is required. There are, however, a number of additional considerations for developing a transient analysis model.

- The location (including elevation) of hydraulic devices (pumps, control valves, check valves, regulating valves, etc.) is required for the model.

- A transient model should carry out calculations at all local high and low points because the pressure extremes often occur at these locations.

- Cavitation must be modeled for transient analysis. If cavitation occurs at any location in the distribution system, it can greatly affect the transient analysis results (Jung et al. 2009b).

196 COMPUTER MODELING OF WATER DISTRIBUTION SYSTEMS

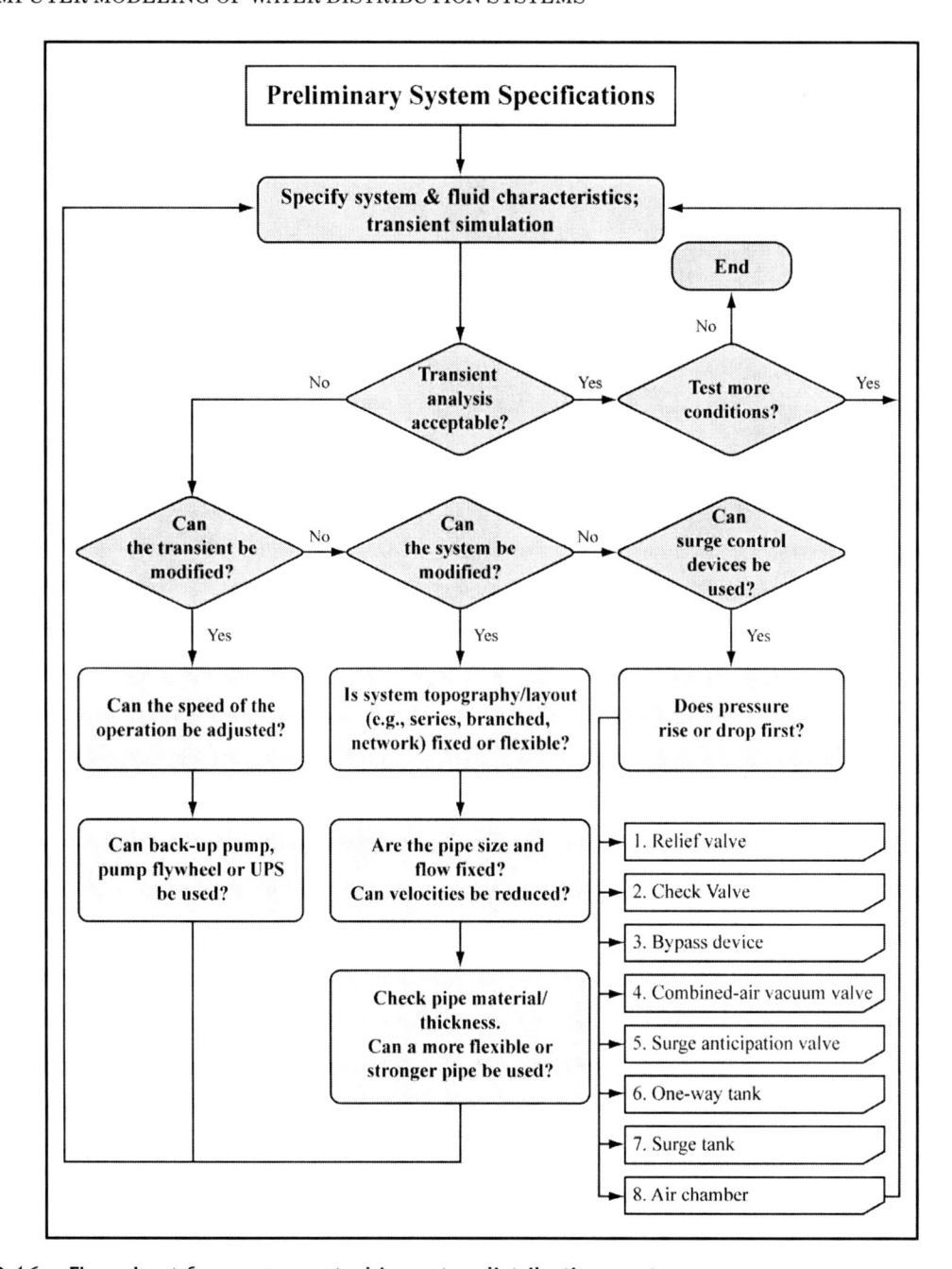

Figure 8-16 Flowchart for surge control in water distribution systems

- Skeletonization guidelines are significantly different than those for steady-state analysis. Dead-end pipes, for example, will have a very significant effect on a transient analysis while having no effect on the steady-state analysis. Jung et al. (2007) performed a detailed study of the issues associated with water distribution model skeletonization for surge analysis. They concluded that skeletonization can introduce some significant error in estimating pressure extremes and can overlook water-column separation and subsequent collapse at vulnerable locations in the distribution system.

- Series and parallel pumps with similar characteristics may be modeled as a single equivalent pump if none of the individual pumps to be combined are subjected to a pump trip, startup, or a speed change.

- Pressure (positive or negative) surges can drastically alter the local pressure and affect the demand magnitude that can be extracted at any given node in the system. Pressure-sensitive demands versus fixed (constant or pressure independent) nodal outflows should be used to assess the impact of pressure changes and produce more accurate transient results (Jung et al. 2009a).

- It is good practice to allow a transient model to operate initially at steady-state for a short period before the transient is initiated. This provides additional assurance that the transient model is operating correctly.

- For transient analysis it is necessary to use a computational time interval such that the pressure wave travel times for all pipe segments will be a multiple of this time increment and this integer multiple will be calculated for each line segment. Some adjustment is required to obtain a time interval that is not unreasonably small. This will amount to actually analyzing a model of the piping system with pipe lengths (or wave speeds) different than the actual values and a sensitivity check should be made to assure that the time interval chosen is acceptable. The time interval required for accurate transient analysis may be quite small (0.01 to 0.001 seconds). For example if the wave speed is 4,000 ft/sec (1,220 m/sec), a time increment of 0.01 seconds represents a travel distance (and therefore, length accuracy) of 40 ft (12.2 m) for the model. A time increment of 0.001 seconds represents a travel distance (and length accuracy) of 4 ft (1.2 m) for the model. The length accuracy of the model (maximum difference between actual and model pipe lengths) must be sufficient to generate an accurate solution. However, increasing the accuracy will require a longer computational time. It should be noted that the shortest pipe in the network model plays a dominant role in determining the computational time interval. Consideration should be given to merging very short pipes into longer pipe segments or simply removing them from the network model but only when their resulting effect on the system transient behavior is negligible.

8.9. DATA REQUIREMENTS

The data requirements for surge analysis include all data necessary to do a steady-state analysis. A steady-state analysis for the starting conditions is required. Initial conditions of flow in all pipe segments and static pressure head (or pressure) at all junctions and components must be calculated before initiating the transient analysis. Note that the required initial pressure head (or pressure) conditions are static heads (P/γ) and not total heads or hydraulic grade lines (elevation + P/γ). The static pressure is the pressure that a pressure gage attached to the line at that position would read. This requirement is necessary to address cavitation during transient conditions.

In addition to the usual pipe data (length, diameter, and roughness), the wave speed must be calculated for each pipe. This requires the wall thickness of each pipe, material, Poisson's ratio and elastic (Young's) modulus, and the bulk modulus and density of the fluid. In addition, information on the degree of restraint of the pipes is required. Because many of these parameters may not be directly available, estimates for typical pipe materials are often used.

Characteristic data (or an equation) that relate the pressure change and flow to the operating point must be used for all hydraulic components and for the full range of operating conditions during the transient analysis. All system components (pumps, valves, etc.) should be initially in a balanced state (the initial pressure change across the component and flow through the component should be compatible with the characteristic relationship for that component). The characteristic data for a component must account for changes in the set point of the component (valve opening, pump speed, etc.) that occur during the transient analysis. Normally, the steady-state characteristics are used. Figure 8-17 shows typical valve closure characteristics for a linear stem movement for several types of valves (Boulos et al. 2006; Wood et al. 2005a). It is assumed that the valves will operate on these curves during transient conditions.

Pump steady-state head-flow-speed curves are often used to model pump operation. However, if during the transient the pump operates in abnormal zones (turbining, etc.), it is necessary to use more detailed four quadrant pump operating characteristics (i.e., ± speed and ± flow). Of the many methods developed for this purpose, the Suter curve is the most widely used (Thorley 1991; Suter 1966). A typical four quadrant pump characteristic curve is shown in Figure 8-18. In this figure, the x-axis represents the four quadrants of pump operation, and the y-axis represents the corresponding head and torque characteristics. These data are normally not provided by pump manufacturers. However, many of these four quadrant curves are available for a range of pump specific speeds. Typically, one of these curves is selected by matching the specific speed of the pump to the available curves. The additional pump data required for transient analysis include the efficiency and the total inertia (pump and motor).

A characteristic equation that relates the pressure change and flow to the operating point of all hydraulic surge control devices along with their locations must also be available. All such devices (surge tanks, air valves, etc.) should be initially in a balanced state (the initial pressure change across the component and flow through the component should be compatible with the initial characteristic relationship for that component). The characteristic equation for a surge control device must account for changes in the set point of the device (relief valve, nonslam air valve, etc.) that occur during the transient analysis.

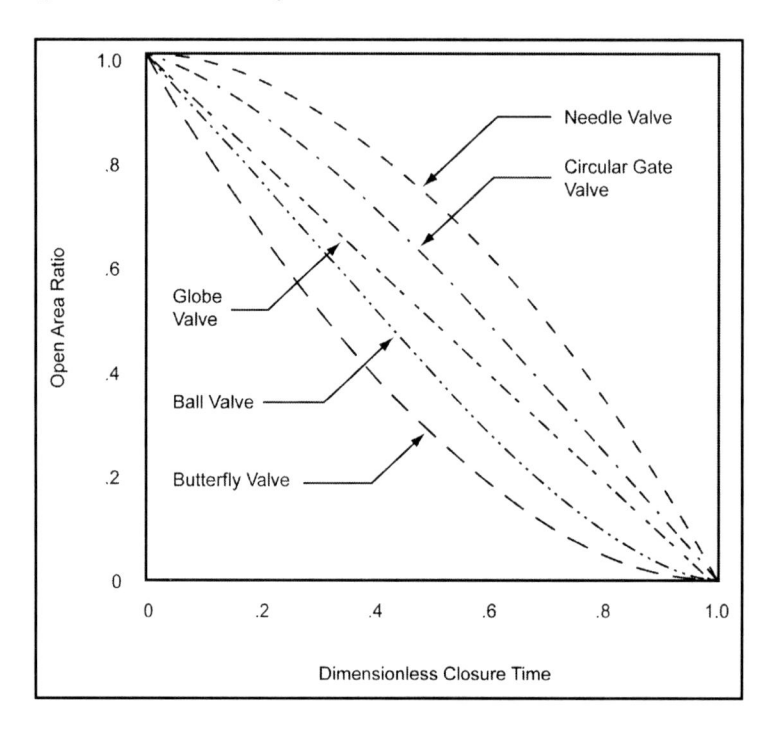

Figure 8-17 Representative valve closure characteristics

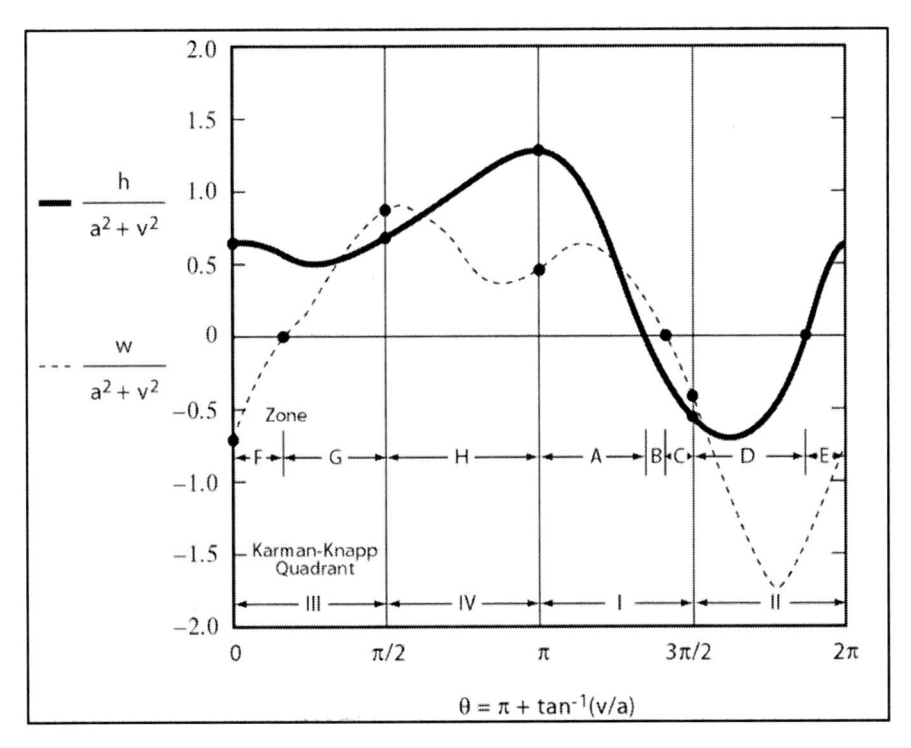

Figure 8-18 Typical pump four quadrant characteristics (Suter curve)

Finally, the exact nature of the disturbance and its intended timescale must be specified. The disturbance can be either a change in the open area ratio for a valve, the speed ratio for a pump, or any other change in operation (check valve closing, pipe rupture, etc.) that results in a change in flow.

8.10. SUMMARY

Hydraulic transient, also called *pressure surge* or *water hammer*, is the means by which a change in steady-state flow and pressure is achieved. When flow conditions in the water pipeline system are changed, such as by closing a pump or a valve or by starting a pump, a series of pressure waves is generated. These disturbances propagate with the velocity of sound within the medium until dissipated down to the level of the new steady-state by the action of some form of damping or friction. In the case of flow in a water distribution network, these transients produce velocity and pressure changes. When sudden changes take place, however, the results can be dramatic because pressure waves of considerable magnitude can occur and are quite capable of destroying equipment and pipelines. Only if flow regulation occurs very slowly is it possible to go smoothly from one steady-state to another without large fluctuations in pressure head or flow velocity.

Clearly, flow control actions can be extremely important, and these actions have implications not only for the design of the hydraulic system but also for other aspects of system operation and protection. Problems such as selecting the pipe layout and profile, locating control elements within the system, and formulating operating rules as well as the ongoing challenges of system management are all influenced by the details of the control system. A rational and economic operation requires accurate data, carefully calibrated (static- or extended-period simulation) models, ongoing predictions of future demands and the response of the system to transient loadings, and correct selection of both individual components and remedial strategies. These design decisions cannot be considered an after-thought to be appended to a nearly complete design. Transient analysis is a fundamental and challenging part of rational system design.

Transient analysis is essential to good design and operation of piping systems. Transient modeling provides the most effective and viable means of predicting potentially negative impacts of hydraulic transients under a number of worst-case scenarios, identifying weak spots, and evaluating how they may possibly be avoided and controlled. The basis of surge modeling is the numerical solution of conservation of mass and linear momentum equations. A number of widely used computer codes based on Eulerian (MOC) and Lagrangian (WCM) numerical solution schemes are currently available and have been successfully validated against field data and exact analytical solutions. However, surge analysis computer models can only be effective and reliable when used in conjunction with properly constructed and well-calibrated hydraulic network models. Poorly defined and calibrated hydraulic network models may result in poor prediction of pressure surges and locations of vapor cavity formation and, thus, defeat the whole purpose of the surge modeling process.

Looped water distribution systems comprising short lengths of pipes may be less vulnerable to problems associated with hydraulic transient than a single long pipe system. This is because wave reflections (e.g., at tanks, reservoirs, junctions) will tend to limit further changes in pressure and counteract the initial transient effects. For networks with long pipelines and irrespective of whichever numerical basis is used, a good transient model will have nodes along those pipes defining the important high and low points to ensure accurate calculations are made at those critical locations. An important consideration is dead ends (which may be caused by closure of control or check valves) that lock pressure waves into the system in a cumulative fashion (wave reflections will double both positive and negative surge pressures). As a result, the effects of dead ends need to be carefully evaluated in transient analysis.

Proper selection of components for surge control and suppression in water distribution systems requires a detailed surge analysis to be effective and reliable. In addition, good maintenance, pressure management, and routine monitoring programs are essential components of transient protection. With these capabilities, water utility engineers can greatly enhance their ability to better understand and estimate the effects of hydraulic transients and to conceive and evaluate efficient and reliable water supply management strategies, safeguard their systems and public health with maximum effectiveness, and forge closer ties to their customers. It is understanding complexity through simplicity.

8.11. GLOSSARY OF NOTATIONS

A cross-sectional area [L^2]

A, B, C coefficients of control element characteristic equation

c sonic wave speed [$L\ T^{-1}$]

D pipe diameter [L]

E_c elastic modulus of the pipe [$M\ L^{-1}T^{-2}$]

E_f elastic modulus of the fluid [$M\ L^{-1}T^{-2}$]

f pipe friction factor

$f(Q)$ pipe resistance (nonlinear) term that is a function of flow rate [L]

g acceleration of gravity [$L\ T^{-2}$]

H head [L]

K_r coefficient of restraint for longitudinal pipe movement

L pipe length [L]

P pressure [M L^{-1} T^{-2}]

Q volumetric flow rate [L^3 T^{-1}]

R surge wave reflection coefficient at pipe junction

T time [T]

T surge wave transmission coefficient at pipe junction

t_l pipe thickness [L]

V flow velocity [L T^{-1}]

Δt time interval [T]

Δx spatial grid size [L]

μ_P Poisson's ratio for the pipe material

ρ liquid density [M L^{-3}]

8.12. REFERENCES

Besner, M.C., Lavoie, J., Morissette, C., and Prevost, M. 2008. Effect of Water Main Repairs on Water Quality. *Jour. AWWA*, 100(7):95–109.

Besner, M.C., Lavoie, J., Payment, P., and Prevost, M. 2007. Assessment of Transient Negative Pressures at the Site of the Payment's Epidemiological Studies. In *Proceedings of the American Water Works Association WQTC, Charlotte, N.C.* Denver, Colo.: AWWA.

Boulos, P.F., Lansey, K.E., and Karney, B.W. 2006. *Comprehensive Water Distribution Systems Analysis Handbook for Engineers and Planners*. 2nd edition. Broomfield, Colo.: MWH Soft.

Boulos, P.F., Karney, B.W., Wood, D.J., and Lingireddy, S. 2005. Hydraulic Transient Guidelines for Protecting Water Distribution Systems. *Jour. AWWA*, 97(5):111–124.

Boulos, P.F., Wood, D.J., and Funk, J.E. 1990. A Comparison of Numerical and Exact Solutions for Pressure Surge Analysis. In *Proc. of the 6th International BHRA Conf. on Pressure Surges*, A.R.D. Thorley editor. Cambridge, UK.

Boyd, G.R., Wang, H., Britton, M.D., Howie, D.C., Wood, D.J., Funk, J.E., and Friedman, M.J. 2004. Intrusion Within a Simulated Water Distribution System Due to Hydraulic Transients. 1: Description of Test Rig and Chemical Tracer Method. *Journal of Environmental Engineers*, ASCE, 130(7):774–783.

Chaudhry, M.H. 1979. *Applied Hydraulic Transients*. New York: Van Nostrand Reinhold Co.

Fleming, K.K., Gullick, R.W., Dugandzic, J.P., and LeChevallier, M.W. 2006. *Susceptibility of Distribution Systems to Negative Pressure Transients*. Denver, Colo.: AwwaRF.

Friedman, M., Radder, L., Harrison, S., Howie, D.C., Britton, M.D., Boyd, G.R., Wang, H., Gullick, R.W., Wood, D.J., and Funk, J.E. 2004. *Verification and Control of Pressure Transients and Intrusion in Distribution Systems*. Denver, Colo.: AwwaRF.

Gullick, R.W., LeChevallier, M.W., Svindland, R.C., and Friedman, M.J. 2004. Occurrence of Transient Low and Negative Pressures in Distribution Systems. *Jour. AWWA*, 96(11):52–66.

Hooper, S.M., Moe, C.L., Uber, J.G., and Nilsson, K.A. 2006. Assessment of Microbiological Water Quality After Low Pressure Events in a Distribution System. In *Proceedings of the 8th Annual International Water Distribution Systems Analysis Symposium*, Cincinnati, Ohio.

Jung, B.S., Boulos, P.F., and Wood, D.J. 2009a. Effect of Pressure Sensitive Demand on Surge Analysis. *Jour. AWWA*, 101(4):100–111.

Jung, B.S., Boulos, P.F., Wood, D.J., and Bros, C.M. 2009b. A Lagrangian Wave Characteristic Method for Simulating Transient Water Column Separation. *Jour. AWWA*, 101(6):64–73.

Jung, B.S., Karney, B.W., Boulos, P.F., and Wood, D.J. 2007a. The Need for Comprehensive Transient Analysis of Water Distribution Systems. *Jour. AWWA,* 99(1):112–123.

Jung, B.S., Boulos, P.F., and Wood, D.J. 2007b. Pitfalls of Water Distribution Model Skeletonization for Surge Analysis. *Jour. AWWA,* 99(12):87–98.

Karim, M.R., Abbaszadegan, M., and LeChevallier, M.W. 2003. Potential for Pathogen Intrusion During Pressure Transient. *Jour. AWWA,* 95(5):134–146.

Karney, B.W., and McInnis, D.A. 1990. Transient Analysis of Water Distribution Systems. *Jour. AWWA,* 82(7):62–70.

Kirmeyer, G.J., Friedman, M., Martel, K., Howie, D., LeChevallier, M., Abbaszadegan, M., Karim, M., Funk, J., and Harbour, J. 2001. *Pathogen Intrusion Into the Distribution System.* Denver. Colo.: AWWA and AwwaRF.

LeChevallier, M.W., Gullick, R.W., Karim, M.R., Friedman, M., and Funk, J.E. 2003. The Potential for Health Risks From Intrusion of Contaminants Into Distribution Systems From Pressure Transients, *Journal Water Health,* 1(1):3–14.

McInnis, D.A., and Karney, B.W. 1995. Transients in Distribution Networks: Field Tests and Demand Models. *Journal of Hydraulic Engineering,* ASCE, 121(3): 218–231.

National Research Council. 2006. *Drinking Water Distribution Systems: Assessing and Reducing Risks.* Washington, D.C.: National Academies Press.

Ndambuki, J.M. 2006. Water Quality Variation Within a Distribution System: A Case Study of Eldoret Municipality, Kenya. In *Proceedings of the Environmentally Sound Technology in Water Resources Management.* Gaborone, Bostwana.

Ramalingam; D., Lingireddy, S., and Wood, D.J. 2009. Using the WCM for Transient Modeling of Water Distribution Networks. *Jour. AWWA,* 101(2):75–89.

Suter, P. 1966. Representation of Pump Characteristics for Calculation of Water Hammer. *Sulzer Tech. Rev.,* 66:45–48.

Thorley, A.R.D. 1991. *Fluid Transients in Pipeline Systems.* Herts, UK: D&L George.

Tullis, J.P. 1989. *Hydraulics of Pipelines.* New York: John Wiley & Sons.

Walski, T.M. 2009. Not So Fast! Close Hydrants Slowly. *Opflow,* 35(2):14–16.

Walski, T.M., and Lutes, T. 1994. Low Pressure Problems Caused by Hydraulic Transients. *Jour. AWWA,* 86(12):24–32.

Walski, T.M., Chase, D.V., Savic, D.A., Grayman, W.M., Beckwith, S. and Koelle, S. 2003. *Advanced Water Distribution Modeling and Management.* Waterbury, Conn.: Haestad Press.

Wood, D.J., Dorsch, R.G., and Lightner, C. 1966. Wave Plan Analysis of Unsteady Flow in Closed Conduits. *Journal of Hydraulics Division,* ASCE, 92:83–110.

Wood, D.J., Lingireddy, S., and Boulos, P.F. 2005a. *Pressure Wave Analysis of Transient Flow in Pipe Distribution Systems.* Broomfield, Colo.: MWH Soft.

Wood, D.J., Lingireddy, S., Boulos, P.F., Karney, B.W., and McPherson, D.L. 2005b. Numerical Methods for Modeling Transient Flow in Distribution Systems. *Jour. AWWA,* 97(7):104–115.

Wylie, E.B., and Streeter, V.L. 1993. *Fluid Transient in Systems.* Englewood Cliffs, N.J.: Prentice Hall.

AWWA MANUAL M32

Chapter 9

Storage Tank Mixing and Water Age

9.1. INTRODUCTION

This chapter provides a summary of various methods of modeling distribution system tanks and reservoirs, and the advantages and disadvantages of each. It also describes the most important concepts of flow mechanics inside tanks and reservoirs. Some systems distinguish tanks from reservoirs by the vertical location of the storage facility (e.g., tanks are elevated while reservoirs are in-ground); by size (e.g., tanks are small and reservoirs are large); or by shape (e.g., tanks are regularly shaped while reservoirs are irregularly shaped). In this chapter, the words *tank* and *reservoir* are used interchangeably because both refer to facilities that are used to store water and have similar purposes in a distribution system.

Distribution system storage tanks (and reservoirs) store a significant portion of the total volume of water in a potable water system. Maintaining water quality in these storage tanks is necessary to reduce the loss of disinfectant residuals, prevent nitrification, minimize disinfection by-product (DBP) formation, and reduce biological growth. Maintaining acceptable water quality in distribution system storage tanks depends on the quality of water that entering a tank, the water mixing characteristics in the tank, and the rate of turnover. The quality of the water entering tanks is primarily determined by treatment-plant process performance, disinfection practices, and design and operation of the distribution system. Operational and design aspects of an individual tank affect mixing characteristics and the change in water quality in the tank. Adequate mixing is needed to prevent the formation of stagnant zones in tanks, which over time can lead to the release of poor quality water to the distribution system supplied by the tanks. In some cases, insufficient mixing leads to thermal stratification and dead zones of water inside storage tanks that may exacerbate water quality deterioration.

9.2. TYPES OF TANKS AND RESERVOIRS

Tanks and reservoirs can generally be classified into two broad types: (1) those that provide storage for fire flow requirements and peak demands, and (2) those that are used for disinfection to achieve desired contact time (CT). In most cases, a tank or reservoir serves one of these purposes, but sometimes it can be used to serve both. The tanks and reservoirs in distribution systems are generally used to serve fire flow and peak flow demands, whereas the tanks and reservoirs at water treatment plants are generally used for disinfection credit (although some of them are also used for peak demands). The primary focus in this chapter is modeling of distribution system tanks and reservoirs that are used for water distribution storage. For these types of tanks and for given water quality entering the tanks, the change in water quality inside the tanks and exiting the tanks is primarily governed by the mixing processes and residence time within the tank.

9.3. BACKGROUND

Physical and mathematical models have been used to study the mixing behavior in distribution system tanks and reservoirs. A summary of some of these studies is provided in an Awwa Research Foundation report (Grayman et al. 2000). Some of the major findings of this report are as follows:

9.3.1. Tank Designs Should Achieve Good Mixing

There are two theoretical schemes that provide the ways in which water may flow through a storage tank:

- Completely mixed state
- Plug flow manner

When comparing two tanks of the same size and inflow-outflow characteristics, the one operating under plug-flow conditions will generally lose more disinfectant than the one operated under mixed-flow conditions, because the rate of disinfectant decay is concentration dependent. Therefore, achieving a completely mixed state should be a primary goal in a water storage tank design.

9.3.2. Fill Time Versus Mixing Time

The typical time required to fill a tank should exceed the time required to achieve complete mixing. For a tank operating in a fill and draw mode, mixing occurs primarily during the fill cycle. As a result, if the tank is relatively well mixed by the end of each fill cycle, mixing problems such as stratification and dead zones are unlikely to occur. For a wide range of tank and tank designs, experimentation has shown that the mixing effectiveness is primarily dependent on:

- Volume of water in the tank at the start of filling
- Inlet diameter
- Filling flow rate
- Filling time

9.3.3. Develop Turbulent Jet to Promote Mixing

Mixing a fluid requires a source of energy. In distribution system storage tanks, this energy is normally introduced during tank refilling. As water enters a tank, jet flow occurs at the inlet, ambient water is entrained into the jet, and circulation patterns are formed that result in mixing. To have efficient mixing, the path of the jet must be long enough to allow for the entrainment and mixing process to develop. Therefore, the inlet jet should not be pointed directly toward nearby barriers such as walls, the tank bottom, or deflectors. The rate and degree of mixing depend primarily on the size of the tank and the momentum of the incoming jet. If the turbulent jet is insufficient to ensure good mixing, more complex inlet configurations or mechanical mixing may be required.

9.3.4. Avoid Conditions That Inhibit Complete Mixing

Conditions that inhibit complete mixing and promote stratification should be avoided. Whenever there is a temperature difference between the contents of a tank and its inflow, the potential for poor mixing and stratification exists. Temperature differences result in what is called a *buoyant jet*. A positively buoyant jet occurs when the density of the incoming water is less than the density of the ambient water (may occur during winter), while a negatively buoyant jet occurs when the density of the incoming water is greater than the density of the ambient water (may occur in elevated storage tanks during the summer). Incoming water with excessive buoyancy (either positive or negative) relative to its momentum will lead to ineffective mixing and could produce stable stratified conditions within the tank. Based on this relationship, tall tanks and tanks with large-diameter inlets have a greater tendency toward stratification. If significant temperature differences are experienced, then increasing the inflow velocity is an effective strategy for reducing the likelihood of stratification. Standpipes, where the water depth exceeds the diameter of the tank, are typically more susceptible to stratification.

9.4. FACTORS AFFECTING WATER QUALITY

9.4.1. Hydraulic Residence Time (HRT)

Hydraulic residence time (HRT) is the amount of time that water spends within a storage tank. This is frequently referred to as the tank *water age*. However, HRT is a more precise term because it differentiates between the total water age within the water system (i.e., the combination of residence time within the distribution system and in tanks) and the water age just associated with the time spent in the tank. Excessive HRT in a distribution system storage tank can make it difficult to consistently maintain an adequate disinfectant residual inside the tank and can lead to excessive DBPs, regardless of mixing conditions. This can result in inadequate residuals and elevated DBPs downstream of the tank when the water flows into the distribution system. Acceptable values for HRT depend on both the inherent water quality composition of the water (i.e., water with higher amounts of organic material result in faster consumption of disinfectant) and the relationship of the tank to other parts of the distribution system (i.e., if a tank feeds into other tanks, the total hydraulic residence time through the system can quickly become excessive).

Acceptable HRT is system-specific because of the differences in water quality entering the distribution system, disinfection strategy, and water temperature among different systems. A bulk water disinfectant decay test that indicates disinfectant residual as a function of time can be conducted to determine the criteria for desirable, acceptable, and undesirable categories.

The HRT can be calculated as follows (Mahmood et al. 2005):

$$\text{Theoretical average residence time} = \frac{V_{max}}{V_{max} - V_{min}} \cdot \frac{1}{N}$$

Where:

V_{min} = typical minimum cycle volume
V_{max} = typical maximum cycle volume
N = number of fill/draw cycles per day

9.4.2. Fill Time

Mixing primarily occurs during the filling cycle. The ability of a water storage tank to achieve complete mixing can be assessed by comparing the actual average fill time to the mixing time theoretically required for a complete mix to occur. For good mixing, actual average fill times should exceed the theoretically required time for complete mixing. For a given filling flow rate, this can be expressed in terms of fill time or volume turnover per fill cycle. Based on detailed laboratory tests, the fill time required to achieve complete mixing can be defined by the following empirical equation (Grayman et al. 2000; Roberts et al. 2006):

$$\text{Mixing time (hours)} = \frac{k}{3600} \cdot V_{min}^{2/3} / M^{1/2}$$

Where:

M = momentum = U · Q
Q = average fill rate (cfs)
U = average inflow velocity (ft/sec)
V_{min} = average minimum daily volume (ft^3)
k = mixing coefficient
 = 10 (for H/D <1)
 = 10 + 3.5 (H/D −1) (for H/D > 1)
H = tank height (ft)
D = tank diameter (or equivalent diameter for no cylindrical tanks) (ft)

The mixing time equation assumes isothermal conditions (i.e., the temperature of the inflow is the same as the temperature of the ambient water in the tank). As a result, if the water temperature of the inflow significantly differs from the water temperature in the tank, the mixing equation may not be applicable and mixing may be inhibited and lead to stratification. The mixing equation applies to tanks that operate in either simultaneous inflow-outflow mode (tank inflow and outflow through different pipes) and in fill-and-draw mode (tank fills and empties through same pipe). However, for the case of simultaneous inflow-outflow mode, there is not a representative fill time that can be used to compare to the mixing time. Therefore professional judgment must be applied to determine the adequacy of the calculated mixing time.

9.4.3. Inlet Momentum

Inlet momentum is a key factor for mixing in storage tanks. Generally, the higher the inlet momentum, the better the mixing characteristics in a storage tank. Inlet momentum is defined as:

$$\text{Momentum} = \text{velocity} \cdot \text{flow rate}$$

Average inlet velocity and inlet momentum are calculated for a given tank based on the average filling rates. Higher momentum can be achieved by increasing the inflow rate or by increasing the velocity. In many situations, it is not practical to increase the flow rate into tanks because of system hydraulic limitations. Pumping water into the tanks may increase the flow rate, but this approach is typically not desirable due to increased operational complexity and cost. It is generally more feasible to increase the inlet velocity by decreasing the inlet diameter. Prior to decreasing inlet diameter, the resulting increase in head loss needs to be considered.

9.4.4. Inlet Pipe Location and Orientation and Tank/Reservoir Geometry

The location and orientation of the inlet pipe relative to the tank walls can have a significant impact on mixing characteristics. For example, when the height of a tank is much larger than the diameter or width, a horizontal inlet pipe at the bottom of a tank is more likely to result in incomplete mixing of the water in the tank than a vertical inlet. In this type of tank, the horizontal orientation of the inlet pipe would most likely result in the water jet hitting the opposite vertical wall of the tank before sufficient time has elapsed for mixing, resulting in a loss of inlet momentum.

The mixing characteristics of a tank also depend on the initial water depth in the tank. When the inlet orientation is horizontal and the initial water depth is very high, the inlet momentum may not be able to mix all the water in the tank completely during the fill cycle.

9.4.5. Volume Turnover

Turnover is the percentage of water that is exchanged on a daily basis between the tank and the distribution system. Increasing the daily turnover can result in lowering the residence time in the tank and improve mixing. There are no firm standards on the amount of turnover that is needed although a typical goal is one-third of the total volume each day.

9.4.6. Seasonal Variation

Several factors that influence the water quality in tanks are affected by seasonal variation. These include: changes in water usage and systems operation, changes in water and air temperature that may lead to stratification or affect mixing, and changes in temperature that may affect reactive processes in the tank. To account for seasonal variation, water quality within tanks should be assessed under different seasonal conditions.

9.5. TYPES OF MODELING

9.5.1. Physical Scale Modeling

Tanks and reservoirs can be modeled by building a geometrically scaled model and observing water movement and flow characteristics. However, a physical model has to be scaled properly, and dimensional analysis is used to achieve this goal. The scale range for most studies is typically between 1:10 and 1:100, and generally three conditions need to be satisfied according to the dimensional analysis:

- Geometric similarity—the model should be the same shape as the actual tank but scaled by a constant scale factor.

- Kinematic similarity—the velocity at any location in the model and the actual tank should be proportional to each other by a constant factor.
- Dynamic similarity—all forces in the model flow should be proportional to all forces in the actual tank flow.

To achieve dynamic similarity, it is necessary to establish dimensionless parameters and to match these parameters between the modeled tank and actual tank. The common dimensionless parameters used in tank hydraulics are

- Froude number, $F_r = V^2/gL$—ratio of inertial force to gravitational force.
- Reynolds number, $Re = VL/\nu$—ratio of inertial force to viscous force.
- Weber number, $W = \rho V^2 L /\sigma$—ratio of inertial force to surface tension force.

Where:

V = velocity of water
g = gravitational acceleration
L = characteristics length such as diameter of tank or depth of water
ν = viscosity of water
σ = surface tension of water
ρ = density of water

Sometimes it is difficult to match all of the dimensionless parameters between a physical model and the actual tank. In this case, emphasis should be placed on matching the parameters that correspond to the predominant forces, and in the case of tanks, the most important dimensionless parameter is the Froude number. For inlet and outlet pipes, the most important dimensionless parameter is the Reynolds number. When tanks and reservoirs are modeled according to Froude number similitude, the following relationships govern:

- Geometric similarity: Length = L, Area = L^2, Volume = L^3
- Kinematic similarity: Velocity = $L^{0.5}$, Time = $L^{0.5}$, Flow = $L^{2.5}$
- Dynamic similarity: Force = L^3, Mass = L^3

Roberts (et al. 2006) applied a unique tank scale modeling technique called *three-dimensional laser induced fluorescence* (3DLIF). Two fast scanning mirrors drive a laser beam from an argon-ion laser through the flow in a programmed pattern. A small amount of a fluorescent dye is added to the effluent; the laser causes the dye to fluoresce, and the emitted light is captured by a video camera. Mirrors are used to sweep the laser beam and to create multiple vertical slices through the flow. After computation of tracer concentrations, the data can be converted, by image processing techniques, into three-dimensional images of the flow field. The 3DLIF system can obtain millions of sampling points leading to great insight into the hydrodynamics of turbulent mixing in tanks.

Physical scale models can provide accurate information about water flow mechanics inside a tank. However, depending on the geometry of the tank and inlet/outlet, it may become challenging to achieve similarity of the dimensionless parameters. Larger scale models are preferable, but building large physical models is more expensive. Building a model may also be impractical for many desiring to study mixing within a water storage tank. Therefore, other methods of modeling need to be explored.

9.5.2. Empirical Modeling

Tank and reservoir mixing can be studied using empirical models that can be used as stand-alone tank models or integrated into network simulations. There are several types of empirical models that have been developed for tanks and reservoirs:

- Completely mixed
- Plug flow (first in/first out)
- Short-circuiting (last in/first out)
- Compartment

Figure 9-1 illustrates the various types of empirical models previously listed. Though each of these theoretical modes of behavior are idealistic representations of flow behavior in tanks that are not actually fully achieved in practice, a properly applied empirical model based on specific tank design and operational conditions may provide a good representation of the mixing behavior of a tank. These types of empirical models are also referred to as *systems models*, *input-output models*, and *blackbox models*.

Completely mixed models assume that the water in the tank mixes instantaneously and completely, and remains in this state indefinitely. The concentration throughout the tank is a blend of the tank's current bulk water and the newer water entering the tank. The concentration at the outflow of the tank is equal to the concentration in the completely mixed tank. This is the most commonly used empirical model and is the default mechanism used by most network modeling software. Depending on the tank geometry, inlet/outlet configuration, and flow rates, it is possible to approximately model many distribution system tanks as completely mixed. Generally, tanks with high momentum jets and high volume turnover may achieve predominantly mixed flow behavior.

Plug-flow models assume that the water does not mix at all and moves as parcels through the tank from the inlet to the outlet. This type of model represents first in–first out (FIFO) behavior for water flow, meaning the first parcel of water to enter the tank is also the first parcel of water to leave the tank. Tanks that operate in simultaneous inflow-outflow mode with baffles to guide the flow of water through the tank are most likely to achieve the plug flow FIFO behavior. Treatment plant clearwells and contact chambers most commonly follow this type of behavior. Most tanks in the distribution system do not fit this type of model because they are not typically baffled and usually operate in fill-and-draw mode.

Short-circuiting models assume that water entering a tank is bypassing a significant portion of the tank volume by travelling toward the outlet preferentially. This type of model represents LIFO behavior for water flow. Depending on the tank geometry, inlet/outlet configuration, and flow rates, it is appropriate to model some distribution system tanks as short-circuiting type. Generally, tanks with low flow rates, stratified conditions, low volume turnover, and a common inlet/outlet or an outlet close to the inlet may exhibit short-circuiting behavior. This type of model may be applicable to tall and narrow standpipes in the distribution system.

Compartment models are hybrid representations that assume that the entire volume of the tank does not fit into one of the model types previously indicated but may be represented by two or more compartments that are each completely mixed and interact with other compartments. The volume of each compartment may vary or can remain fixed (for example, to represent stratified conditions). New water entering the tank completely mixes with the stored water in the first compartment, and when it fills, excess water is sent to another compartment where the water completely mixes

210 COMPUTER MODELING OF WATER DISTRIBUTION SYSTEMS

with the stored water in that compartment. There has been only limited application of compartment models due to the lack of guidance and experience on the selection and parameterization of compartment models.

Empirical models are most frequently used to model water age or disinfectant concentrations in a tank over an extended period of time. Figure 9-2 illustrates the output from an empirical model used to model water age in a tank for a period of one year.

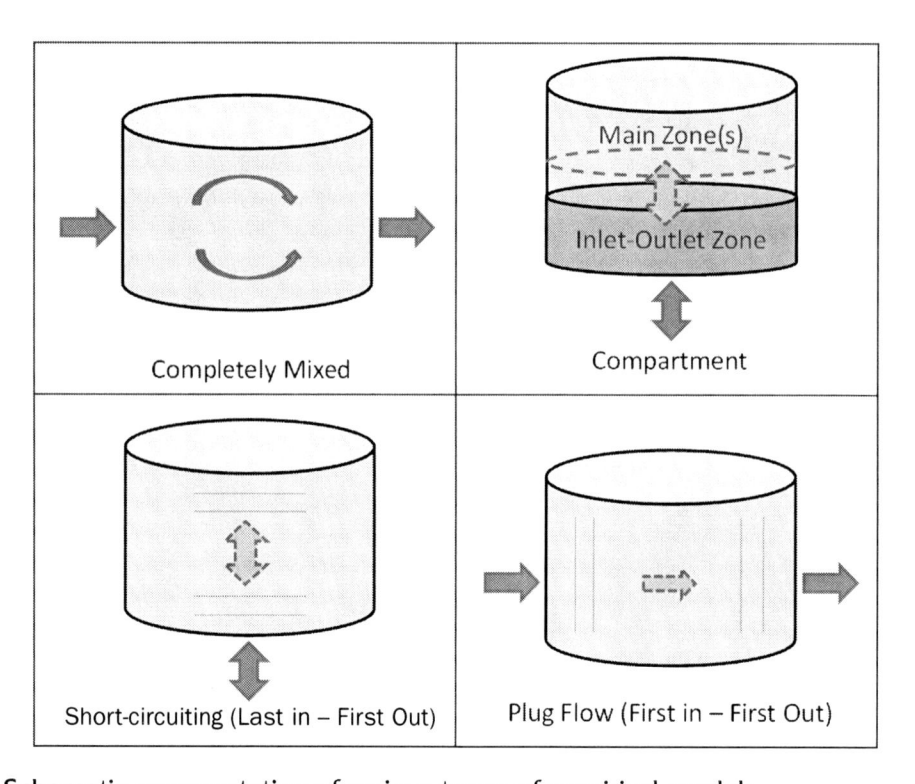

Figure 9-1 Schematic representation of various types of empirical models

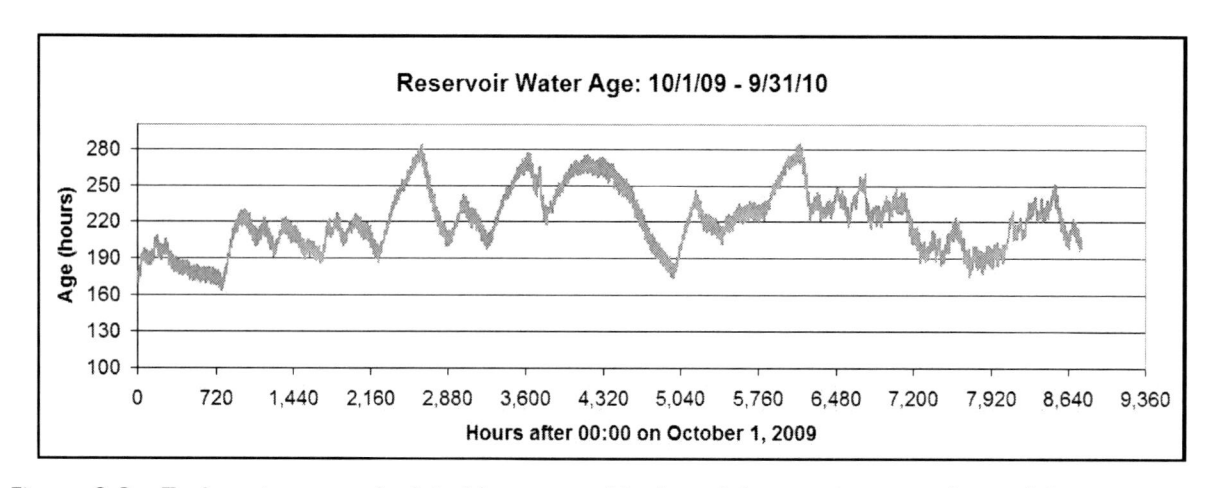

Figure 9-2 Tank water age calculated by an empirical model assuming complete mixing

Empirical models offer a very fast computational solution for modeling water age and disinfectant residuals in the *outflow* from a tank, and can therefore be easily used in network simulations. A properly applied empirical model based on specific tank design and operational conditions may provide a good representation of the mixing behavior of a tank. However, these theoretical modes of behavior are idealistic representations of actual flow behavior in tanks that are not actually fully achieved in practice. They do not predict the actual internal flow behavior within a tank and, therefore, cannot be used to assess the mixing characteristics or distribution of water age *within* the tank. Field data can be used to determine which type of empirical model best represents a tank, but care must be taken with field data collection and the additional cost of field sampling and analysis should be considered.

9.5.3. Computational Fluid Dynamics (CFD) Modeling

Mixing in tanks and reservoirs can be studied using computational fluid dynamics (CFD) modeling techniques that use the fundamental equations of fluid mechanics as governing equations. These fundamental equations are:

- Conservation of mass
- Conservation momentum
- Conservation of energy

Within the framework of a CFD model, a tank's geometry, including inlet and outlet features, is represented by a grid or mesh consisting of cells or control volumes with which the equations are solved. Gridding may be characterized by both the density (number of grid cells and size of the individual cells) and the quality of the gridding. There are various ways to generate grids. The resulting quality of the gridding, the type of grids, and the grid density can influence both the results and the computational time associated with solving the equations.

Appropriate boundary conditions need to be specified for the walls, air/water interface, inlets, and outlets to obtain accurate CFD solutions. For example, tank walls can be specified as having zero velocity both normal to the wall and tangent to the walls. However, if the water is allowed to move along the surface, the tank water surface can be specified as a slip wall with zero stress to allow tangential movement of water along the surface. The inlets and outlets can be specified with known velocity or known pressure crossing the inlet/outlet. If reverse flow is observed at the outlet boundaries, the computational domain needs to be extended outside of the confines of the tank.

Appropriate solution algorithms should be selected based on the type of problem. For example, if the water surface remains relatively constant, a tank can be modeled using a single-phase approach involving modeling of the water only. However, if the water surface level varies significantly, a multiphase approach involving the air/water interface is more appropriate. If the flow is turbulent, an appropriate turbulence closure scheme also needs to be used.

As a CFD simulation progresses, the user should monitor the convergence of the solutions, defined as *residuals*. Convergence occurs when the residuals are small. The residuals should decrease with time, and the magnitude of various residuals at any given time provides information about how well the solution is approximating the actual tank hydraulic behavior as defined by the Navier-Stokes equations.

Because CFD modeling is capable of generating millions of data points representing the spatial and temporal behavior of flow within the tank, graphical methods for visualizing the output data are important. Figures 9-3 through 9-6 provide examples of how CFD modeling results can be displayed to evaluate water movement and water age inside a tank for various types of scenarios. Figure 9-3 (tall tank) and Figure 9-4 (short, large-diameter tank) illustrate the effect of thermal differences using tracer and velocity values and section views. Figure 9-5 illustrates the effect of operational or design changes using velocity values and isometric views. Figure 9-6 illustrates different methods of displaying water age distribution inside a tank using pathlines; one showing pathlines by water age values and the other showing percent of the total tank volume with specific water age ranges.

CFD modeling, if applied properly, can provide accurate information with regard to the hydraulic behavior of tanks and reservoirs, insights for the design of mixing mechanisms and inlet piping, and improvement of operations. In addition, three-dimensional images of important parameters of interest such as velocity, temperature, pressure, constituent concentration, and water age can be produced by model results. However, CFD modeling is computationally intensive and can require significant training for the modeler.

9.6. MODEL VERIFICATION

Several types of data can be collected and used to verify the accuracy of modeling results for tanks and reservoirs:

- Tracer data
- Temperature data
- Water quality data
- Hydraulic data

Field data may be collected as part of a special study or may be routinely measured in a tank using permanent monitors reporting via a SCADA system. The procedure for data collection to verify the results from tank modeling is similar to the procedure for data collection to calibrate network models. The collection of tracer, water quality, and hydraulic data has been discussed in earlier chapters.

Tracer studies using conservative tracers such as fluoride or calcium chloride can be used to both verify mixing models and to directly calculate the hydraulic residence time distribution of the water inside a tank. The tracer tests can be simulated using physical scale models and CFD models and modeling results compared to the field data to verify or refine the model.

Temperature measurements may be taken at different levels within a tank to determine whether the tank is stratified. These measurements can also be used as a means of verifying or refining physical scale models or CFD models of the tank.

Water quality parameters can also be used to verify results from physical, empirical, or CFD modeling. However, many water quality parameters such as chlorine residual are not just impacted by tank hydraulics but also by biological or chemical reactions. Hence, care must be taken to choose the appropriate water quality parameter for model verification and accurately represent the reactive processes.

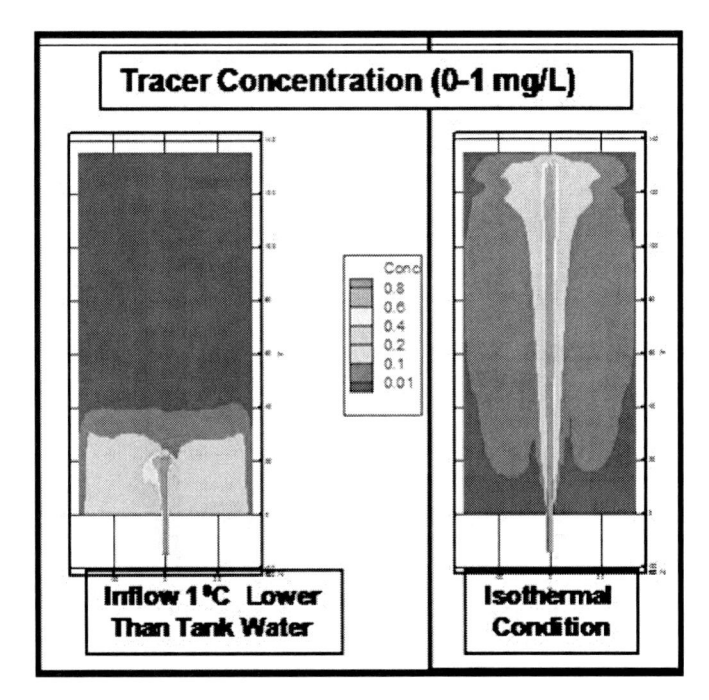

Figure 9-3 Effect of thermal differences for tall tank

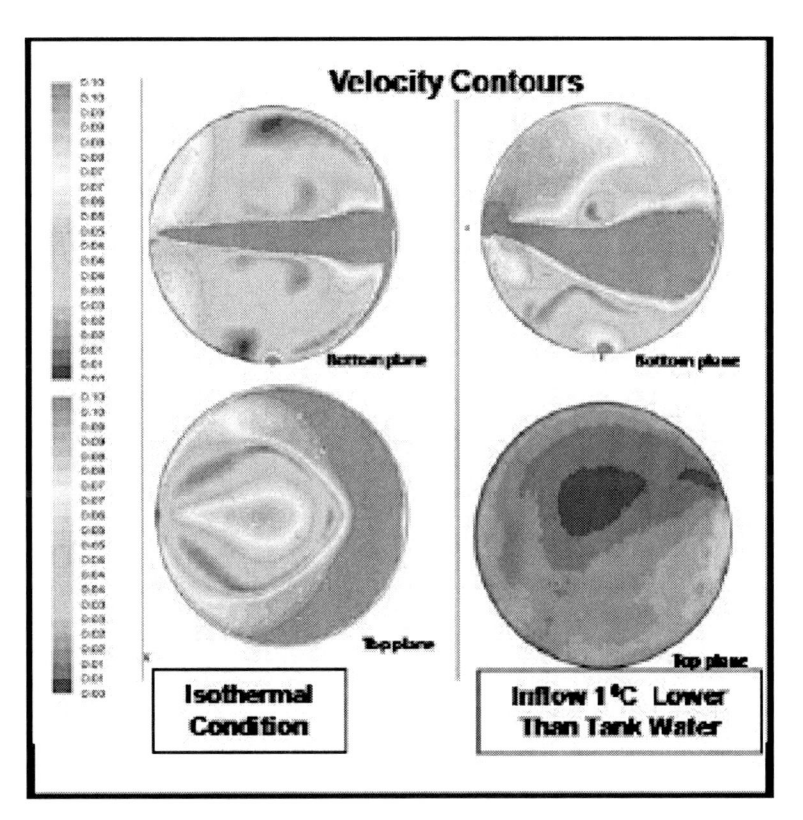

Figure 9-4 Effect of thermal differences for short tank

214 COMPUTER MODELING OF WATER DISTRIBUTION SYSTEMS

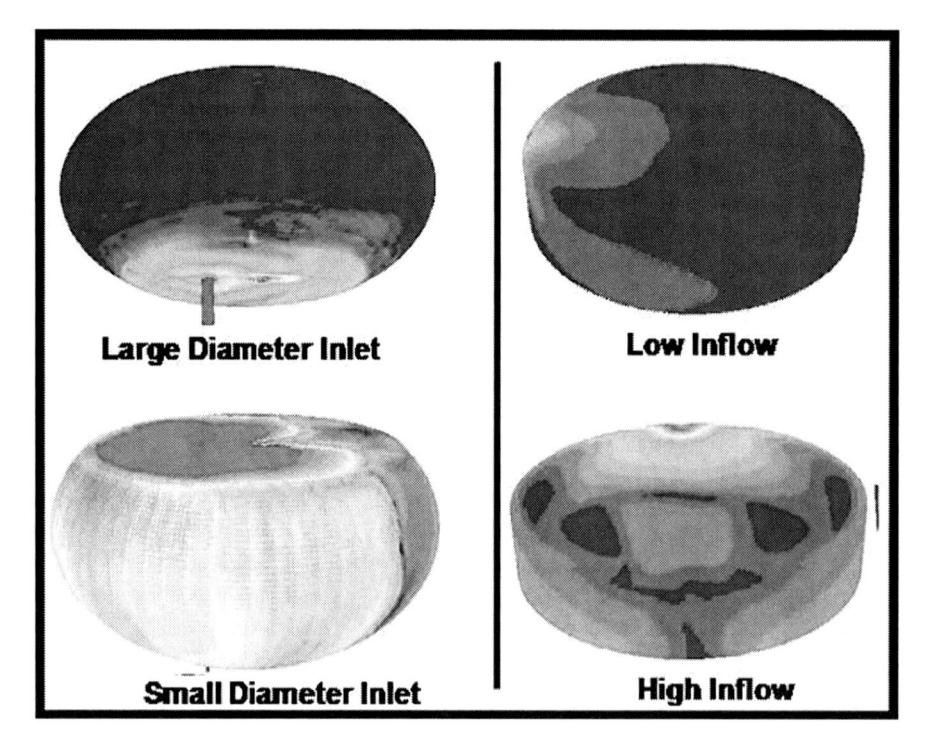

Figure 9-5 Effect of operational and design changes

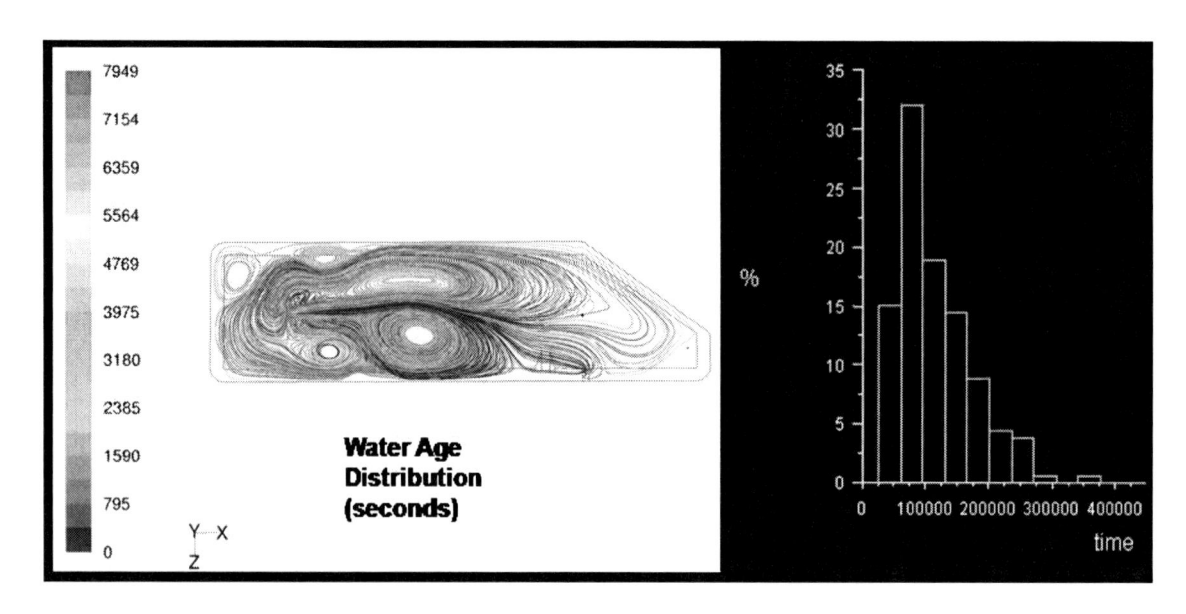

Figure 9-6 Water age distribution

9.7. STRATEGIES TO PROMOTE MIXING AND REDUCE WATER AGE

9.7.1. Improving Mixing Within Tanks

The consistent achievement of well-mixed conditions in distribution system storage tanks will minimize the potential for development of stratified conditions and dead (stagnant) zones. Stratification and stagnant zones within distribution system storage tanks can lead to wide ranges of water age, DBP levels, and microbial activity within the tanks. Slugs of poorer quality water present in the higher stratified layers or stagnant zones within a tank can be released into the distribution system when larger than normal drawdowns occur in the tank, resulting in transient occurrences of low residuals and high DBPs within the distribution system.

The goal of achieving acceptable mixing within distribution system storage tanks is relevant to all systemwide water quality management strategies. Table 9-1 identifies typical alternatives for improving the mixing and turnover characteristics of distribution system storage tanks. These improvements are characterized as physical or operational modifications. The table also identifies the relative ease with which these modifications can typically be made.

The physical modifications for improving mixing within the tanks may include:

- Changing the orientation of the inlet pipe and/or
- Decreasing the inlet diameter to increase the jetting action
- Separating inlet and outlets
- Adding mechanical mixing devices

As mentioned previously, the inlet characteristics are important for mixing, while the outflow characteristics do not significantly influence mixing. When the inlet/outlet is a common pipe, the ability to reduce the inlet diameter to achieve a higher inflow velocity and better jetting action may be constrained by the need to maintain an outflow capacity adequate to satisfy system operational and fire flow requirements. Generally, the headloss for modified piping in distribution system tanks does not exceed more than a few feet, even at relatively high velocities, which is small compared to typical system pressure head. Nonetheless, a reduction in the inlet diameter should be checked hydraulically to assure the hydraulics are not significantly impaired.

Table 9-1 Example modifications to improve tank mixing characteristics

Variable Impacting Mixing and Water Quality	Type of Improvement	Desirable Characteristic	General Ability to Control/Implement
Inlet diameter	Physical	Smaller diameter for higher velocity	Good
Inlet orientation	Physical	Toward maximum length of water	Good
Separate inlet and outlet	Physical	Avoid short circuiting	Good
Active mixer	Physical	Increase mixing	Good
Booster chlorination	Physical/Operational	Boost residual	Good
Initial water level	Operational	Lower water level at beginning of fill cycle	Moderate
Filling flow rate	Operational	Higher flow rate during fill cycle	Moderate
Temperature difference	Physical/Operational	Inlet temperature $\geq$ tank temperature	Low/No independent control

9.7.2. Reducing Water Age

The ability to reduce the average HRT of stored water typically requires significant modifications to the operations of the system. These may include changes to pressure gradients, inventory management, and operating set points (e.g., pump station operation and level set points). These modifications can result in increased operational complexity. Potential impacts to customers and the assurance of the adequacy of reserves for fire flow and emergencies are important considerations to address as part of any implementation strategy.

In addition to the example physical modifications to tanks listed in Table 9-1, operational changes are often needed to reduce HRT in distribution system storage tanks. Possible operational changes include:

- Increasing the daily drawdown in the tank
- Increasing the number of fill-and-draw cycles each day
- Manually draining the tank at set intervals or as needed to force refilling with freshwater
- Installing tank booster pumps to pump water out of "locked out" tanks and back into the distribution system
- Installing new water mains to eliminate hydraulic bottlenecks that prevent a tank from contributing to the supply of water during peak hourly demand times
- Demolishing tanks that are too low, too small, too old, or otherwise no longer needed
- Seasonal modifications to tank usage including lower water levels in the winter or taking selected tanks out of service in the winter.

It should, however, be recognized that the ability to implement operational modifications may be limited by the following considerations:

- Control of flow rates during tank filling may be needed to minimize the potential for low pressure in the distribution system
- Changes in operating protocol for booster stations and other tanks to achieve turnover while maintaining adequate pressure systemwide
- Additional energy costs associated with extended operation of pumps

9.7.3. Sampling Locations Within Tanks

While improved mixing and reduced HRT may improve water quality and reduce DBP formation within storage tanks, sampling within storage tanks can help provide early warning of impending water quality problems within the tanks if the sampling ports are placed at appropriate locations. Water quality parameters such as disinfectant residual and temperature can provide information about mixing and water quality conditions within a tank. Tools such as CFD modeling that provide three-dimensional flow patterns inside storage tanks are useful in locating poorly mixed and well-mixed zones, and thus can help determine appropriate sampling locations.

9.8. REFERENCES

Grayman, W.M., Rossman, L.A., Arnold, C., Deininger, R.A., Smith, C., Smith, J.F., and Schnipke, R. 2000. *Water Quality Modeling of Distribution System Storage Facilities*. Denver, Colo.: AwwaRF.

Hwang, N.H.C., and Houghtalen, R.J. 1996. *Fundamentals of Hydraulic Engineering Systems*. Upper Saddle River, N.J.: Prentice Hall.

Mahmood, F., Pimblett, J.G., Grace, N., and Grayman, W.M. 2005. Evaluation of Water Mixing Characteristics in Distribution System Storage Tanks. *Jour. AWWA*, 97(3):74–88.

Roberts, P.J.W., Tian, X., Sotiropoulos, F., and Duer, M. 2006. *Physical Modeling of Mixing in Water Storage Tanks*. Denver, Colo.: AwwaRF.

This page intentionally blank.

Index

NOTE: *f.* indicates a figure; *t.* indicates a table.

Computer Modeling of Water Distribution
    Systems (M32)
Index
Note: *f.* indicates figure; *t.* indicates table.

All-mains models, 12, 13
All-pipes models, 20, 21*f.*, 22
All-pipes reduced models, 20, 21*f.*
American Water Works Association
    Calibration Guidelines for Water Distribution
        System Modeling (ECAC Calibration
        Guidelines Report), 88
    Engineering Modeling Applications
        Committee, 88
ArcView, 3
Area isolation, 7
Asset management systems (AMS), 2, 11
AutoCAD, 3
Automated meter reading (AMR), 13

Best-fit, defined, 86. *See also* Calibration
Blackbox models. *See* Empirical modeling
Buoyant jet, 205

C-factors (roughness), 33, 33*t.*–34*t.*
    adjustments to (EPS models), 101
    calibration of, 94
    See also Hazen-Williams C-factor tests
CADD. *See* Computer-aided design and drafting
Calibration, 85
    accuracy reporting, 89–90
    and adjustments to C-factors (EPS
        models), 101
    and adjustments to closed valve data (EPS
        models), 101
    and adjustments to control logic (EPS
        models), 100–101
    and adjustments to diurnal demand patterns
        (EPS models), 101
    and adjustments to pump curves (EPS
        models), 100
    and adjustments to valve position data (EPS
        models), 100
    automated, 89
    AWWA ECAC guidelines, 88
    and C-values (pipe roughness coefficients), 94
    and control valve data, 92–93
    and data from flow recording devices, 91–92
    and data from pressure recording devices, 91
    and data from system maps, 90
    and data on demand-peaking factors, 93–94
    and data on minor losses, 94
    and data on unmetered flows, 93
    data sources for, 90–94
    defined, 85
    and demand data, 90
    and demand peaking factors (EPS models), 99–100
    and elevation data, 90–91
    and errors in head loss measurement, 83
    of extended-period simulation (EPS) models, 87, 98–102
    of fire flow (all-pipes models), 96
    of fire flow (EPS models), 100
    of fire flow (steady-state models), 96, 97*f.*
    goal of, 87–88
    guidelines, 88–89
    of HGL (steady-state models), 95, 97*f.*
    hydraulic, 87
    and identifying closed valves (steady-state models), 96
    and information system connectivity, 94
    macrocalibration, 87
    and mass balance, 94
    of maximum day (EPS models), 98–100
    and metered sales data, 93
    microcalibration, 87
    and nonrevenue water data, 93
    and operational data, 90
    and operations management, 7
    of peak hour (steady-state models), 95
    and physical facilities data, 90
    presenting results of (for EPS models), 101–102, 101*f.*
    presenting results of (for steady-state models), 96–97
    and pump controls (EPS models), 98
    and pump curve data, 92
    reconciling discrepancies, 86–87

and related terms, 86
of SCADA data, 91
and SCADA errors (EPS models), 101
sources of discrepancies between model results and field data, 85–86
and Stage 2 DBP Rule, 89
of steady-state models, 87, 95–97, 97f.
and storage facility water levels (EPS models), 99
types of, 87
United Kingdom (WRc) guidelines, 89
and valve controls (EPS models), 99
and water demand, 51
of water quality, 87
Capital improvement programs, 5, 13
Chloramines, 152, 153
Chlorine
decay bottle test, 164, 165f.
decay coefficients, 152–153
levels and decay, 8
See also Combined chlorine; Free chlorine; Total chlorine
CIS. *See* Customer information systems
CMMS. *See* Computerized maintenance management systems
Combined chlorine, 152
Compartment models (tank and reservoir mixing), 209–210, 210f.
Completely mixed models (tank and reservoir mixing), 209, 210f.
Computational fluid dynamics (CFD) modeling, 211–212
display showing effect of operational and design changes, 212, 214f.
display showing effect of thermal differences for short tank, 212, 213f.
display showing effect of thermal differences for tall tank, 212, 213f.
display showing water age distribution, 212, 214f.
and Navier-Stokes equations, 211
residuals, 211
Computer-aided design and drafting (CADD) systems, 2, 10
as data sources for model development, 28, 29
Computerized maintenance management systems (CMMS), 2, 10–11
as data sources for model development, 28, 29
and model design, 18–19
Contaminant tracking, 7

Continuity equations, 8
Customer information systems (CIS), 2, 10
and model design, 18–19

Darcy-Weisbach formula, 32–33, 35, 36f.
Data
confidence level, 26
from electronic records, 28–29
on facilities, 24–25
geographical, 24, 26
operational, 25–26, 27
from paper records, 28
on physical facilities, 27
from physical inspections, 29–30
quality assurance, 83–84
sources, 24–26, 27–30
topographic or elevation, 26–27
on water consumption, 25
on water demand, 25, 27, 45
on water production, 25
Databases, enterprise, 12, 28, 29
Demand. *See* Water demand
Digital terrain models (DTMs), 28, 29
Distribution system modeling, 1–2, 15
analysis types, 8–9
benefits of, 4
data considerations, 8
data import and export, 9
engineering design applications, 5–6
equations, 8
generating multiple scenarios, 9
historical development of, 2–4
in-house modeling vs. outside consultants, 11
links, 8
model developer vs. decision maker, 12
modelers vs. rest of utility, 12
nodes, 8
one-time vs. long-term use, 11
planning applications, 4
process, 2, 3f.
and related software systems, 2, 9–11, 13
selective reporting of results, 9
skeletonized vs. all-mains models, 12
software categories, 4–8
system operations applications, 6–7
trends in, 12–14
water quality improvement applications, 7–8
See also Extended-period simulation; Hydraulic modeling; Model construction and development; Steady-state simulation
Distribution System Requirements for Fire Protection (M31), 110, 116

Electrical power measurement, 81
Emergency planning, 14
    operations scenarios, 6
Emergency Planning for Water Utility Management (M19), 117
Empirical modeling (tank and reservoir mixing), 209–211
Energy
    analysis, 9, 13
    equations, 8
Energy cost management, 7
Enterprise databases, 12
EPA. *See* US Environmental Protection Agency
EPANET, 3, 18
Extended-period simulation (EPS), 8–9, 13, 103, 125–126
    analysis and sizing of storage tanks, 143
    analysis of additional storage capacity needs, 145–146
    analysis of booster station and well needs, 145
    analysis of emergency system operations, 138–140, 141f.
    analysis of operational improvements, 144
    analysis of potential energy optimization, 140–142, 142f.
    analysis of pressure-regulating station needs, 144
    analysis of storage vs. production, 136–138, 137f., 138t., 139f.
    analysis of vulnerability, 138, 139f.
    and average day conditions, 130
    calibration, 87, 98–102, 136
    case study (Fullerton, Calif.), 143–146, 145f.
    and control valve data, 128–129, 132
    and demand data from different customer types, 133–135
    and diurnal demand curves, 127, 133–135, 134f., 135f.
    in identification of existing inefficiencies, 144
    initial conditions, 129
    input data, 126–129
    and minimum day conditions, 130
    and miscellaneous supply sources, 128
    and nonelectric pump drivers, 141–142, 142f.
    objective, 129
    and operational philosophy, 129–130
    in operator training, 143
    and peak day conditions, 130
    and peak week conditions, 130
    and pump performance curves for booster stations and wells, 127–128
    and pump station information, 131
    and reservoir data, 128, 130–131
    with SCADA data, 132–133, 133f.
    without SCADA data, 133
    and seasonal variations, 130
    and service storage data, 130–131
    setup, 129–135
    and system operating conditions, 129
    and tank data, 128
    time interval, establishing, 127
    and time-of-day electric rates, 141, 142f.
    and time-of-day variations, 130
    types of, 136–143
    and variable speed pumps, 142
    verification of system operation, 126–127
    and well data, 131

Field tests, planning, 66
    equipment preparation, 66
    flow monitoring and mass balance, 66
    and maintaining system operability during tests, 67
    map of distribution network, 66
    and timing of tests, 67
Fire flow
    automated calculation, 9
    calibration of, 96, 97f., 100
    steady-state simulation, 108–110
    studies, 5
    tests, 76–77, 77f.
Flow
    calculations, 6
    throttling and inducing, 80
Flow-measuring equipment, 68
    calibration of data from, 91–92
    hydrant Pitot gauges, 68–70, 69f., 70f., 71f.
    magnetic meters, 74, 75f.
    master meters, 73–74, 80
    meter calibration test, 75
    Pitot tubes, 70–71, 72f.
    propeller meters, 73, 73f.
    readings at time corresponding to pressure measurements, 84
    strap-on meters, 71, 72f.
    turbine meters, 73, 73f.
    Venturi meters, 74, 74f., 75f.
Flowmeters in Water Supply, 73
FORTRAN, 3
Free chlorine, 152–153
    bulk decay coefficients, 153–154
Froud number, 208
Fullerton (California) water system, 143, 145f.

analysis of additional storage capacity needs, 145–146
analysis of booster station and well needs, 145
analysis of operational improvements, 144
analysis of pressure-regulating station needs, 144
EPS case study, 143–146
identification of existing inefficiencies, 144

Geographic information systems (GIS), 2, 3, 10
communications between GIS databases and model results, 59–62
coordination of GIS and modeling departments, 60–61
as data sources for model development, 28, 29
geometric network concept, and connectivity with hydraulic models, 61–62
integration with hydraulic models, 59, 60
interaction with hydraulic models, 18, 19, 21–22
interchange with hydraulic models, 59
interface with hydraulic models, 59–60
and model reconstruction, 58–59
and model topology, 61–62
relating GIS IDs to model IDs, 61
tracking and translating GIS IDs to model elements, 61
in tracking land-use and zoning data, 49
Geometric network, 61–62

Haloacetic acids (HAAs), simulation of, 155
Hardy-Cross method, 2
Hazen-Williams C-factor tests, 77–78
formula, 78
and head loss measurements, 78
miscellaneous methods, 80
parallel hose method, 79, 79f.
and pipe diameters, 78
two-gauge method, 79–80, 79f.
Hazen-Williams formula, 32–33, 34–35, 76, 77
Head
measurements, 83
See also Total dynamic head; Velocity head
Head loss
and data collection for calibration, 83
See also Hazen-Williams C-factor tests
HRT. See Hydraulic residence time
Hydrant Pitot gauges, 68–70, 69f., 70f.
diffusers, 70, 71f.
flow equation, 68–69
and hydrant outlets and discharge coefficient, 69, 71f.
Hydrants, 42
and transients, 178, 179f.
Hydraulic gradient line (HGL)
calibration of, 95, 97f.
and Hazen-Williams C-factor tests, 78, 79
and hydraulic gradient tests, 81, 82
slope adjustment caused by demand and pipe roughness adjustments, 83–84
Hydraulic gradient tests, 81
flow conditions, 82
plots, 82, 82f.
pressure measurements, 81–82, 81f.
Hydraulic modeling
conservative water quality tracers in calibration of, 162–163
geometric network concept, and connectivity with GIS, 61–62
integration with GIS, 59, 60
interaction with GIS, 18, 19, 21–22
interchange with GIS, 59
interface with GIS, 59–60
performance similar to water system operations, 51
See also Distribution system modeling; Extended-period simulation; Model construction and development; Steady-state simulation
Hydraulic residence time (HRT), 205–206. *See also* Water age
Hydraulic tests and measures, 65
and overall effects on water system, 83
miscellaneous, 83
and SCADA data, 75
See also Field tests, planning; Fire flow tests; Flow-measuring equipment; Hydraulic gradient tests; Pressure measurements; *and* tests *under* Pumps and pumping
Hydraulic transients. *See* Transient analysis; Transients

Information systems integration, 13
Input-output models. *See* Empirical modeling

Laboratory information systems (LIMS), 10
Lines, 23–24
Links, 8, 31
defined, 30
minor loss coefficients, 37, 37t.
pipe data and development, 31–32

pipe direction, and check valves, 37
and pipe roughness coefficients, 31, 32–36

Mains. *See* Water mains
Maintenance. *See* Model maintenance
Mass balance, 66
    calibration of, 94
Maximum day, 98–100
Mechanical pressure gauges, 67
Meters and metering. *See* Automated meter reading; Flow-measuring equipment
Model construction and development, 17
    and data availability and quality, 22
    data confidence level, 26
    data sources, 24–26, 27–30
    data types, 26–27
    demand development, 45–53
    detailed models, 21
    and geographic coordinates, 28
    and GIS, 18, 19, 21–22
    and GIS categorization of nodes, lines, and polygons, 23–24
    and hardware selection, 18
    hybrid models, 21
    level of detail, 19–20
    and links, 31–37
    and long-term sustainability, 22–23, 23*f.*
    model maintenance, 57–62
    model topology and GIS connectivity, 61–62
    and nodes, 30–31, 38–43
    objectives of model, 19
    operational data, 53–57
    and other enterprise applications, 18–19
    planning process, 19–24
    quality assurance (topology and connectivity review), 43–45
    relating model IDs to GIS IDs, 61
    relating model IDs to SCADA IDs, 56–57
    software selection, 18
    standards, 22
    structure of model, 20–22, 21*f.*
    tracking GIS IDs, 61
Model maintenance, 57
    communications between model results and GIS databases, 59–62
    coordination of modeling and GIS departments, 60–61
    frequency of, 58–59
    model reconstruction, 58–59
    periodic, 58
    types of, 57
Moody Diagram, 35, 36*f.*

Multi Species eXtension (MSX) modeling, 154, 155–156, 170

Navier-Stokes equations, 211
Newton-Raphson method, 188
Nodes, 8, 23–24, 38
    in close proximity, 43, 43*f.*
    control or facility elevations, 31
    defined, 30
    disconnected, 44, 44*f.*
    elevation data and quality assurance, 31
    functions and requirements, 30–31
    ground elevations, 31
    hydrants, 42
    junction type, 31
    pumping stations, 38–40
    and pumping water level, 38
    reservoirs, 38
    storage facilities, 38
    tanks, 38
    turnouts, 42
    valves, 40–42
    wells, 42–43
North Marin (Calif.) Water District, water quality modeling and testing (case study), 165, 169–170
    average percent of Stafford Lake water in district, 168, 169*f.*
    district description, 165–166
    hydraulic validation, 166–167
    observed and modeled chlorine residual, 169, 170*f.*
    observed and modeled sodium concentrations, 166–167, 168*f.*
    sampling study, 166
    skeletonized representation of network, 166, 167*f.*
    water quality calibration, 167–169

Operational data, 53
    and automated controls, 53
    calibration, 90
    from charts, 55
    and manual controls, 53
    from operations staff, 53
    required for modeling (by facility/equipment type), 53, 54*t.*
    from SCADA systems, 55–57
    from written records, 54
Operations
    modeling applications, 6–7
    real-time modeling for operators, 14

Personnel training, 6
Physical scale modeling (tank and reservoir mixing), 207–208
Pipes and piping
 C-factors (roughness), 33, 33t.–34t.
 cavitation in, 174, 182, 191
 Darcy-Weisbach formula (roughness), 32–33, 35, 36f.
 data for model development, 31–32
 dead-end pipelines and transients, 182, 182f., 183f., 186
 diameter discrepancies, 44
 direction, and check valves, 37
 disconnected, 44, 45f.
 equivalent sand grain roughness, 33, 34t.
 friction factor (Moody Diagram), 35, 36f.
 Hazen-Williams formula (roughness), 32–33, 34–35
 intersecting pipes, 44, 44f.
 as links, 30, 31
 materials, 32
 minor loss coefficients, 37, 37t.
 parallel (duplicate), 44, 45f.
 physical properties of common materials for, 185, 185t.
 pipe-split candidates, 43, 43f.
 pipeline profiles and transients, 180, 181f.
 pipeline ruptures and transients, 178
 quality assurance (topology and connectivity review), 43–44, 43f., 44f., 45f.
 restraints, 184–185
 roughness coefficients, 32–36
 system design criteria, 112–113
 water quality modeling of reactions within, 151
 See also All-mains models; All-pipes models; All-pipes reduced models; Transient analysis; Transients
Pitot tubes, 70–71, 72f.
Plug-flow models (tank and reservoir mixing), 209, 210f.
Polygons, 23–24
Pressure
 calculations, 6
 maximum, 111
 measurements, 67–68
 minimum during fire flows, 112
 minimum during peak hour, 111–112
 negative, 112
Pressure logger and chart, 67, 68f.

Pressure recording devices, calibration of data from, 91
Pressure-reducing valves (PRVs), 66
Pumps and pumping
 adjustments to pump curves (EPS models), 100
 bypass lines, 192
 calibration of controls, 98
 calibration of pump curve data, 92
 efficiency, 81
 efficiency curves, 114, 116f.
 electrical power measurement, 81
 and induced flow, 80
 in-line control valves, 193
 multiple pump rating curves, 114, 115f.
 pump curves, 39–40, 40f.
 pump curves, tested vs. assumed, 80
 pump rating curve vs. system head curve, 114, 115f.
 sizing, 6
 station sizing, 6
 stations, 38–40
 stations (GIS detail vs. model detail), 38, 39f.
 system design criteria, 113–114
 system head curves for pump stations, 114
 tests, 80–81, 81f.
 and throttling of flow, 80
 total dynamic head (TDH), 80, 81f.
 and transients, 176, 177, 177f., 178f.
 velocity head, 81

Real-time modeling, for system operators, 14
Reconciliation, defined, 86. *See also* Calibration
Reduction, defined, 20
Reservoirs, 203
 data in EPS models, 128, 130–131
 field studies and water quality modeling, 156, 163–164, 164f.
 as nodes, 38
 siting, 5
 sizing, 5–6
 See also Storage facilities; Tank and reservoir mixing; Tanks
Reynolds number, 208
Roughness testing. *See* Hazen-Williams C-factor tests

SCADA. *See* Supervisory control and data acquisition
Security applications, 14

Short-circuiting models (tank and reservoir mixing), 209, 210*f.*
Skeletonization, defined, 20
Skeletonized models, 12, 20, 21*f.*
Skeletonized reduced models, 20–21, 21*f.*
Source management, 6–7
Stage 2 DBP Rule, Initial Distribution System Evaluation (IDSE), 89
Steady-state simulation, 8, 103–104
    average day demand, 105, 108
    calibration, 87, 95–97, 97*f.*
    and control valve set points, 122
    and development of system improvements, 120–122
    and effective storage, 117
    and elevated storage, 118, 119*f.*
    and emergency storage, 116–117
    and equalization storage, 116, 117*f.*
    fire flow analysis, 108–110
    and fire storage, 116
    and ground storage with pumping, 118, 119*f.*
    and high-level ground storage, 118, 119*f.*
    limiting demand conditions, 104–105
    and local impact analysis, 122
    maintenance of model for continued use, 123
    and master planning, 120–121
    maximum day demand, 104, 105, 105*f.*, 106–107
    maximum day demand plus fire flow, 105
    maximum hour demand, 104, 105, 105*f.*, 107–108
    and maximum pressure, 111
    and minimum effective elevation, 117
    minimum hour demand, 104, 105*f.*
    minimum hour demand of average day, 105, 110
    and minimum pressure during fire flows, 112
    and minimum pressure during peak hour, 111–112
    and multiple pump rating curves, 114, 115*f.*
    and negative pressures, 112
    and number and location of storage facilities, 119–120
    and outage planning, 122
    and piping system design criteria, 112–113
    and pressure design criteria, 111–112
    and pump efficiency curves, 114, 116*f.*
    pump rating curve vs. system head curve, 114, 115*f.*
    and pumping systems design criteria, 113–114
    and rehabilitation of neighborhood distribution mains, 121
    and reliability during planned maintenance activities, 110
    and reliability in emergency conditions, 110–111
    and standpipes, 118
    and storage allocation, 118, 118*f.*
    and storage facilities design criteria, 115–120
    and storage types, 118–119, 119*f.*
    and subdivision planning, 121
    and system design criteria, 111–120
    and system head curves for pump stations, 114
    system performance analyses, 104–106, 106*t.*
    types of problems analyzed by, 103
Storage facilities, 203
    analysis and sizing of tanks (EPS models), 143
    analysis of additional capacity needs (case study), 145–146
    calibration of water levels, 99
    data in EPS models, 130–131
    design criteria, 115–120
    for disinfection contact time, 204
    effective storage, 117
    elevated storage, 118, 119*f.*
    emergency storage, 116–117
    empirical modeling, 209–211, 210*f.*
    equalization storage, 116, 117*f.*
    for fire flow and peak demands, 204
    fire storage, 116
    ground storage with pumping, 118, 119*f.*
    high-level ground storage, 118, 119*f.*
    and minimum effective elevation, 117
    as nodes, 38
    physical scale modeling, 207–208
    reactions within pipes and storage tanks (water quality modeling), 151
    standpipes, 118
    and storage allocation, 118, 118*f.*
    See also Reservoirs; Tank and reservoir mixing; Tanks
Supervisory control and data acquisition (SCADA) systems, 2, 10
    calibration data from, 91
    data errors in, 56
    data format considerations, 56

as data sources for model development, 28, 29
EPS models with and without data from, 132–133, 133f.
errors in, and EPS model calibration, 101
importance to modeling, 55
and measured vs. calculated values, 56
and model design, 18–19
and model input data, 55–56
relating SCADA IDs to model IDs, 56–57
representation in model software, 53
as source of data for modeling, 55–57
transient or surge data in, 56
units for data, 56
and verification or reference data, 56
Surge protection devices, 191–193
Surge protection strategies, 193–195, 195f.
Surges. *See* Transient analysis; Transients
Systems models. *See* Empirical modeling

Tank and reservoir mixing, 14, 151, 203
    avoiding stratification, 205
    blackbox models, 209
    and buoyant jet, 205
    compartment models, 209–210, 210f.
    completely mixed models, 209, 210f.
    completely mixed state, 204
    computational fluid dynamics (CFD) modeling, 211–212, 213f., 214f.
    and conservative tracers, 163
    developing turbulent jet flow to promote, 204–205
    effect of operational and design changes, 212, 214f.
    effect of thermal differences for short tank, 212, 213f.
    effect of thermal differences for tall tank, 212, 213f.
    and empirical modeling, 209–211
    and fill time, 206
    and hydraulic residence time (HRT), 205–206
    improvement strategies, 215, 215t.
    and inlet momentum, 206–207
    and inlet pipe location and orientation, 207
    input-output models, 209
    model verification, 212
    modeling methods, 207–212
    and physical scale modeling, 207–208
    plug flow manner, 204
    plug-flow models (first in/first out), 209, 210f.
    sampling locations for monitoring of, 216
    and seasonal variation, 207
    short-circuiting models (last in/first out), 209, 210f.
    systems models, 209
    and tank or reservoir geometry, 207
    and temperature variations, 163–164, 164f.
    and volume turnover, 207
    in water quality modeling, 151
    See also Reservoirs; Storage facilities; Tanks; Water age
Tanks, 203
    analysis and sizing of, 143
    data in EPS modeling, 128
    field studies and water quality modeling, 156, 163–164, 164f.
    fill time vs. mixing time, 204
    as nodes, 38
    water quality modeling of reactions within, 151
    See also Storage facilities; Tank and reservoir mixing
Three-dimensional laser induced fluorescence, 208
Topology, defined, 61
Total chlorine, 152–153
Total dynamic head (TDH), 80, 81f.
Transient analysis, 9, 14, 173–174, 199–200
    and cavitation, 174, 182, 191, 195
    considerations, 195–197
    continuity equation, 189–190
    control element head-flow equation, 187–188
    data requirements, 197–199
    and four quadrant pump characteristics, 198, 199f.
    friction factor, 189
    fundamental equation of water hammer, 184
    glossary of notations, 200–201
    governing equations,189–190
    head rise equation, 184
    Joukowsky relation, 184
    method of characteristics (MOC), 190
    momentum (dynamic) equation, 189–190
    momentum principle, 184
    negative wave reflection at open ends or connections to reservoirs, 186
    and Newton-Raphson method, 188
    and passive resistance elements, 187
    and physical properties of common pipe materials, 185, 185t.
    pipe data needed, 197
    and pipe length and time interval, 197
    and pipeline restraint, 184–185

and Poisson's ratio, 185
positive wave reflection at dead ends, 186
pressure change and flow relation to surge
    control devices, 198
and pressure-sensitive demands, 197
propagation speed of pressure wave, 184
and quadratic formula, 188
relationship between pressure change and
    flow change, 183–184, 183*f.*, 197–198
resistance, defined, 188
and resistive control elements, 188
resistive parameters, 188
and series and parallel pumps, 197
simplifications and assumptions in, 195
skeletonization, 196
specification of disturbance and intended
    timescale, 199
and steady-state modeling, 195, 196, 197
and surge protection devices, 191–193
and surge protection strategies, 193–195,
    195*f.*, 196*f.*
transmission coefficient, 185–186
and valve closure characteristics, 198, 198*f.*
wave action at control elements,
    186–189, 187*f.*
wave action at pipe junctions, 185–186, 186*f.*
wave characteristic method (WCM), 190–191
wave propagation with friction,
    188–189, 189*f.*
wave speed, 184–185
Transients, 173–174
    and air release valves, 192
    and bladder tanks, 191
    and broken air admission valve, 178, 180*f.*
    causes of, 175, 176–182
    cavitation, 174, 182, 191
    and check valves, 192
    and closed surge tanks, 191
    control methods, 191–195
    and damaged pump bowl, 178, 180*f.*
    and dead-end pipelines, 182, 182*f.*, 183*f.*, 186
    defined, 176, 176*f.*
    and downsurge, 191
    and feed tanks (one-way surge tanks), 192
    and float-operated air valves vs. surge
        resistant air valves, 194
    and flywheels, 192–193, 193*f.*
    and hybrid tanks, 191
    and hydrants opening or closing too quickly,
        178, 179*f.*
    identifying systems vulnerable to, 181–182
    and in-line pump control valves, 193

mitigation techniques, 178–180
and normal pump operation protection vs.
    power failure protection, 194–195
and pipeline profiles (good to bad), 180, 181*f.*
and pipeline ruptures, 178
possible consequences of, 174–175
and pressure relief valves, 192
pressure surge fluctuations following pump
    shutdown, 182, 183*f.*
and pressure waves, 176
and pump bypass lines, 192
and pump shutdown or trip, 176, 177, 177*f.*
and pump startup, 177, 178*f.*
and pump station and downstream pipeline
    protection, 194
and SCADA data, 56
and simple surge tanks (open), 191
and surge anticipation valves, 192
and surge vessels (air chambers), 191
and upsurge, 191
and vacuum breaking valves, 192
and valves opening or closing too rapidly,
    176, 177, 179*f.*
Transport equations, 8
Trihalomethanes (THMs)
    formation coefficients, 155
    laboratory formation test, 164
Troubleshooting, 6

Validation, defined, 86. *See also* Calibration
Valves, 40
    adjustments to closed valve data (EPS
        models), 101
    adjustments to position data (EPS
        models), 100
    altitude type, 41
    calibration of control valve data, 92–93
    calibration of controls, 99
    check type, 37, 40
    closed, identifying in model calibration, 96
    control, 40–41
    control valve set points, 122
    flow control type, 41
    isolation (zone) type, 40
    modeling tips, 41–42
    pressure-reducing, 41
    PSVs, 41
    sizing, 5
    throttle control type, 41
    and transients, 176, 177, 178, 179*f.*, 180*f.*, 192
Velocity head, 81

Water age
    calculated by empirical model assuming complete mixing, 210, 210f.
    distribution, 212, 214f.
    reduction strategies, 216
    studies, 149
    tracking, 7
    See also Hydraulic residence time; Tank and reservoir mixing
Water Audit and Loss Control Program (M36), 73
Water conservation impact studies, 5
Water demand
    adjustments to diurnal demand patterns (EPS models), 101
    allocation process, 49–50
    allocation to model, 13, 45
    assigning to appropriate junctions of model, 49
    average day demand, 105, 108
    base demand, defined, 49
    base demands, adjusting, 51–52
    base demands, CIS as source for allocating, 50
    calibration of data, 51, 90
    commercial, 46
    consumption data from meters via CIS, 50
    data, 25, 27, 45
    data from customer information systems (CISs), 47–48
    data from land-use classifications, 49
    data from population counts, 49
    data on water production by pressure zone, 48–49
    data sources, 47–49
    diurnal curves, 52–53, 52f.
    diurnal curves and EPS modeling, 127, 133–135, 134f., 135f.
    and geographic variations, 51
    industrial, 46
    for irrigation, 46
    maximum day demand, 104, 105, 105f., 106–107
    maximum day demand plus fire flow, 105
    maximum hour demand, 104, 105, 105f., 107–108
    minimum hour demand of average day, 105, 110
    minimum hour demand, 104, 105f.
    and nonservice junctions (nodes), 50
    residential, 45–46
    and seasonal variations, 51–52
    and service junctions (nodes), 50
    steps in allocating demands to model, 45
    types of, 45–47
    and water consumption data from meters, 47
    water loss as source of, 46–47
    wholesale (sales to other utilities), 46
Water Distribution System Analysis: Field Studies, Modeling, and Management, 59
Water hammer. *See* Transient analysis; Transients
Water loss, 46–47
    allocation of, 47
    calculations, 6
    sources of, 47
Water mains
    flushing programs, 7
    rehabilitation programs, 5
    See also All-mains models
Water Meters—Selection, Installation, Testing, and Maintenance, 73
Water quality
    analysis, 9, 14
    analysis procedures, 160
    automated analysis in field, 157
    automated grab samples, 157
    calibration and review of analytical instruments for surveys, 161
    calibration of, 87
    changes in distribution system, 147
    chlorine decay bottle test, 164, 165f.
    collection of ancillary data, 159
    communications and coordination (surveys), 161
    contingency plans for surveys, 161
    continuous sample collection, 157
    data recording (surveys), 160
    equipment and supply needs (surveys), 160
    in-situ measurements, 157
    laboratory analysis, 157
    logistical arrangements for surveys, 160
    manual analysis in field, 157
    manual grab samples, 157
    monitoring and sampling principles, 156–158
    monitoring locations, 8, 158
    monitoring parameters, 157–158
    preparation of sampling sites, 159
    preparation of survey data report, 161
    safety issues in surveys, 160
    sample analysis, 157
    sample collection, 157
    sample collection procedures, 159

sampling frequency, 159
sampling locations, 158–159
surveys, 156, 158–161
surveys, steps in, 158
and system operation, 159
THM formation test, 164
tracer studies, 156, 162–163
training requirements (surveys), 161
Water quality modeling, 147–148
advances in, 149
and advective transport of mass within pipes, 149–150
bulk decay coefficients, 153–154
case study (North Marin, Calif., Water District), 165–170, 167f., 168f., 169f., 170f.
and chlorine decay bottle test, 164, 165f.
chlorine decay coefficients, 152–153
computational methods (extended-period models), 151–152
computational methods (steady-state models), 151
and conservation of constitutent mass, as basis for studies, 149
and conservative constituents, 151
and conservative tracers in calibration of hydraulic model, 162–163
constituent grow and decay studies, 149
and disinfection by-products, 155
estimation of growth or decay of multiple constituents simultaneously, 149
and field testing of water quality, 156
governing principles, 149–151
historical data, use of, 156, 162
hydraulic data requirements, 152
input data requirements, 126
and laboratory kinetic studies, 156, 164, 165f.
and mixing of mass at pipe junctions, 150
and mixing of mass within storage tanks, 151
model equilibration, 152, 153f.
and monitoring of water quality, 156–158
Multi Species eXtension (MSX) modeling, 154, 155–156, 170
prediction of spatial and temporal distribution of constituents, 148
reaction rate data, 152–155
and reactions within pipes and storage tanks, 151
source and contaminant tracing, 149
and tank and reservoir field studies, 156, 163–164, 164f.
THM formation coefficients, 155
and THM formation test, 164
and tracer studies, 156, 162–163
types of studies performed, 148
wall demand coefficients, 154
water age studies, 149
water quality data requirements, 152
Weber number, 208

Zone boundary selection, 6

This page intentionally blank.

AWWA Manuals

M1, *Principles of Water Rates, Fees, and Charges*, Sixth Edition, 2012, #30001

M2, *Instrumentation and Control*, Third Edition, 2001, #30002

M3, *Safety Practices for Water Utilities*, Sixth Edition, 2002, #30003

M4, *Water Fluoridation Principles and Practices*, Fifth Edition, 2004, #30004

M5, *Water Utility Management*, Second Edition, 2004, #30005

M6, *Water Meters—Selection, Installation, Testing, and Maintenance*, Fifth Edition, 2012, #30006

M7, *Problem Organisms in Water: Identification and Treatment*, Third Edition, 2004, #30007

M9, *Concrete Pressure Pipe*, Third Edition, 2008, #30009

M11, *Steel Pipe—A Guide for Design and Installation*, Fifth Edition, 2004, #30011

M12, *Simplified Procedures for Water Examination*, Fifth Edition, 2002, #30012

M14, *Recommended Practice for Backflow Prevention and Cross-Connection Control*, Third Edition, 2004, #30014

M17, *Installation, Field Testing, and Maintenance of Fire Hydrants*, Fourth Edition, 2006, #30017

M19, *Emergency Planning for Water Utilities*, Fourth Edition, 2001, #30019

M20, *Water Chlorination/Chloramination Practices and Principles*, Second Edition, 2006, #30020

M21, *Groundwater*, Third Edition, 2003, #30021

M22, *Sizing Water Service Lines and Meters*, Second Edition, 2004, #30022

M23, *PVC Pipe—Design and Installation*, Second Edition, 2003, #30023

M24, *Dual Water Systems*, Third Edition, 2009, #30024

M25, *Flexible-Membrane Covers and Linings for Potable-Water Reservoirs*, Third Edition, 2000, #30025

M27, *External Corrosion: Introduction to Chemistry and Control*, Second Edition, 2004, #30027

M28, *Rehabilitation of Water Mains*, Second Edition, 2001, #30028

M29, *Fundamentals of Water Utility Capital Financing*, Third Edition, 2008, #30029

M30, *Precoat Filtration*, Second Edition, 1995, #30030

M31, *Distribution System Requirements for Fire Protection*, Fourth Edition, 2008, #30031

M32, *Computer Modeling of Water Distribution Systems*, Third Edition, 2012, #30032

M33, *Flowmeters in Water Supply*, Second Edition, 2006, #30033

M36, *Water Audits and Loss Control Programs*, Third Edition, 2009, #30036

M37, *Operational Control of Coagulation and Filtration Processes*, Third Edition, 2011, #30037

M38, *Electrodialysis and Electrodialysis Reversal*, First Edition, 1995, #30038

M41, *Ductile-Iron Pipe and Fittings*, Third Edition, 2009, #30041

M42, *Steel Water-Storage Tanks*, First Edition, 1998, #30042

M44, *Distribution Valves: Selection, Installation, Field Testing, and Maintenance*, Second Edition, 2006, #30044

M45, *Fiberglass Pipe Design*, Second Edition, 2005, #30045

M46, *Reverse Osmosis and Nanofiltration*, Second Edition, 2007, #30046

M47, *Capital Project Delivery*, Second Edition, 2010, #30047

M48, *Waterborne Pathogens*, Second Edition, 2006, #30048

M49, *Butterfly Valves: Torque, Head Loss, and Cavitation Analysis*, Second Edition, 2012, #30049

M50, *Water Resources Planning*, Second Edition, 2007, #30050

M51, *Air-Release, Air/Vacuum, and Combination Air Valves*, First Edition, 2001, #30051

M52, *Water Conservation Programs—A Planning Manual*, First Edition, 2006, #30052

M53, *Microfiltration and Ultrafiltration Membranes for Drinking Water*, First Edition, 2005, #30053

M54, *Developing Rates for Small Systems*, First Edition, 2004, #30054

M55, *PE Pipe—Design and Installation*, First Edition, 2006, #30055

M56, *Fundamentals and Control of Nitrification in Chloraminated Drinking Water Distribution Systems*, First Edition, 2006, #30056

M57, *Algae: Source to Treatment,* First Edition, 2010, #30057

M58, *Internal Corrosion Control in Water Distribution Systems,* First Edition, 2011, #30058

M60, *Drought Preparedness and Response,* First Edition, 2011, #30060

M61, *Desalination of Seawater,* First Edition, 2011, #30061